Christianity and Critical Realism

One of the key achievements of critical realism has been to expose the modernist myth of universal reason, which holds that authentic knowledge claims must be objectively 'pure', uncontaminated by the subjectivity of local place, specific time and particular culture.

Wright aims to address the lack of any substantial and sustained engagement between critical realism and theological critical realism with particular regard to: (1) the distinctive ontological claims of Christianity; (2) their epistemic warrant and intellectual legitimacy; and (3) scrutiny of the primary source of the ontological claims of Christianity, namely the historical figure of Jesus of Nazareth. As such, it functions as a prolegomena to a much-needed wider debate, guided by the under-labouring services of critical realism, between Christianity and various other religious and secular worldviews. This important new text will help stimulate a debate that has yet to get out of first gear.

This book will appeal to academics, graduate and postgraduate students especially, but also to Christian clergy, ministers and informed laity, and members of the general public concerned with the nature of religion and its place in contemporary society.

Andrew Wright is Professor of religious and theological education at King's College London, UK.

New studies in critical realism and spirituality

Other titles in this series:

Critical Realism and Spirituality
Mervyn Hartwig and Jamie Morgan

Christianity and Critical Realism
Ambiguity, truth and theological literacy
Andrew Wright

Christianity and Critical Realism

Ambiguity, truth and theological literacy

Andrew Wright

Routledge
Taylor & Francis Group
LONDON AND NEW YORK

First published 2013
by Routledge
2 Park Square, Milton Park, Abingdon, Oxon OX14 4RN

Simultaneously published in the USA and Canada
by Routledge
711 Third Avenue, New York, NY 10017

Routledge is an imprint of the Taylor & Francis Group, an informa business

British Library Cataloguing in Publication Data
A catalogue record for this book is available from the British Library

Library of Congress Cataloging-in-Publication Data
Wright, Andrew, 1958–
Christianity and critical realism : ambiguity, truth, and theological literacy / Andrew Wright.
p. cm. – (New studies in critical realism and spirituality)
Includes bibliographical references and index.
1. Christian philosophy. 2. Christianity–Philosophy. 3. Critical realism. 4. Philosophy and religion. 5. Religion–Philosophy. I. Title.
BR100.W65 2013
230.01–dc23 2012014951

ISBN: 978-0-415-53957-9 (hbk)
ISBN: 978-0-203-08513-4 (ebk)

Typeset in Times New Roman
by Wearset Ltd, Boldon, Tyne and Wear

For Elina, Juliana and Mariana

Contents

Acknowledgements

I am, once again, grateful to the ongoing support of colleagues and friends in the Centre for Theology, Religions and Culture at King's College London, in particular Professor Alister McGrath, Dr Philip Barnes, Ann-Marie Brandom and my sister Angela Wright. Thanks also to my doctoral students for their enthusiasm, friendship and intellectual stimulation: Dave Aldridge (Oxford Brookes University), Sheryl Anderson, Sungwoo Baek, Diane Craven, Ciro Genovese (Canterbury Christchurch University), Frances Mackenney-Jeffs, Benji McNair Scott, Laurence Moscrop, Marvin Oxenham (London School of Theology), Tim Parsons (London Metropolitan University), and Adrian Smith, Phil Wall and Johanna Woodcock (Oxford Brookes University). Many of the arguments contained in this book were developed and honed at meetings of the Forum for Religious and Spiritual Education in the Department of Education at King's College London: I am especially grateful to the Forums' convener Dr Tony Wenman, and to Christina Davis, Angela Goodman, Will Griffith and Tom Hibberd. Particular thanks are due to Professor Roy Bhaskar for his friendship, support and advice. My greatest intellectual debt is owed to the wisdom, insight and enthusiasm of my wife, Dr Elina Wright (Regent's Park College, University of Oxford).

Before taking on a life of its own, this book was initially intended as a chapter in a monograph on Critical Realism and Religious Education. I am appreciative of the patience and understanding of the editorial staff at Routledge in facilitating a late revision of my publishing plans. I am grateful to the respective publishers for permission to quote from T.F. Torrance's *Theological Science* (© T. & T. Clark 1969, with permission from Bloomsbury Publishing Plc), and N.T. Wright's *Christian Origins and the Question of God, Volume One: The New Testament and the People of God* (© SPCK 1992). Unless otherwise stated, biblical quotations are taken from the New Revised Standard Version (Anglicized edition). On a few occasions I have slightly amended the NRSV rendering of the New Testament in the light of my own reading of the original Greek text.

Beyond the realm of academia, the congregation of St Philip's Church, Burwash Weald enabled me to put my academic efforts into proper perspective *sub specie aeternitatis*, as did my daughters Becky and Liz, my parents Ruth and Bill, my sisters Heather and Angela, my brother Jonathan and my extended family. Above all, I was surrounded throughout by the love and care of Elina, Juliana and Mariana, to whom this book is dedicated.

Introduction

This book began life as a chapter in a projected monograph on critical realism and religious education. The contemporary practice of religious education in the UK tends to occlude questions of realistic truth: Do the generic truth claims of religion in general, or of the truth claims of specific religious traditions in particular, enjoy any epistemic purchase on the ultimate ontological order-of-things? Or do any of the alternative secular accounts of ultimate reality provide us with a more comprehensive, powerful and truthful explanatory model? The reasons for the occlusion of such questions are not difficult to identify: the Enlightenment's rejection of the epistemic warrant of religious truth claims reduced religious belief to a private affair predicated on an irrational or post-rational leap of faith that precluded informed debate in the public sphere. This, coupled with the Enlightenment's cultivation of the twin liberal values of freedom of belief and tolerance of the beliefs of others, made a realistically oriented religious education vulnerable to the charge of indoctrination. In some quarters the subject was seen as an agent of Christian confessionalism, striving to impose a set of ideological beliefs upon pupils in a closed and distinctly uncritical manner. The 1970s saw a fundamental paradigm shift in the theory and practice of religious education in state-funded schools in the UK. An open multi-faith approach was adopted, which sought to attend to a range of different religious and secular belief systems without bias or prejudice. In this new situation the charge of indoctrination was avoided by presenting pupils with neutral descriptions of the truth claims of various religious traditions in a manner that tended to avoid discussion of their veracity. As a result, contemporary religious education often leaves pupils free to express their own beliefs and opinions, provided that in doing so they acknowledge and tolerate the beliefs and opinions of others. The frequent absence of any sustained attempt to employ judgemental rationality to critically assess the various ontologically incommensurable truth claims presented in the religious education classroom means that in many instances religious education functions as an instrument for the transmission of liberal ideology, in which an uncritical freedom of expression equates with a thoroughgoing epistemic relativism: 'You are free to believe whatever you like, regardless of its epistemic warrant and ontological veracity, provided that in doing so you respect and acknowledge the freedom of others to do likewise.' The problem with this ideological representation of religions is threefold: first, it ignores critical ontological questions of truth and truthful living; second, it imposes a premature and unwarranted epistemic closure by treating all beliefs as equally valid; and third, it perpetuates a widespread religious, spiritual and theological illiteracy among religious believers, secular sceptics, and interested/disinterested agnostics alike.

Since the mid-1990s I have worked closely with colleagues in the Centre for Theology, Religions and Education, located in the Department of Education and Professional Studies at King's College London, to develop a model of 'critical religious education' that employs critical realism in an under-labouring role. It draws on critical realism's triumvirate of ontological realism, epistemic relativism and judgemental rationality in a manner designed to: (1) place questions of truth and truthful living at the heart of contemporary religious education (ontological realism); (2) recognize the contested nature of religious and secular accounts of the ultimate-order-of-things (epistemic relativism); and (3) promote appropriate levels of religious, spiritual and theological literacy (judgemental rationality). The projected book, which will be published in Routledge's New Studies in Critical Realism and Spirituality series as *The Spiritual Turn in Critical Realism: Critical Religious Education and Spiritual Literacy*, set out to provide a fuller account of the relationship between critical religious education and critical realism than had hitherto been possible in my previous publications.

The so-called 'spiritual turn' in critical realism, which we will consider in detail in the first part of the current book, made it clear that the internal logic of critical realism opens up, and indeed demands, rational engagement with religious and secular questions about ultimate truth and the ultimate order-of-things. The hermeneutical circle between parts and the whole requires that the interpreter makes sense of the parts of a text in the light of the text as a whole, and of the whole of the text in the light of its individual parts, in an ongoing interpretative dialectic. By extension, in seeking to make sense of reality the critical realist strives to understand the parts in terms of the whole, and the whole in terms of its constituent parts. This necessitates attempts to explain the ultimate nature of the totality of reality, regardless of whether the proffered explanations are religious/theological or secular/naturalistic. In acknowledging the intellectual imperative to attempt to make sense of the *totality* of experiences, objects/events and causal mechanisms, the spiritual turn in critical realism resonates strongly with the largely independent tradition of theological critical realism generated and nurtured by Christian theologians since the late 1950s. In the chapter in the projected monograph from which this current book emerged, I set out to use Christianity as a case study of the way in which critical realism might underlabour for a form of critical religious education committed to the pursuit of truth and truthful living through the cultivation of religious, spiritual and theological literacy.

However, as I began to draft the chapter it quickly became clear that what had, on the surface, promised to be a relatively simple exercise was in actual fact replete with problems – the major one being the relative paucity of intellectual engagement between critical realism and Christian theology. With the notable exception of the work of my colleague Alister McGrath, Christian theologians, despite having developed their own version of critical realism as a tool for theological investigation, have largely failed to engage with the work of Bhaskar and his colleagues. Where conversations have taken place, they have largely been in the specialist sphere of the interface between Christian theology and the natural sciences. Prior to the publication of this book, there has been little substantial engagement between critical realism and theological critical realism with regard to: (1) the distinctive ontological claims of Christianity; (2) their epistemic warrant and intellectual legitimacy; and (3) scrutiny of the primary source of the ontological claims of Christianity, namely the historical figure of Jesus of Nazareth. This failure on the part of theological critical realism is replicated in discussions of Christianity within the critical realist community. This is most clearly visible in the work of Sean Creaven, whose critique of Christianity singularly fails to engage in any viable intellectual depth with the particularities of Christian ontology and epistemology (Creaven 2010; Wright 2011). This problem is compounded by Bhaskar's own reading of Christianity, which suggests that institutional religion tends to occlude spiritual well-being, that spiritual insight is grounded in pre-linguistic experience of transcendence rather than discursive doctrinal formulations, and that consequently Christian doctrines do not offer viable explanations of ultimate reality.

This book is offered as a prolegomena to a much-needed debate, guided by the under-labouring services of critical realism, between Christianity and various other religious and secular worldviews. It is written in the conviction that for such a debate to be effective, a more substantial understanding of the ontological claims and epistemic commitments of Christianity than that currently available within critical realist circles needs to be tabled for discussion. If critical realism is correct in suggesting that we make sense of the world by developing explanatory models that we are justified in embracing until such time as they are trumped by more powerful and comprehensive models, then any evaluation of the Christian retroductive explanation of the ultimate order-of-things must necessarily be cognisant of its ontological substance and epistemic ground. One of the ongoing consequences of the Enlightenment's occlusion of the discipline of theology as a valid academic endeavour is a tendency of critics of Christianity to assume that theological questions do not require the same level of careful academic scrutiny demanded and required in other fields of knowledge. Thus, for example, the anti-Christian polemic of Richard Dawkins is notable for its almost complete lack of engagement with academic Christian theology – a criticism which can also be laid at the feet of Sean Creaven (Dawkins 2006; McGrath and Collicutt 2007). The modern occlusion of theology is rooted in empiricist and idealist assumptions about the epistemic warrant of religious belief systems that critical realism has shown to be woefully inadequate. This being the case, any informed debate needs to engage with Christian ontology and epistemology with the same intellectual rigour and thoroughness granted any other field of intellectual endeavour.

One of the key achievements of critical realism has been to expose the modernist myth of universal reason, which holds that authentic knowledge claims must be objectively 'pure', uncontaminated by the subjectivity of local place, specific time and particular culture. The principle of epistemic relativism recognises that all knowledge claims are necessarily 'contaminated' by the perspective of the knower and the knowing community. This book is no exception: I write as a Christian theologian from within the Anglican Communion into which I was baptised as an adult. As such, I write in the spiritual and intellectual conviction that the Christian account of the Trinitarian God, and of that God's salvific engagement with his creation, above all in the person of Jesus of Nazareth, constitutes the most powerful and comprehensive account of the ontological structures of the ultimate order-of-things currently available to us. This is not to suggest that my primary aim in writing is apologetic, or that my core concern is to convince the reader of the truth of my particular beliefs, though inevitably something of that agenda will be visible beneath the surface of the text. Rather my concern is primarily academic: to help stimulate a debate that has yet to get out of first gear. It is imperative that any discussion of the ontological substance and epistemic warrant of the Christian account of ultimate reality proceeds from an appropriately informed understanding of Christian ontology and epistemology. Hence the primary aim of this book is simply to make available, without any attempt to be either definitive or authoritative, an outline sketch of some key features of Christian onto-theology and epistemology informed by their ongoing engagement with critical realism.

In Part I, 'Critical realism and transcendence', I place the ensuing discussion of Christianity and critical realism in context. Chapter 1, 'Critical realism and dialectical critical realism', provides an overview of the main tenets of critical realism that will already be familiar to some readers. Chapter 2, 'The spiritual turn: transcendence and meta-reality', offers an overview of recent developments in Bhaskar's thought that will again be familiar to some readers, though it does contain an attempt to unpack the provenance of Bhaskar's distinctive spiritual vision which may shed new light on his emergent position. Chapter 3, 'Christianity and critical realism', considers the response of critical realists to Bhaskar's spiritual turn before exploring the independent tradition of theological critical realism. With regard to the latter, though some readers will already be familiar with the critically realistic informed debate between Christian theology and natural science, the account of theological critical realism's concern for the ontological substance and epistemic ground of Christian doctrine, as well as for the application of a critical realist historiography to the life of Jesus of Nazareth, may open up dimensions of Christian thought not previously available in critical realist circles.

In Part II, 'Epistemic relativism: the ambiguity of Christianity', I consider Christianity through the lens of critical realism's commitment to epistemic relativism, focusing on the critical question of the extent to which Christian doctrine can be recognised as epistemically fallible without damage to the integrity of the Christian belief that it is a product of divine revelation. Chapter 4, 'The identity of

Christianity', functions as a prolegomena to that discussion: it utilizes critical realism in an under-labouring role in an attempt to make sense of the significantly under-determined label 'Christianity', which frequently acts as an umbrella term for a diverse range of ontologically incommensurable religious beliefs and truth claims. I suggest that critical realism provides a more powerful tool for making sense of the broad phenomenon of 'Christianity' than either essentialism (idealism) or nominalism (empiricism), and identify 'classical' or 'orthodox' Trinitarian theology as the most appropriate retroductive model (among the many gathered under the 'Christian' umbrella) to focus on in the current context. Chapter 5, 'Trinitarian theology and epistemic relativism', argues that, contrary to popular perceptions of premature and unwarranted epistemic closure engendered by modernist (and frequently fundamentalist) appeals to biblical inerrancy and papal infallibility, Trinitarian theologians have consistently acknowledged the contingency and fallibility of their retroductive models, and have regularly subjected them to iterative development and revision. Chapter 6, 'The "problem" of Christian exclusivism', argues, against Bhaskar, that to affirm the exclusive truth of Trinitarian doctrine is to do no more than avow the hypothesis that the Christian account of the ultimate order-of-things constitutes the most powerful and comprehensive explanatory model currently available. In acknowledging the contingent nature of Christian doctrines, Christian theologians open themselves up to both internal and external critique. Critical realism finds nothing intrinsically irrational or morally questionable in defending a claim that one particular explanatory model is more powerful and comprehensive than any other; on the contrary, failure to do so leads directly to the occlusion of judgemental rationality. This being the case, to affirm and defend the exclusive truth of Christian doctrine is neither intrinsically irrational nor morally questionable.

In Part III, 'Ontological realism: Christianity and truth', attention shifts from epistemic relativism to ontological realism. My primary concern in these chapters is to provide an overview of the material substance of Christian ontology. In Chapter 7, 'Classical theism and the Triune God', I seek to establish the groundwork for an account of Christian onto-theology by distinguishing between classical theism and Trinitarian theology, and arguing that the 'God of the philosophers' is ontologically incommensurate with the 'God of Abraham'. The emergence of modern atheism is predicated, in the main, on critiques of classical theism in general, and seventeenth- and eighteenth-century deism in particular; as such it is concerned only tangentially with the Trinitarian God of Christian ontology. In Chapter 8, 'The Triune God: Father, Son and Holy Spirit', I offer an account of the distinctive *sui generis* Christian understanding of God as Trinity. In Chapter 9, 'The economy of salvation', I seek to relate this account of the Trinitarian God of orthodox Christian faith to the Christian narrative of God's salvific engagement with his creation, paying particular attention to the Christian doctrine of creation, Christian anthropology, and Christian accounts of the theological significance of the person and work of Jesus of Nazareth.

In Part IV, 'Judgemental rationality and Christian epistemology', I attend to the question of the epistemic warrant and intellectual justification of the truth clams embodied in the account of Christian ontology presented in Part III. Chapter 10, 'The epistemology of divine revelation', argues that the epistemic grounding of Christian truth claims in the notion of 'divine revelation', rather than constituting an irrational leap of faith, is entirely consistent with a critical realist epistemology that, in insisting on the primacy of ontology over epistemology, itself requires an epistemology of 'general revelation'. Chapter 11, 'Theological science', offers an overview of the Scottish theologian Thomas Torrance's critical realist account of Christian epistemology, which seeks to identify and justify Christian theology as a legitimate science alongside the natural and social sciences. Chapter 12, 'Radical Orthodoxy: beyond secular reason', seeks to test the hypothesis that Christian theology's claim to possess epistemic warrant and intellectual justification cannot be simply dismissed out of hand as inherently irrational, by way of a case study of 'Radical Orthodoxy', a contemporary movement within Trinitarian theology that proceeds by rejecting the occidental reason valorised by the Enlightenment.

In Part V, 'Judgemental rationality and the historical Jesus', I continue to explore questions of judgemental rationality, and the epistemic warrant and intellectual justification of Christian theology, by focusing on historical investigation of the person and work of Jesus of Nazareth. Chapter 13, 'Towards a critical realist historiography', rejects positivist and idealistic approaches to historical interpretation in favour of a critical realist historiography grounded in the key principles of ontological realism, epistemic relativism and judgemental rationality. Chapter14, 'The quest for the historical Jesus', offers an account of the ways in which investigation of the historical Jesus in the nineteenth and twentieth centuries became moribund through a range of empiricist and idealist presuppositions. Chapter 15, 'Jesus Christ: a critical realist reading', provides an overview of the account of the history and theological significance of Jesus of Nazareth by the Christian theologian, historian and critical realist Tom Wright, which supports the thesis of a continuity between the Jesus of history and the Christ of Christian faith.

Part I

Critical realism and transcendence

1 Critical realism and dialectical critical realism

Critical realism is primarily associated with the philosophical movement instigated by Roy Bhaskar (Collier 1994; Archer *et al.* 1998; Lopez and Potter 2001; Hartwig 2007). It seeks to map a path beyond the extremes of modern certainty and postmodern scepticism via a triumvirate of core philosophical principles: ontological realism, epistemic relativism and judgemental rationality. Ontological realism asserts that reality exists for the most part independently of human perception, epistemic relativism asserts that our knowledge of reality is limited and contingent, and judgemental rationality asserts that it is nevertheless possible to judge between conflicting truth claims while recognising that all such judgements necessarily remain open to further adjudication. Critical realism has had a significant impact across a range of disciplinary fields, from the natural and human sciences through to the arts and humanities. Like any major philosophical movement it has its antecedents. Albert Einstein's observation that 'the belief in an external world independent of the perceiving subject is the basis of all natural science' reflects the critically realistic assumptions underlying natural science (Torrance 2001, p. 2). Similarly, as Alan Norrie points out, it is difficult to make sense of Karl Marx's philosophy unless we read him as a proto-critical realist (Norrie 2010, p. 67). Indeed, insofar as reflective engagement with a world that transcends our ability to fully comprehend it constitutes a near-universal human experience, critical realism may be read as a highly sophisticated philosophy of universal 'common sense'. The term 'critical realism' was first used in the 1920s by a group of American scholars opposed to forms of idealism, pragmatism and naïve realism (McGrath 2002, p. 203f.; cf. Drake 1920). A broad tradition of Christian critical realism, concerned with the interface between natural science and theology, theological epistemology and biblical hermeneutics, first emerged in the late 1950s. Until recently this theological tradition developed almost entirely independently of the work of Bhaskar and his colleagues. Bhaskar's own philosophical development has passed through three phases: (1) *critical realism* was primarily concerned with the ontology and epistemology of the natural and social sciences; (2) *dialectical critical realism* built on critical realism to address issues of human emancipation via a critical conversation with the Western dialectical tradition; and (3) the *philosophy of meta-reality*, the result of Bhaskar's so-called 'spiritual turn', addressed issues of the ultimate nature of reality and the meaning and purpose of life. Here we will examine the contours of critical realism and dialectical critical realism, reserving consideration of the philosophy of meta-reality for the following chapter.

Critical realism

Between 1975 and 1991 Bhaskar published a series of books that established his philosophy of critical realism: *A Realist Theory of Science*; *The Possibility of Naturalism*; *Scientific Realism and Human Emancipation*; *Reclaiming Reality*; and *Philosophy and the Idea of Freedom* (Bhaskar 1997 [first published 1975], 1998 [first published 1979], 1986, 1989, 1991). In highlighting the limitations of modern idealism and empiricism, and of postmodern pragmatism and anti-realism, he sought to provide the natural and human sciences with a sound theoretical and practical basis, and thereby enhance the project of human flourishing and liberation. In insisting that critical realism functions as an under-labourer to humankind's intellectual and emancipatory endeavours, he shares, albeit with a significantly different set of presuppositions, Wittgenstein's philosophical aim of showing 'the fly the way out of the fly-bottle' (Wittgenstein 1968, §1: 309). The development of Bhaskar's original vision has become a communal affair, and the following exposition, which makes no claim to be definitive, will seek to map Bhaskar's work within the broad contours of this shared project.

Since the Enlightenment, Western philosophy has tended to misconstrue the relationship between ontology and epistemology. Modernity tends to restrict reality to our *knowledge of* reality, founded on either idealised concepts or empirical sense data: we know an object because we have a clear and distinct idea of it, or because we can experience it directly through our five senses. Nothing can be deemed 'real' unless it conforms to one or other of these two criteria. Postmodernity, as the flip-side of this modernist coin, tends to deny that our language and sense experiences possess any substantial epistemic purchase on ontological reality: in its soft form it affirms a thoroughgoing scepticism about the possibility of knowledge of external reality, while in its hard form it asserts a systematic anti-realism that denies the existence of any reality beyond the language employed by individuals' sceptical and fractured consciousnesses. Despite their differences, modernity and postmodernity both commit the epistemic fallacy of reducing ontology to epistemology and restricting reality to the extent and limits – whether greater or lesser – of human knowledge: modernity by conflating and colonizing ontology within epistemology; postmodernity by cutting ontology adrift from epistemology and then sinking it. Both positions seek to fulfil the Cartesian-driven quest for epistemic closure, which holds that knowledge of reality is either absolutely certain (modernity) or absolutely uncertain (postmodernity). Each affirms the sovereignty of epistemology and promises emancipation from the supposed hegemony of ontology, so that modernists are empowered to assert control over reality and postmodernists are freed to turn their backs on it.

However, the exercise of such freedom generates tyrannous tendencies that undermine human flourishing. The misplaced epistemic confidence of modernity breeds totalitarian regimes such as those overseen by Hitler and Stalin, whilst the similarly misplaced epistemic scepticism of postmodernity refuses the intellectual resources necessary to challenge an unjust status quo, and thereby buttresses by default the oppressive forces of consumer capitalism. According to Bhaskar, ontology and epistemology are neither separate nor inseparable, but rather discrete-yet-related: contra modernity, because we participate in a reality that is greater than our capacity to fully know it, there can be no epistemic certainty; contra postmodernity, because we nevertheless know a great deal about reality – enough, at least, to perform open heart surgery, recognize genocide as evil and value the music of Mozart – there can be no thoroughgoing epistemic scepticism. This being the case, ontology can neither be colonised by epistemology nor split from it: 'human intellectual enterprises are necessarily fallible, but not for that matter, necessarily mistaken' (Gunton 1983, p. 145). Once we accept the contingency of knowledge and reject the possibility of any premature epistemic closure, we can hold fast to both ontological realism and epistemic relativism without contradiction, and in doing so open up the possibility of a judgemental rationality that promises to provide more powerful, comprehensive and truthful ways of engaging with reality without any descent into positivist, idealist, pragmatic or relativistic forms of totalitarianism.

(1) Ontological Realism. According to ontological realism, objects exist and events occur in reality whether we are aware of them or not. A primary condition for human knowledge is the ontologically grounded distinction between the intransitive realm of real objects and events and the transitive realm of our contingent knowledge of them. If our (epistemic) knowledge of dinosaurs is accurate, then dinosaurs must have existed (ontologically) prior to our establishing knowledge of them. To deny this is to slip into the epistemic fallacy of reducing reality to our knowledge of reality. It is certainly true that we construct accounts of dinosaurs, but palpably untrue that in doing so we construct the dinosaurs themselves. The move to affirm ontological realism and reject the epistemic fallacy is justified because it possesses significantly greater explanatory power than any alternative. The fact that natural scientists are able to construct accounts of the natural order-of-things that enable us to develop technologies that empower us to walk on the moon demands rational explanation. The anti-realist claim that such accounts are mere language games that have no substantial purchase on external reality is significantly less powerful than the realist claim that they constitute relatively accurate, though necessarily incomplete and partially fallible, descriptive explanations that enjoy a measure of transitive epistemic purchase on the intransitive ontological order-of-things.

Once the distinction between ontology and epistemology is established, it becomes possible to develop a rich account of the contours of reality. The fact that we are able to provide multiple explanatory accounts of the same object suggests that reality itself is pluriform and stratified. The physicist and chemist, psychologist and sociologist, historian and geographer, artist and poet, philosopher and theologian are able to provide complementary descriptions of a given object or event. This realistic notion of ontological stratification directly challenges the Kantian claim that, rather than being intrinsic to reality itself, our experience of a stratified world is dependent on epistemic categories, embedded in the cognitive apparatus of the mind, through which we structure and organise knowledge. It is not merely that we *perceive* individual persons as biological, social and moral beings; rather they are, in themselves, *intrinsically* and *simultaneously* biological, social and moral. On this reading, genocide is evil, not merely because a majority perceive it to be evil (epistemology), but because it is an intrinsic aberration of the moral order-of-things (ontology). In the same way, the ontological reality of a beautiful sunset imposes itself on human observers, rather than human observers forcing the category 'beauty' onto the event in an expression of subjective taste: beauty is not wholly in the eye of the beholder. To suggest otherwise is to invoke a fact–value divide generated not by reality itself (ontology), but by the susceptibility of some strata of reality to more 'objective' verification procedures than others (epistemology). On this view the physicality of an object is deemed more 'real' than its aesthetic beauty purely on the basis that the measurement of its physicality is more secure than any assessment of its intrinsic beauty – this despite the fact the intrinsic beauty of a newborn child cradled in his mother's arms is palpably more real, to any but the most crassly insensitive observer, than his atoms, molecules and DNA. Reality, that is to say, is ontologically value-laden: the fact that we cannot weigh and measure goodness and beauty with scales and stopwatches does not make them any less real. Though each emergent stratum is dependent on the preceding ones (there can be no morality without psychology, and no psychology without biology), they are not reducible to them (morality cannot be reduced to psychology, and psychology cannot be reduced to biology). If this were not the case, then the physicist (or, perhaps, the mathematician) would be in a position to pronounce definitively upon every aspect of reality, knowing more about society than the sociologist, more about music than the musician, more about poetry than the poet, more about theology than the theologian. Such a misplaced view is broadly reflected in the logical positivist's distinction between objective, verifiable scientific knowledge, and subjective, unverifiable aesthetic, moral and spiritual taste.

Reality is open and dynamic. The fact that we are justified in judging that our epistemically contingent expressions of causal laws possess universal status does not imply that reality is a closed system. Thus, for example, identifying the mechanisms that cause matches to light when struck in particular circumstances does not negate either counterfactual or transfactual statements. Counterfactual statements envisage alternative courses of events in order to better understand causal mechanisms: for example, that a match may not light when struck because the matchbox is damp and consequently the necessary causal conditions (the 'particular circumstances') are not in place. Transfactual statements draw attention to the fact that causal mechanisms do not always produce expected results even when the necessary conditions are in place: to say that matches *tend* to light when struck is to accept the reality that matches do not *always* do so. Counterfactual and transfactual statements are significant because they show that causal mechanisms operate in open systems that are capable of generating a range of different events without undermining any particular universal law. Since causal mechanisms do not invariably generate particular events, causal laws function to identify potential states of affairs that may or may not be realised. Crucially, the failure of a causal mechanism to realise its potential on a particular occasion does not, in itself, undermine the accuracy of our account of that mechanism. The fact that the operations of a range of interactive causal mechanisms are compatible with a high level of contingency in the world supports the conclusion that reality is open and dynamic, generating higher-order strata of reality whose emergence was not predetermined or necessary, and whose future development is contingent and unpredictable. Thus, for example, discussion of anthropic phenomena among natural scientists and theologians proceeds from the counterfactual recognition that the fundamental causal mechanisms governing the physical universe do not necessitate the emergence of sentient life, and that consequently the apparent 'fine-tuning' of our actual life-bearing universe demands rational explanation (McGrath. 2009a, pp. 86ff., 111ff., 120ff.).

(2) Epistemic relativism. In the wake of the Enlightenment, epistemology became dominated by the binary opposites of the modern pursuit of epistemic certainty and the postmodern insistence on epistemic scepticism. As we noted above, such binary thinking disguises a singular drive to epistemic closure, since both parties claim secure understanding (whether positive or negative) of the extent and limitations of knowledge. Critical realism, in affirming epistemic relativism, seeks a route beyond this impasse.

Epistemic relativism asserts the priority of ontology over epistemology: reality precedes knowledge of reality, so that we cannot know something unless there is first something to know. Epistemic certainty and scepticism constitute forms of the epistemic fallacy, which effectively reverses this order of priority by reducing reality to our ability (or inability) to know it, so that unless we know something with relative certainty it cannot be deemed to exist. In its positive form the epistemic fallacy manifests itself whenever reality is forced into the strait-jacket of fixed epistemic criteria. Thus the logical positivist's equation of knowledge with verifiable sense data, and attendant reduction of statements of aesthetic, moral and spiritual value to expressions of subjective preference devoid of any purchase on reality, generates a fact–value divide in which reality is limited to facts and stripped of all value. In this situation the principle of verification functions as a means of distinguishing between objective fact and subjective value, and thereby of providing the knower with epistemic closure. In its negative form postmodernism embraces the epistemic fallacy in two interrelated ways: (1) epistemic scepticism, in which substantial *knowledge* of any reality external to the knower is denied on epistemic grounds; and (2) anti-realism, in which the *existence* of any reality external to the knower is denied. In both cases knowledge is limited to a set of relativistic linguistic constructions that have no purchase on the external world: the soft epistemic assumption that we cannot access any reality beyond our linguistic constructs generates the hard ontological assumption that there is not actually any external reality to access. Claims to the contrary are seen as evidence of the interplay of power structures that have a strong totalitarian tendency, and draw the conclusion that in order to avoid the hegemonic imposition of power-disguised-as-knowledge it is necessary to deconstruct all truth claims. This has the supposed effect of emancipating individuals from economies of power and enabling them to live brokerless lives ordered by a pragmatism rooted in personal desire, taste and preference. However, such emancipation is illusory, since reacting against epistemic certainty by embracing epistemic scepticism does nothing to remove the hegemony of the epistemic fallacy itself. Postmodern claims that we have no access to external reality, or that external reality does not actually exist, are themselves *truth* claims that, insofar as they are deemed to be self-evident and non-negotiable, take on precisely the same hegemonic and totalitarian features they were designed to avoid.

To commit the epistemic fallacy is to allow epistemic criteria to set the boundaries of reality, despite the fact that the priority of ontology means that it is reality that should progressively shape epistemic criteria. In affirming that knowledge is constituted by the relationship between the knower and the object of knowledge, critical realism rejects the extremes of epistemic certainty and epistemic scepticism, together with the epistemic fallacy that nurtures and supports them. Since we are clearly fallible creatures lacking the omniscient capacity to know everything infinitely, our knowledge is necessary limited and epistemically relative. However, this rejection of epistemic certainty does not necessitate a move to a thoroughgoing epistemic scepticism. Few, if any, of our beliefs about the world are entirely false: though the child who responds to the question 'What is 2+2?' with the answer 'Blue' makes a fundamental categorical error, she has nevertheless recognised that a question has been asked; similarly, the psychiatric patient who believes that his doctor is Napoleon Bonaparte is still living in and responding – however inadequately – to the same world inhabited by his doctor. Some beliefs may be profoundly mistaken: the world is not flat, and alchemy and astrology are predicated on erroneous understandings of the natural order-of-things. All beliefs fall short of absolute truth: Einsteinian physics constitutes an advance on Newtonian physics, but still falls short of a complete understanding of the natural world.

Since knowledge lies between the extremes of absolute certainty and thoroughgoing scepticism, we have a rational warrant to embrace and act on our beliefs, so long as we have good reason to hold them to be true, and until such time as we encounter good reasons for rejecting them. Our knowledge thus takes the medieval scholastic form of 'faith seeking understanding', rather than the modern form of 'understanding seeking faith'. Theists and atheists subscribe to incompatible fundamental beliefs: both groups can claim to possess reasonable epistemic warrant for their beliefs (the naïve assumption that theistic belief is intrinsically irrational can only be made out of ignorance of the fact that many highly intelligent people subscribe to theistic beliefs and have set out their reasons for doing so in the public arena) – yet both may be mistaken. In such circumstances neither group can, in practice, avoid acting on their beliefs as if they are true; and both are intellectually justified in doing so, provided they are willing to subject their beliefs to rational scrutiny and if necessary revise or abandon them. The affirmation of epistemic relativism acknowledges the limits of our knowledge, but does not deny either the actuality of genuine knowledge or the possibility of establishing better knowledge in the future.

(3) Judgemental rationality. Accounts of reality are not all of equal value: it is possible to judge some to be more truthful than others. If this were not the case we would be faced with a relativistic tyranny, prematurely imposing epistemic closure on the pursuit of truth, undermining the possibility of emancipatory praxis, and sounding the death-knell of intellectual debate.

There are no secure foundations upon which to construct knowledge, and no protected procedures with which to adjudicate between conflicting truth claims. The empiricist appeal to verifiable sense data, the idealist appeal to logical coherence, and the romantic appeal to intuitive sensibility – together with its close relation the postmodern appeal to personal preference and desire – have all been found wanting. To stipulate in advance a secure foundation of knowledge, or fixed criteria for exercising judgemental rationality, is to slip into the epistemic fallacy. The priority of ontology means that we must adapt our epistemic tools in response to the objective demands of reality, rather than adjust reality to bring it into conformity with our epistemic tools.

Epistemology is contextual rather than foundational. Contra Descartes, we do utilise a hermeneutic of scepticism to strip away current knowledge and expose secure foundations on which to build a new edifice; rather, we employ a critical hermeneutic of faith, starting out from our already established beliefs and knowledge relationships, and seeking to progressively refine, revise and test them. Because we have no access to a god's-eye 'view from nowhere', there is no vantage point 'from which we can view and evaluate our beliefs' other than 'our already existing beliefs' (Nagel 1986; McGrath 2002, p. 35). The brittle tools of thoroughgoing scepticism and naïve incredulity are ill equipped to engage truthfully with reality, and consequently tend to generate alienating ideologies. Between these extremes stands the wisdom of discernment, summed up in the Jesuit theologian Bernard Lonergan's transcendental precepts: be attentive, be intelligent, be reasonable and be responsible (Lonergan 1973, p. 55). He argues that 'the basic form of alienation is man's disregard of the transcendental precepts', and that 'the basic form of ideology is a doctrine that justifies such alienation' (p. 55). Critical realist epistemology follows the path of inference in pursuit of the best possible explanation (Harman 1965; Lipton 2004). We make sense of the world by constructing theoretical models designed to provide powerful and comprehensive explanations of the objects and events we seek to understand. This constructive process proceeds from abduction through retroduction to iteration. Abduction entails the intuitive generation of novel insights in response to encounters with new facts or the revisiting of old problems (McGrath 2009a, p. 44). Abductive insight paves the way for the production of retroductive explanatory hypotheses designed to identify the underlying causal mechanisms that enable the object under investigation to be and to behave in the way that it does. Once a hypothesis is established, it is then subject to iterative testing and revision in the light of further insights gleaned from the investigator's ongoing interaction with the object (McGrath 2009a, p. 32).

If we wish to reject an explanatory model we must replace it with an alternative model possessing greater explanatory power. Because knowledge is contextual, there can be no fixed universal criteria for adjudicating between competing models. This means that any judgemental criteria we might employ must itself be open to debate as part of the ongoing epistemic process. The choice between competing hypotheses is initially tacit, intuitive and instinctive: we know more than we can say, and embrace the power of a hypothesis – e.g. that genocide is evil or that the music of Mozart is sublime – long before we can give a full rational explanation for doing so. We may recognise that Shakespeare provides us with deeper insights into the human condition than, say – and with all due respect – the novels of Jeffrey Archer, even though we might struggle to provide an adequate explanation as to why this is the case. Explanations, that is to say, tend to lend rational support to our prior beliefs rather than establish or ground them. The retroductive process means that we engage in judgemental rationality *a posteriori*, in the light of and in response to our already established and ongoing knowledge relationships. In doing so, we strive to allow the intrinsic nature of the objects under investigation to mould and shape our understanding, and guard against the danger of imposing our preferences and prejudices upon them. Judgemental rationality cannot be reduced to the application of a single assessment criterion. Rather, as in a court of law, we seek to attend to a raft of evidence offered by a range of different witnesses, and to arrive at the best possible available explanation in the light of our weighing of both the evidence and the integrity of the witnesses.

Dialectical critical realism

Dialectic: The Pulse of Freedom, together with *Plato Etc.: The Problems of Philosophy and Their Resolution*, first published in 1993 and 1994 respectively, marked Bhaskar's turn from critical realism to dialectical critical realism (Bhaskar 1993, 1994). His determination to enrich critical realism by introducing a dialectical dimension was driven: (1) by an ethical concern to establish the ontological grounds for the emergence of a genuinely eudemonic and emancipated society; and (2) by an intellectual concern to offer a fundamental metacritique of mainstream Western philosophy from the pre-Socratic philosophers through to the present day. With regard to the former, the deeply rooted influence of Karl Marx on critical realism came to the fore: the central aim of philosophy is to change society for the better rather than merely understand it. In identifying the structural and agential mechanisms that work together to configure society in different ways Bhaskar established the ground for a dual shift: in critical realism from description to explanation, and in dialectical critical realism from explanation to emancipatory practice. To achieve the latter, dialectical critical realism needed to move beyond a static-spatial account of reality and engage more fully with the dynamic-temporal process of change. Understood in spatio-temporal terms, being emerges as a process of becoming. Bhaskar drew on the dialectical philosophies of Hegel and Marx as his primary conversation partners: where for Hegel dialectic was essentially a process of idealised abstract argument, Marx grounded his dialectics firmly in the material reality of the historical process itself. Bhaskar's reworking of Hegelian dialectic sought to undermine its totalitarian thrust by affirming an essential openness to the future, and identifying 'absence' rather than 'presence' as the driving force of dialectical philosophy. This move, Bhaskar argued, directly contradicts the mainstream of Western philosophy, which has been dominated by a metaphysic of presence. Hegelian dialectic invoked a three-stage process of thesis > antithesis > synthesis, in which the apparent differences between a thesis and its antithesis are synthesised into a seamless whole. Thus art and its antithesis religion are synthesised into philosophy; generic religion and its antithesis particular religions are synthesised into the absolute religion (Christianity); the logical idea-in-itself and its antithesis nature (the idea-outside-itself) are synthesised into Absolute Spirit (Geist). This served to establish the essential unity of all things, identify Absolute Spirit as the underlying essence and immediate presence of being qua being, affirm Hegel's system as the culmination of philosophy, and assert the status quo of Western bourgeois culture as the culmination of the process through which Absolute Spirit achieves self-consciousness and self-actualisation. In sharp contrast, Bhaskar invokes a four-stage process of non-identity > absence > totality > transformative praxis, which serves to identify the actuality of fractured relationships and the absence of a genuinely harmonious and eudemonic order-of-being, and thereby affirm both the possibility and necessity of transformative emancipatory praxis.

(1) Non-identity. The starting point of dialectical critical realism, described by Bhaskar as the 'First Moment' (1M), is the recognition of the non-identity of objects in the real world. There is, ingrained in the ontological structures of reality, an irreducible difference between things. I am not you, and you are not me: we are distinct from one another, and our unique existential identities cannot be eclipsed by subsuming them under the essential generic category 'human being'. In taking non-identity as his starting point Bhaskar affirms the priority of existence over essence, in direct contradiction of Hegel's affirmation of the priority of essence over existence. Bhaskar's decision to begin by affirming non-identity over identity has obvious affinities with nominalism and existentialism, as well as with post-structuralism's concern to deconstruct Hegel's onto-theological metaphysic of presence in order to create space for the alterity of the 'other'. However, where post-structuralism took its cue from epistemology in asserting the impossibility of establishing a logo-centric ontology, Bhaskar affirms non-identity on the ontological ground that differentiated transcendental realism is a necessary condition for knowledge. It is important to recognise that non-identity is merely the starting point of Bhaskar's dialectic. If the Hegelian

affirmation of identity leads inexorably towards a totalitarian eclipse of personal autonomy, then the post-structural affirmation of non-identity opens the door to an anarchic eclipse of authentic relationship. Both extremes constitute forms of alienation that Bhaskar's dialectic seeks to overcome.

(2) Absence. The second stage of dialectical critical realism, described by Bhaskar as the 'Second Edge' (2E), is oriented towards, and continually cuts into, the future: reality is in the process of change, being is in the process of becoming. In a further departure from Hegel, Bhaskar affirms a dispositional ontology in which the temporal movement from that-which-is to that-which-will-be can only be properly understood in terms of absence or negativity. Bhaskar does not use 'absence' to refer to either metaphysical nothingness (the absence of Being qua Being) or logical operations (the negation of a proposition). Rather, he argues that every object and event must be understood, not merely in terms of that which is present, but more fundamentally in terms of that which is absent. Thus the absence of a person at a family gathering – whether because they are elsewhere, deceased, not yet born or never existed (say, the long-hoped-for spouse of an elderly bachelor) – may be just as significant as the actual presence of other family members. Bhaskar argues that real determinate absence is a causally significant and ontologically necessary feature of reality: there can be no change without negation, whether this be the absenting of that which was previously present, or the coming into presence of that which was previously absent.

If ontological polyvalence asserts the significance of both absence and presence, then ontological monovalence restricts meaning and value to that which is positively present. Bhaskar argues that the dominance of ontological monovalence in Western philosophy can be traced back to the pre-Socratic philosophers. Parmenides' assertion of the permanency and indivisibility of a singular reality (the 'One') left no room for change and diversity (the 'Many'), and in doing so effectively negated the ontological significance of absence. In similar vein, Plato interpreted non-being as merely 'different' to being rather than as its positive negation, while Augustine interpreted evil as merely the privation of goodness. Ontological monovalence functions to obscure both the dynamic nature of reality and the contingent nature of our knowledge of reality, and as a consequence possesses a totalitarian tendency. The ethical implications of this state of affairs are central to Bhaskar's concerns. It is possible to valorise the status quo by affirming that which is present and rejecting or ignoring that which is absent. To do so is to embrace what Bhaskar refers to as the 'TINA Syndrome' ('There Is No Alternative'). However it is also possible to undermine the status quo by negating that which is present and/or bringing into presence that which is absent. In an unjust society ontological monovalence encourages stagnation in the status quo, while ontological polyvalence foregrounds the absence of justice. Absence is ontologically real, causally significant, and a necessary feature of temporal change. The negation and absenting of what is, and its replacement by what is to come, constitute the necessary conditions for emancipatory change and transformative praxis.

(3) Totality. The third stage of dialectical critical realism, described by Bhaskar as the 'Third Level' (3L), is that of totality. Even if objects are irreducibly distinctive, non-identical and marked by their differences from other absent objects, we cannot avoid the question of their relationship with one another. Bhaskar affirms the interrelatedness of the totality of all objects and events, and in doing so seeks to transcend the extremes of nominalism and essentialism: though reality cannot be reduced to an essential undifferentiated whole, it is nevertheless greater than the nominal sum of its constituent atomistic parts.

Contra essentialism, in invoking the concept of 'totality' Bhaskar does not embark on a retrograde return to Hegelian idealism, and continues to reject Hegel's claim that individual objects can be subsumed within a greater totalitarian whole. Hegel arrives at his position by presenting reality as an evolving rational complex that culminates once 'reason has ultimately worked its way through the world, making explicit the sense of a rational whole that was implicit in it' (Norrie 2010, p. 16). The Hegelianism system claims to achieve epistemic and ontological closure: 'the work of dialectical philosophy is done, for the totality is present and complete' (p. 16). For Bhaskar, however, the appeal to totality is not a means of overcoming non-identity and absence; rather it serves to highlight the persistence of non-identity and absence in the totality of all things. Because objects are not yet related together as they should be, and because the conditions necessary for a eudemonic society remain absent, any epistemic or ontological closure must necessarily be premature.

Contra nominalism, Bhaskar rejects any reduction of reality to a series of atomistic objects randomly interacting with one another. The empiricist notion that we perceive atomistic packets of sense data, from which we construct nominal wholes on the basis of observation of the regular conjunction of objects and events, generates an understanding of reality in which the interrelationship between objects is purely arbitrary, and in which the ultimate order-of-things is reduced to the sum total of its constituent parts. Such a view has ethical implications, since it supports a reductive account of society as a nominal entity ultimately reducible to the sum total of the activities of individual agents. This in turn supports a liberal understanding of society as a collection of isolated individuals, each pursuing 'some variant of that most invisible, because it is the most pervasive, of all modern goods, unconstrained freedom' (Taylor 1992, p. 489). John Milbank suggests that the freedom of individuals to pursue their own version of the good life results in an inevitable clash of interests, and that as a result liberal society is predicated on an 'ontology of violence' (Milbank 1990a, pp. 278ff.).

Bhaskar's attempt to transcend essentialism and nominalism is predicated on a necessary relationality between objects. This is a direct result of the ontological structure of reality itself, which is constituted by the complex constellational connections between objects and events, and can be accounted for by identifying and explaining the network of causal relationships and mechanisms that comprise and configure reality. The different strata of reality are simultaneously united by their interdependency and distinguished by their irreducibility: our moral being is relationally dependent on our biological being, but cannot be reduced to it. There are no ontological parts without an ontological whole, and there is no ontological whole without its ontological parts. Bhaskar understands the relationship between parts and whole in terms of the temporal edge of becoming and absenting: as an object changes in time, so it absents what it was and makes present what it has become. Reality is thus an emergent process in which being is constantly in the process of becoming. This process of becoming is neither entirely random (contra atomistic nominalism) nor entirely determined (contra totalitarian idealism); rather reality is emergent and progressive, driven towards an ordered-yet-open future by a dialectical dynamic of absenting and presenting.

(4) Transformative praxis. The final stage of dialectical critical realism, described by Bhaskar as the 'Fourth Dimension' (4D), is concerned with the role of agents in transformative praxis. Reality is marked not by a dislocated atomism, nor by an undifferentiated totality, but by dynamic relationality. We are the people we are, not because we are autonomous existential individuals, nor because we are largely inconsequential emanations from some greater whole, but insofar as we relate to ourselves, to others-in-community, to the natural world and to the presence/absence (existence/non-existence) of a transcendent order or divine reality. The causal interaction between the different parts of reality can transform the nature of the whole. Causal agents are simultaneously limited by social structures and yet capable of transforming them. The fact that the ultimate order-of-things is irreducible to either a chaotic nominalism or a deterministic essentialism makes possible ontologically grounded attempts to transform the world through human agency. The moral order cannot be reduced to the capitalist pursuit of individual self-interest (the Many), nor the Stalinist structural imposition of a collective social ideal (the One); rather, it requires individuals-in-community to work together to transform society in a manner in which the flourishing of each individual (the Many) and of the common good (the One) are indistinguishable – from each according to their ability, to each according to their need. Currently, reality is marked by the absence of a eudaimonic society: the reality of brokered master/slave and client/patron relationships undermines a social order in which the freedom and flourishing of each individual is dependent on the freedom and flourishing of society as a whole. At the same time, however, the eudaimonic society already exists as a current aspiration and future possibility. Ethical agency, operating in an ordered-yet-open world, opens up the possibility of intentional actions, both individual and communal, designed to absent-the-absence of the eudaimonic society via both personal transformation and structural change. Since social structures cannot be transformed apart from human agency, the possibility of transformative praxis is wedded to the willingness and ability of agents to bring about change. The failure of Stalinism was to assume that human nature could be changed simply by altering social structures. The recognition of the need to transform fractured human nature in order to renovate splintered social structures was one of the factors that led to the 'spiritual turn' in Bhaskar's thought.

2 The spiritual turn

Transcendence and meta-Reality

Bhaskar's developing interest in spirituality, God, and to a lesser extent religion, was revealed in four books that together constituted the first fruits of what became known as the 'spiritual turn' in critical realism: *From East to West: Odyssey of a Soul*; *From Science to Emancipation: Alienation and the Actuality of Enlightenment*; *Reflections on Meta-Reality: Transcendence, Emancipation and Everyday Life*; and *Meta-Reality: Creativity, Love and Freedom* (*The Philosophy of Meta-Reality, Volume 1*) (Bhaskar 2000, 2002a, 2002b, 2002c). Initially, Bhaskar referred to this new philosophical outlook as transcendental dialectical critical realism, and argued that 'dialectical critical realism must develop into a philosophy of (universal) self-realisation' (Bhaskar 2000, p. x). Partly, perhaps, due to the negative reaction to this 'spiritual turn' among some – though by no means all – critical realists, Bhaskar moved on to offer a revised version of his emergent philosophy, now dubbed the 'philosophy of meta-Reality'. Though the differences between the two stages of the spiritual turn are not insignificant, ultimately the continuities outweigh the dissimilarities. The following sections will offer an overview of the philosophy of transcendental dialectical critical realism presented in *From East to West*, trace its spiritual roots and lineage, explore the transition to the philosophy of meta-Reality, and finally outline the contours of the philosophy of meta-Reality itself.

Transcendental dialectical critical realism

From East to West: Odyssey of a Soul combines a philosophical account of transcendental dialectical critical realism with an apparently semi-autobiographical novella outlining the successive reincarnations of a soul. The latter was read by many as evidence of a descent into esoteric New Age mysticism. Our primary concern here is with Bhaskar's philosophy; with regard to the novella we must content ourselves for the present with drawing attention to the deeply rooted tradition of utilising the aesthetic categories of poetry and myth as a means of accounting for spiritual realities that rational philosophical categories struggle to grasp (Balthasar 1989, pp. 43ff.). According to critical realism we discover the alethic truths of reality through intuitive moments of abductive insight, from which we generate retroductive explanatory hypotheses that we then iteratively test and revise in the light of alternative explanatory models (Bhaskar 1986, p. 68). Previously, Bhaskar had argued that critical realism offers a more powerful retroductive explanation of our engagement with reality than those provided by idealism, positivism and pragmatism, and that dialectical critical realism constitutes an iterative development and expansion of critical realism. He now argues that 'the dialectic of critical realism at once prepares the ground for and necessitates its development into *transcendental* dialectical critical realism or philosophy of (universal) Self-realisation (and ultimately of God-realisation)' (Bhaskar 2000, p. 21; emphasis in original).

Critical realism identified a stratified universe in which higher strata were dependent upon, but not reducible to, lower strata. Where modern naturalism affirms physics as the basic stratum of reality, classical philosophy and theology had sought to identify a deeper metaphysical and/or theological substratum responsible for generating and sustaining the natural order, and providing it with its intrinsic order and rationality. Dialectical critical realism identified a dynamic forward momentum in reality: being is in process of becoming. This makes possible the emergence of higher strata of reality, including sentient life capable of the conscious discernment of moral, aesthetic and spiritual dimensions of reality. The recognition that these interrelated strata, when taken as a whole, constitute a non-dualistic totality resurrected the classical philosophical and theological questions of the existence of being qua being in the face of nothingness, the ultimate source, ground and nature of reality, and the ultimate meaning and purpose of life. The Christian theologian Thomas Torrance points out that explanatory accounts of the hierarchical strata of reality combine to 'constitute a vast semantic focus of meaning' (Torrance 1980a, p. 38):

> While theoretically the number of these ascending levels might be conceived to be infinite, in the actual universe which is finite, the number is necessarily limited. So far as natural science is concerned, this must mean that the intelligibility or meaning of the whole semantic fabric demands and rests upon some ultimate self-sufficient ground of intelligibility as its sufficient reason, otherwise the whole of natural science would be finally pointless for its alleged intelligibility would run out into nothing.

The modern occlusion of such questions, most evident in logical positivism's rejection of metaphysics and theology, led to the assumption that naturalism offers, by default, a self-evidently truthful explanation of the ultimate ground of reality. Torrance suggests that it is far from self-evident that the natural order is self-generating, self-rationalising and self-sustaining, and proposes instead that 'natural science through its remarkable intelligibility cries aloud for a proper doctrine of creation' (p. 38). *From East to West* may be read both as Bhaskar's recognition that such 'ultimate' questions can no longer be legitimately occluded from rational debate, and as his initial attempt to provide a more powerful retroductive explanation of the totality of things than that proffered by naturalism and classical theism.

Bhaskar's account of the deep ontological substratum of reality proceeds via the affirmation of categorical and dispositional realism. *Categorical realism* asserts that categories are not merely human constructs, but rather substantial ontological realities inherent in the fabric of the universe. Bhaskar identifies 'God' as the ultimate 'categorical structure of the world', the 'causally and taxonomically irreducible' ground and alethic truth of being, 'on which the rest of being is unilaterally existentially dependent' (Bhaskar 2000, p. 40). As unbounded absolute, God creates and bounds, contains and unifies, and categorically defines the totality of reality (p. 42). *Dispositional realism* asserts the ontic, epistemic and logical priority of possibility over actuality (actual objects and events were once potential), being over action (forces must exist in order to cause), and self over agency (persons must exist in order to act). This means that the ultimate structure of the universe, and hence – given the fact that categorical *values* possess ontological status – the ultimate meaning and purpose of life, exists as a potential awaiting future actualisation and realisation. The openness of reality to the future, within which the human capacity for self-transcendence is ingrained, constitutes a potentiality and possibility that Bhaskar also identifies as 'God'. As defined by Bhaskar, 'God' is thus both the ultimate categorical structure of being and the ultimate potentiality of being-as-becoming. As the ultimate categorical structure of the world, God is ingrained in reality, but is neither saturated by it nor exhaustive of it. As the ultimate future disposition of the world, God as being-in-becoming transcends the world as that-which-the-universe-will-ultimately-become. In classical theological terms, Bhaskar's account of God appears to constitute a form of panentheism that needs to be carefully distinguished from forms of theism, deism and pantheism. Whereas theism and deism insist on a fundamental distinction between God and the universe, and pantheism views God as synonymous with the universe, panentheism maintains that God is simultaneously immanently ingrained within the structure of the universe and transcendent of it.

According to Bhaskar, God is ontologically immanent as the inner-God-within-man, and ontologically transcendent as the outer-God-without-man. God is ontologically immanent within man but not saturated by man, and ontologically transcendent of man and but not exhaustive of man. In his deep categorical structure, 'man is essentially God, already essentially free, even now already enlightened' (p. 41). However, this reality is occluded and overlain by illusion, structural faults and wrong actions which dislocate man from his divine essence. Man's intrinsic nature manifests itself as an inner urge to realise God through a de-alienating dialectic of self-transcendence and self-realisation. Since God is essentially love, such dialectic requires the realisation of that unconditional love which constitutes the ultimate unifying, totalising and liberating power of the universe. To achieve these goals one must absent the presence of illusion, wrong action and duality, and make present the absence of transcendent moments of enlightened cosmic consciousness. If our consciousness of God is necessarily bounded, relative and subjective, it is nevertheless experience of God as unbounded beauty, love and power. At the transcendent moment of cosmic consciousness the distinction between the bounded relative subject and the unbound absolute object collapses into a 'subject-object identity, in which alterity, otherness – and with it the possibility of both referential detachment and emotional attachment – give way' (p. 46).

If dialectical critical realism opens up the possibility of self-transcendence through transformative action, then transcendental dialectical critical realism identifies the ultimate goal of human self-transcendence as the experience of identity with God, the ultimate categorical ground of all being and the ultimate dispositional summation of being-as-becoming. The essential thesis of *From East to West* is that because 'man is essentially God', self-realisation *is* God-realisation (p. ix).

The roots of the spiritual turn

Logical positivism's insistence that metaphysics and theology are illegitimate fields of intellectual enquiry is a classic example of the modern occlusion of the question of transcendence. Prior to Bhaskar's spiritual turn, critical realism's recovery of metaphysics focused chiefly on the natural and social realms, and avoided addressing metaphysical and theological questions regarding the categorical ground and dispositional thrust of the totality of reality. It is to Bhaskar's credit that he recognised the intellectual mandate, as well as the moral and spiritual imperative, to challenge the occlusion of such questions. It is clear that the primary roots of the emergent philosophy of meta-Reality lie within critical realism and dialectical critical realism. However, the roots of Bhaskar's distinctive spiritual vision that was to lend substance to his philosophical reflections are less transparent.

Bhaskar acknowledges a debt to 'both radical libertarian Western thought and mystical Eastern thought', and claims to draw on 'some of the best insights of the New Age and the New Left movements' (p. ix). The odyssey of the soul described in the 'novella' references a range of 'great world belief systems – Ancient Greek, Judaic, Essene and Christian, (Vedic) Hindu, Buddhist, Confucian, Taoist, Zen, (Sufi) Islam and modern materialist thought' (p. 1). Bhaskar's encounters with these belief systems are reported in an eclectic manner that reveals an apparent lack of engagement with the fruits of the academic study of religions. For example, in Life One the soul encounters Moses and teaches 'the children of the chosen people … the esoteric teachings of the perennial wisdom' (p. 13). Yet there is no historical evidence, either in the Mosaic traditions preserved in Hebrew Scripture or elsewhere in antiquity, to suggest any actual historical link between Moses and esoteric teaching or perennial wisdom; such a view only emerges in Gnostic circles in the opening centuries of the Christian era, for example, in the Hermetic literature discussed below. Indeed, Bhaskar acknowledges that the novella 'makes no claim to (though it might be or contain) historical truth' (p. x). He appears to adopt a syncretistic approach to the world religions, removing concepts from their original contexts and utilising them to enhance his own philosophical system. This being the case, our search for the spiritual roots of the philosophy of meta-Reality needs to look beyond the historical insights into the world's religious traditions gleaned by academic scholarship. It appears, by his own reckoning, that Bhaskar's syncretic and selective assimilation of aspects of various religious beliefs and practices is guided by his personal encounter with a range of counter-cultural spiritual traditions.

Bhaskar acknowledges a debt to the New Age movement which, despite its great diversity and rich variety of expression, tends to embrace a common, and relatively transparent, understanding of spirituality. Thus ordinary life is seen as driven by social structures, 'inculcated by parents, the educational system and other institutions', that make us victims 'of unnatural, deterministic and misguided routines' (Heelas 1996, p. 18). To live authentic spiritual lives we must turn away from social convention and rediscover the latent 'divine spark or seed ... innate in the individual human soul' (Braaten 1995, p. 7). 'The inner realm, and the inner realm alone, is held to serve as the source of authentic vitality, creativity, love, tranquillity, wisdom, power, authority and all those other qualities which are held to comprise the perfect life' (Heelas 1996, p. 19). By drawing on a range of alternative spiritual rituals, practices and psycho-technologies we can gain emancipation from the tyranny of society, as well as from the ego responsible for internalising repressive social structures. 'The Self must be liberated; "de-identification" must be effected; the person must drop "ego-attachments" or "games" ' (p. 20).

Bhaskar also acknowledges the influence of theosophy on his thought (Bhaskar and Hartwig 2010, pp. 6ff.). Founded in 1875 by Helena Blavatsky, the Theosophical Society seeks to merge religion, metaphysics and science into a seamless whole. Theosophy claims access to a perennial wisdom in which: (1) knowledge of God is derived from a combination of intuitive experience and scientific exploration that reveals occult principles immanent within the cosmos; (2) reality is constituted by a radical unity, in which all things are inextricably connected and continually evolving towards consciousness; and (3) human souls form part of the universal soul, and through the law of karma progress through a succession of incarnations towards a higher universal consciousness. Such knowledge is potentially accessible to all human beings, provided they have successfully cultivated the requisite capacity for spiritual insight. Since most of the world's religious traditions, together with many secular ones, possess a kernel of this perennial truth, it is possible to discern arcane principles at work beneath their established doctrines and practices. Though human knowledge of the divine source of all being is often distorted and obscured, it can be rediscovered through a pedagogic process of anamnesis, in which human beings achieve enlightenment by recollecting, recovering and contemplating the latent spark of divinity deep within themselves.

The close relationship between the New Age movement and theosophy is well documented (Heelas 1996, pp. 44ff.). Both traditions have roots in nineteenth-century Romanticism, which rejected empiricism and rationalism as primary sources of knowledge in favour of an idealistic sensibility grounded in moral instinct, aesthetic feeling and spiritual intuition. Romanticism took up and nurtured a long-standing tradition of esoteric spirituality that may be traced back through a variety of routes; for example, from William Blake via Emanuel Swedenborg to the Renaissance mystic Jakob Böhme. One loosely knit countercultural strand within the Renaissance related magic, occultism, alchemy, astrology and mysticism to an archaic and perennial spiritual philosophy held to pre-date the world's religious traditions. This perennial wisdom was seen as the primal source of all spiritual and religious traditions, a fact that those religious traditions which claim exclusive possession of spiritual truth – pre-eminently, at the time of the European Renaissance, the Christian tradition – were accused of seeking to suppress. However, the suggestion that the world's religious traditions emerged from a common perennial spiritual wisdom rooted in mystical experience is not supported by the historical evidence. A more powerful retroductive hypothesis, I suggest, is that belief in a perennial wisdom is a product of countercultural attempts to secure spiritual autonomy by resisting the hegemonic structures of organised religion in general and institutional Christianity in particular. It is possible to trace a historical lineage of such attempts from the Renaissance, through nineteenth-century Romanticism, to the contemporary New Age movement. In its modern guise, the notion of a perennial wisdom accessible to all sits comfortably with a democratic liberal economy: if all human beings are free and equal, then they must have free and equal access to spiritual truth; and if social institutions function to undermine freedom and weaken equality, then religious institutions must be held to account for attempting to restrict access to spiritual truth to a privileged elite. If the notion of perennial wisdom is indeed a product of the emergence of modernity in the West, rather than a product of primal mystical encounters that pre-date all religious traditions, then it is vulnerable to the charge of committing the epistemic fallacy by forcing realistic spiritual truth claims into a liberal democratic polity. Three examples must suffice to indicate how this hypothesis might be substantiated.

1 The popular notion that Hinduism is essentially pluralist and inclusive in its attitude towards other religious traditions might suggest that it constitutes a prime example of a religious tradition that has retained an awareness of its perennial roots and successfully avoided the pitfalls of religious exclusivism. However, the perception of Hinduism as intrinsically open and inclusive is largely a product of neo-Hinduism, a movement rooted in the encounter of Hinduism with European culture in general, and Christianity in particular, following colonial and missionary activity on the Indian subcontinent during the nineteenth century (D'Costa 2000, pp. 53ff.). In classical Hinduism *dharma* referred to right action intended to maintain a proper relationship with the cosmic order, and required the performance of cultic rituals and maintenance of religious customs sanctioned by the priestly class of Brahmans. Neo-Hindu scholars such as Vivekānanda, Mahatma Ghandi and Radhakrishnan sought to revise this classical understanding of *dharma*. They played down the established connection between *dharma* and cultic practice, and replaced it with an understanding of *dharma* as ethical

behaviour in conformity with a set of universally applicable principles. The net result was to disconnect Hinduism from its cultic roots and reconceptualise it as the vehicle of a universally valid spiritual ethic. A primary motivation of neo-Hindu revisionism was to resist Christian claims to exclusive possession of the 'true *dharma*' by asserting 'the pre-eminence of "Hinduism" as the eternal, universal, all-encompassing *dharma* within which all specific religions – including Christianity – are subsumed' (Holdrege 2004, pp. 244ff.). Classical Hindu sources have almost nothing to say about the non-Hindu (*mlecchas*), and the few scattered references that have been passed down to us are primarily concerned to warn of the dangers of ritual pollution should the Hindu come into contact with the non-Hindu – a concern diametrically opposed to the suggestion that Hinduism is essentially open and inclusive (D'Costa 2000, pp. 53ff.). The notion that classical Hinduism, by virtue of its essentially pluralist and inclusive outlook, is rooted in perennial wisdom thus appears historically unsustainable. Rather, such a view is best seen as a direct product of Hinduism's encounter with modernity and an attempt to resist Christian colonisation. Bhaskar's appeal to Hinduism in support of transcendental dialectical critical realism appears to draw on a version of neo-Hinduism filtered through the theosophic tradition – a tradition which he describes as 'basically Hinduism for westernised Indians' (Bhaskar and Hartwig 2010, p. 3).

2 In 1460 the nucleus of the ancient *Corpus Hermeticum*, containing a treatise attributed to an ancient Egyptian sage, Hermes Trismegistus, was discovered in Florence (Hermes Trismegistus 1999). According to ancient tradition Hermes' teaching pre-dated Moses and Plato, and provided the underlying inspiration of all consequent religion and philosophy. Such was the interest in the possibility that a source of perennial wisdom pre-dating Hebraic Scripture and Greek philosophy had been unearthed that the Medici family suspended the ongoing translation of the works of Plato in order to hasten the translation of the newly discovered manuscript. However, in 1614 Casaubon demonstrated conclusively that the text was actually written sometime between the first and third centuries of the Christian era. The *Corpus Hermeticum* has affinities with Gnosticism, and has its *Sitz im Leben* in the plethora of esoteric spiritual cults and practices that emerged at the start of the Common Era. Rather than seeking accommodation with Gnostic and quasi-gnostic cults, the emerging Christian Church actively resisted them and sought instead an intellectual engagement with Greek philosophy. The conversion of the Emperor Constantine to Christianity (*c.*312 CE) and subsequent emergence of Christendom forced such esoteric traditions into the counter-cultural underground. The extent to which this process was the result of the greater retroductive explanatory power of Christian theology over esoteric speculation, as opposed to the emergent hegemony of the Christian Church, remains an open question. Though the recognition that the *Corpus Hermeticum* was a historical product of religious syncretism in the early centuries of the Common Era

rather than a source of perennial wisdom pre-dating Moses and Plato led to the demise of interest in the Hermetic tradition, it continued to play a significant role within Freemasonry, and the discovery of Hermetic texts in the Gnostic library uncovered near Nag Hammadi in Egypt in 1945 led to renewed speculation regarding perennial wisdom that was later taken up in New Age circles (Robinson 1996).

3 Despite the fact that the Hermetic literature perpetuated the fiction of its pre-Mosaic and pre-Platonic origins, it is difficult to trace any substantial belief in the existence of perennial wisdom outside of Gnostic circles prior to the Renaissance. Though Bhaskar references religious traditions current in the early Christian era, in particular Gnosticism and the Essene community at Qumran, there is little, if any, evidence to support the suggestion that either the Gnostic sects or the authors of the Dead Sea Scrolls affirmed the existence of a *universally accessible* perennial wisdom *underlying all religious and philosophical systems*. On the contrary, both traditions were marked by a thoroughgoing religious exclusivism. The various Gnostic sects tended to claim elite access to a secret pool of knowledge available only to cult members, and explicitly rejected the presence of spiritual wisdom in other traditions. In particular, a number of Gnostics sects were explicitly anti-Semitic and viewed Hebrew Scripture as the product of a demonic demi-God. Similarly, the Qumran community, 'claiming to be the true heirs of all the promises and the scriptures', regarded not only pagan gentiles but also 'devout Jews of other persuasions as dangerous deceivers' (Wright 1992, p. 203). Ironically, probably the closest we come to the notion of a *universally accessible* perennial wisdom in the Hellenistic era is found within orthodox Christianity itself, specifically in the theology of the second-century Christian Apologists. Thus Clement of Alexandria argues that God prepared the world for Christ's coming by giving the Jews the Torah and the Greeks philosophy, so that Moses and Plato offered Jews and Gentiles respectively a useful-yet-limited *universal* wisdom that was to prepare the way for, and to be brought to fruition by, the incarnate Christ: 'God's plan included the pagans, to whom God gave philosophy as well as the sun, moon and stars to be a path by which they might ascend to God' (Osborn 2008, p. 165). If there is a link between Bhaskar's philosophy and the classical era – albeit a relatively loose one – it is perhaps to be found, not in the existence of an esoteric tradition of perennial wisdom, but in the neo-Platonism of Plotinus' *Enneads* (Plotinus 1991; O'Meara 1993). Plotinus' theory that all things emanate from a single divine source, that individual souls are in process of reuniting with the divine source, and that this return is facilitated by recollection and contemplation of a soul's divine origins has a certain resonance with Bhaskar's notion that human beings are categorically grounded in God, dispositionally oriented to return to God, and that this process is facilitated by experiential awareness of the essential unity of human beings and God made available through a pedagogy of recollection.

It is, then, possible to identify a specific, albeit loosely knit, spiritual tradition underlying Bhaskar's spiritual turn. However, the claim that this tradition is perennial, and that the world's religious traditions emerged from it, is not substantiated by the historical evidence. Rather, it appears that the tradition was generated in the West, especially during the transition to modernity, and to some extent at least in reaction to Christian exclusivism. The lack of historical evidence does not in itself rule out the alethic truth of Bhaskar's spiritual position: to argue otherwise would be to slip into the genetic fallacy of assessing meta-Reality against its (possible) historical origins rather than its ontological claims. However, there can be no getting around the fact that Bhaskar's retroductive hypothesis regarding the role and significance of the world's religions vis-à-vis perennial philosophy makes ontological and empirical claims that require testing in the light of the available evidence and against alternative hypotheses, not least those proffered by Christianity and naturalistic materialism.

One final observation needs to be made by way of conclusion. This attempt to identify the spiritual resources underlying the philosophy of meta-Reality makes no claims to be definitive. In a book devoted to the interface of Christianity and critical realism there is insufficient space for a detailed exploration of Bhaskar's 'spiritual turn', even though the current brief – and to some extent speculative – survey is necessary to contextualise the book as a whole. One of the core arguments of the book is that any account of the interface between critical realism and Christianity must be attentive to the ontological claims and epistemic procedures of Christianity. Precisely the same argument can, and should, be applied to Bhaskar's own position. There is much work still to be done in this area, and those who set themselves against Bhaskar's meta-Reality and Christian theology have a moral and intellectual responsibility to attend closely to their ontological claims and epistemic procedures, and be intelligent, reasonable and responsible when critiquing them.

The transition to meta-Reality

The emergence of the philosophy of meta-Reality was, in part at least, informed by the reception of *From East to West*. MacLennan sums up the negative response of Bhaskar's more vociferous critics succinctly: not only had Bhaskar found God, but, worst of all, 'it was a very down-market god, nothing more than your common or garden New Age variety, the type readily available at any incense saturated shop' (MacLennan 2000). Such expressions of taste are, or course, no substitute for attentive, intelligent, reasonable and responsible debate. Within a critically realistic framework such debate cannot be limited to the mere refutation of a particular retroductive explanatory model; rather, it is necessary to develop an alternative hypothesis capable of demonstrating possession of greater explanatory power. Thus, for example, it is insufficient merely to reject the existence of a creator God on the grounds that there is insufficient evidence to support such a hypothesis. If appeals to a creator God function – among other things – to explain the existence of an inherently rational and meaningful universe, then in rejecting the existence of God it is necessary either to offer a more powerful explanation of the universe's existence, intrinsic intelligibility and inherent meaningfulness, or at the very least to explain why such facts do not require explanation.

Sean Creaven has provided the most sustained attack on Bhaskar's 'spiritual turn' from within the critical realist community (Creaven 2010). He seeks to defend a form of naturalism in which the physical world is self-generating, self-rationalising and self-sustaining, and the evolutionary emergence of sentient human beings from material nature is sufficient to warrant a moral realism informed by a Marxist-inspired politics of universal emancipation from oppression. He claims that the explanatory power of this retroductive model is sufficiently powerful to undermine the alternative models proffered by the philosophy of meta-Reality and theology of classical Christianity. His core criticism of Bhaskar's later philosophy is that it replaces realism with idealism, substitutes the rational pursuit of truth with an intuitive transcendentalism, and seeks human emancipation in subjective experience rather than the material conditions of society. Creaven has been criticised on a number of counts: Mervyn Hartwig claims that that his core arguments are dependent on inductive empiricism rather than critical realism, while I have suggested elsewhere that his attack on Christianity is marked by a radical failure to attend to the available evidence and literature with appropriate academic rigour (Hartwig 2001; Wright 2011).

Hartwig, despite remaining somewhat agnostic towards the philosophy of meta-Reality, has emerged as a key defender of Bhaskar's spiritual turn. He argues that despite concerns about its supposed 'New Age' thrust, all critical realists ought to be able to accept that if human beings are to survive as a species, we will have to come both to see ourselves, and to act, as part of a highly valued cosmic whole (Hartwig 2000). The most positive response to Bhaskar's later work has, perhaps not surprisingly, come from Christians within the critical realist community, most notably Margaret Archer, Andrew Collier and Douglas Porpora (Archer *et al.* 2004). Their baseline position is that, since critical realism restores the ground for rational debate about God, religion and spirituality, both Christian theology and the philosophy of meta-Reality can claim a legitimate place in the public space of academic debate. Porpora has helped establish what appears to be an emergent consensus regarding Bhaskar's later philosophy in critical realist circles. Strictly speaking, he suggests, there has been no 'spiritual turn' in critical realism per se, just a 'spiritual turn' in Bhaskar's emergent philosophy. It remains an open question whether the philosophy of meta-Reality is a necessary logical development out of critical realism and dialectical critical realism, and as such it has the status of a retroductive hypothesis that remains open to further debate (Porpora 2005).

There are a number of ways in which Bhaskar's emergent philosophy of meta-Reality marks a step beyond transcendental dialectical critical realism.

1 Bhaskar clarifies an ambiguity regarding the status of his project by paying greater attention to epistemic relativism. In *From East to West* he claimed to

'show how dialectical critical realism *must* develop into a philosophy of (universal) self-realization', while simultaneously affirming that transcendental dialectical critical realism constitutes '*only one possible development*, of dialectical critical realism' (Bhaskar 2000, pp. ixff.; emphases added). The failure to convince critical realists *en masse* appears to have led to a greater willingness on Bhaskar's part to recognise the contingent status of his spiritual turn. Though he continues to subscribe to the philosophy of meta-Reality and argues consistently and rigorously for its alethic truth, he also acknowledges its contested nature. The philosophy of meta-Reality constitutes, 'like everything else in life, a provisional terminus on a journey' (Bhaskar, 2002c, p. liv). In drawing a distinction between critical realism as 'the best description of the world of duality to date' and meta-Reality as a description of the world of non-duality that sustains the world of duality, Bhaskar implicitly acknowledges that not all critical realists are able to follow him, and in doing so affirms a continuing role for critical realism and dialectical critical realism independent of the philosophy of meta-Reality (p. ix). This fits with Porpora's suggestion that the philosophy of meta-Reality 'is not a spiritual turn in [critical realism] itself but the articulation of one possibility that [critical realism] allows' (Porpora 2005, p. 160).

2 Bhaskar is far more circumscribed in his references to 'God', preferring instead the language of 'ground-state' and 'cosmic envelope' (Bhaskar 2002c, p. xii). It is enticing to read this as an attempt to respond to the hostility his spiritual turn generated among a largely secular audience suspicious of its apparent affinities with both the New Age movement and – albeit to a far lesser extent – institutional religion. However, Bhaskar presents it as an intentional pedagogic move: 'the problem was not so much hostility to, as a lack of any level of intellectual seriousness about, talking about God' (Bhaskar and Hartwig 2010, p. 167). In speaking about God, he constantly found himself faced with religious exclusivists who wanted to know to which particular God he was referring. This, however, missed the point entirely, because he had used the term 'God' to refer to the 'absolute', understood as the categorical ground and dispositional possibility of all emancipatory practices, regardless of whether they were conducted within religious or secular communities. Bhaskar dropped the term 'God' because his philosophy sought to both identify the necessary 'presupposition of all emancipatory practices' and 'make spirituality compatible with secularism'; his previous use of theological language served to obscure that fact that 'all forms of materialism and atheism are on a par with religion in so far as they posit, explicitly or implicitly, some sort of absolute' (pp. 167ff.).

3 Bhaskar steps back from the suggestion, implicit in *From East to West*, that the experience of transcendence and the moment of transcendental identification of the self with the absolute take the form of extraordinary, quasi-mystical, encounters. In the philosophy of meta-Reality, 'transcendence and transcendental identification are no longer confined to the realm of what one might call bliss or peak experiences' (Bhaskar 2002c, p. xii). Rather,

'transcendental identification and transcendental agency are seen to be necessary features of all social interactions', since 'the zone of non-duality is continually accessed and experienced in everyday life' (p. xiii).

4 Bhaskar abandons the division between what might be termed the 'philosophical-theoretical' and 'aesthetic-practical' modes of expression in *From East to West*. As we have already noted, one of the most striking and problematic features of that book was the way in which its philosophical argument is offered as a theoretical introduction to a decidedly non-philosophical 'novella' that describes the journey of a soul towards enlightenment through successive reincarnate lives culminating in what appears to be Bhaskar's own current autobiography. The literary genre of the novella is difficult to categorise. It reads more like a popular spiritual manual than a philosophical exercise, and it is difficult to avoid the impression that Bhaskar is attempting to pass on spiritual wisdom rather than engage in intellectual debate. Indeed, he acknowledges that he is appealing to two different constituencies of readers, and suggests that those without a formal philosophical background may wish to skip the philosophical sections of the book and proceed directly to the novella (Bhaskar 2000, p. 2). In *Meta-Reality: Creativity, Love and Freedom* the two modes of expression are woven into a relatively seamless whole and addressed to a single constituency of readers (Bhaskar 2002c).

5 The philosophy of meta-Reality reveals an increasing emphasis on education and intentional spiritual formation. Though *From East to West* recognises the Bhaskarian soul's vocation 'to be an enlightened and enlightening spiritual teacher', references to education and spiritual formation are only implicit in the text (Bhaskar 2000, p. 73). However, Bhaskar's educational concerns come to the fore in *Meta-Reality: Creativity, Love and Freedom*. Integrated within the core philosophical argument is an emergent philosophy of education as spiritual formation, in which the task of recollecting, accessing and becoming aware of one's ground-state is linked directly to Platonic anamnesis. At times the pedagogic thrust of Bhaskar's argument is expressed with a rhetorical urgency, reflecting his conviction that philosophy must be a serious, consistent and transformative activity:

> Are you really reading this book? Or are you judging it – or me? Do you look at, notice, experience, enjoy and act, have an effect on this world or does the world pass you by? Scaring, scaring you, shuddering down your spine.
>
> (Bhaskar 2002c, p. 168)

The philosophy of meta-Reality

Critical realism operates within a world of duality: a world of unhappiness, oppression, strife and alienation, in which things relate discordantly with one another. The philosophy of meta-Reality is concerned with non-dual states of being: states of freedom, love, creativity and right action, in which things relate in harmony with one another. The world of duality is dependent upon and sustained by the world of non-duality so that

> the very world of misery and destitution we have created itself contains and is sustained by the seeds of a society of abundance, peace and fulfilment, in which we are free to express and fulfil our essential natures.
>
> (Bhaskar 2002c, p. viii)

Bhaskar affirms the ontological priority of identity over non-identity: all things are connected, and the nature of their interconnection presupposes a moment of identification. Though the knower and the object of knowledge are distinct entities, in the knowledge relationship they become one, and it is this essential unity that takes ontological priority over their disunity. Love, the binding force that unites society, has ontological priority over hate, the dislocating force that splits society. The notion of the ontological priority of identity is not to be confused with either 'a punctiform, atomistic point' or 'an abstract blanket whole' (p. xiv). Rather, Bhaskar avows a rich, holistic, differentiated and developing identity, in which it is possible to identify and differentiate particular aspects of complex wholes. By becoming aware of the non-dual world we transform the dual world and emancipate ourselves from its hegemonic structures. If the dual world constitutes a demi-reality in which violence reigns, and the non-dual world constitutes a meta-reality in which love reigns, then we inhabit a relative-reality in which the battle between violence and love is fought out. Whereas realism assumes a split between ontological reality and epistemic descriptions of reality, meta-Reality seeks to transcend this split by affirming the unity of the object of knowledge and the knower. Because of this, meta-Reality 'is not really even a system of thought, but an intervention in the discursive process which is designed to enable agents to reflectively situate their own non-dual being in the context of their growth and development' (p. xxiv).

Bhaskar presents meta-Reality as a development beyond transcendental dialectical critical realism. As we have seen, dialectic critical realism identified four stages or dimensions of the dialectic process: the first moment (1M) of *non-identity*, the second edge (2E) of *absence*, the third level (3l) of *totality*, and the fourth dimension (4D) of *transformative praxis*. Though these are all related to the dualistic world of demi-reality, aspects of the non-dualistic world of meta-Reality, upon which the dual world of demi-reality depends, continually break through. This happens especially at the third level of totality, in which we experience being at one with the whole of reality. Transcendental dialectical critical realism, as described in *From East to West*, adds a fifth aspect (5A) of *spirituality* to this scheme (p. xix). Emancipatory projects in the fourth dimension of transformative praxis cannot succeed merely by transforming the structures of society: they also require the spiritual transformation of human agents. This fifth aspect paves the way for the philosophy of meta-Reality, which is directly concerned with the non-dualistic world, and contains two further dimensions (pp. xixff.). The sixth realm (6R) of *re-enchantment* reconnects the dual world with the non-dual world, and is conventionally conceived as the realm of the paranormal, the supernatural and of religious practice. However, because 'one can be spiritual without a religious practice' and 'observe religious rituals without being spiritual', the realm of re-enchantment points beyond itself to the seventh zone (7Z) of *awakening*, in which we become fully conscious of our place in the totality of reality and thereby fully realise ourselves, regardless of our religious or secular beliefs and commitments (p. xxi). Building on his earlier notion of God as the categorical structure of the world, and ultimate ground and alethic truth of being, Bhaskar now drops the theological terminology in identifying three modalities of non-duality, three ways in which the world of non-duality sustains and grounds the world of duality and thereby establishes the necessary condition for all human life.

1 The *ground-state* is 'the ultimate ingredient of all other states of being, activity and consciousness' (p. x). When a person is in their ground-state they inhabit the world of non-duality and are in harmony with the totality of all things. Human beings have ground-state qualities by virtue of their implicit potential for creativity, love and right action. The connectivity of all ground-states constitutes the *cosmic envelope* that binds all beings and objects into a unified whole. The cosmic envelop is the ultimatum, the ultimate ingredient of all things in the inter-connective unity of their ground states.
2 Certain aspects of *transcendence* ground everyday life. Here Bhaskar distinguishes between relative and absolute transcendence. Relative transcendence refers to the resolving, surpassing and transcending of a particular problem or state of affairs. Absolute transcendence refers to the ultimate resolving, surpassing and transcending of *all* dual states of being, so that in 'its absolute sense transcendence just is non-duality or unity in or with a total context' (p. x). The phenomenon of the emergence of new states of affairs 'out of the blue' provides a bridge between relative and absolute transcendence. Previously Bhaskar had identified four principal forms of the transcendence of consciousness: the transcendence of subjectivity, in which the subject loses itself in an external object; the transcendence of objectivity, in which the subject turns inward and becomes one with itself; the transcendence of consciousness, in which the consciousness of the subject is absorbed in a particular action or agency; and the transcendence of consciousness, in which the consciousness of the subject is absorbed in communal actions marked by holistic teamwork (Bhaskar 2002b, pp. 208ff.). He now adds three further dimensions of transcendence: self-transcendence, in which the conscious subject becomes aware of their ground-state; transcendence of false consciousness, in which the conscious subject becomes aware of the unity of all things; and transcendence of dualism, in which the consciousness subject becomes aware of their ingredient participation in the cosmic envelope. As we have already noted, Bhaskar stresses that transcendental consciousness of being in the ground-state, of being ingredient

in the totality of all things, is not 'confined to the realm of what one might call bliss or peak experiences', but is rather a necessary feature of all perceptions and actions, so that 'the zone of non-duality is continually accessed and experienced in everyday life' (Bhaskar 2002c, pp. xiiff.).

3 All being and consciousness possesses a *fine structure*. If a person focuses attentively and single-mindedly on an object they will experience its 'deep interior' and become aware of qualities such as bliss, emptiness and pure, unbounded love. These qualities constitute the ground-state of creation that is ingredient in all other states and from which all other states emerge.

> If you go deeply enough into anything, whether it be a broom or the corner of a room, a sound or your breath you will eventually reach some ground-state, some aspect or part of the cosmic envelope from which it derives its energy and being, in the sense of being part of a single universe, one cosmos.
>
> (Bhaskar 2002c, p. xiii)

Moving on, Bhaskar identifies three mechanisms by which relations of identity come into being: (1) *Transcendental identification*. This occurs when a person is absorbed by and becomes one with an object or action, perhaps the book they are reading, the television programme they are viewing, or the simple act of making a cup of coffee. (2) *Reciprocity*. Whenever we come into contact with an object we enter into a reciprocal relationship with it. In this relationship both subject and object necessarily share something in common, whether they are aware of that fact or not; if this were not the case, it would be impossible for them to be interrelated. A reciprocal relationship is partly constitutive of the identity of both subject and object, so that I am the person I am by virtue of my relationship with myself and with that which is other than me. In the demi-world of duality such relationships remain fractured: the relationship between the murderer and his victim is real, and such that in the act of killing the two are united. This foregrounds the moral, aesthetic and spiritual task of restoring discordant reciprocal relationships in the dual world to their proper harmony in the world of non-duality. (3) *Co-presence*. At the deepest level objects are enfolded within, implicit within, and co-present within one another. Thus the totality of reality is implicitly enfolded and co-present within myself, and I am implicitly enfolded and co-present within the totality of reality. Bhaskar argues that this is why, despite the objective constraints of hegemonic structures in the world of duality, 'we subjectively feel and experience a commitment to the project of universal self-realisation, that is the fulfilment and flourishing of all beings in the universe' (p. xviii).

Transcendental consciousness is the consciousness of being united with our ground-state. Since the ground-state is ingredient in all being, transcendental consciousness is ultimately constitutive of all states of consciousness. However, in the dual world consciousness is in bondage to 'systems of thought, ideology, received opinion, organised religions, codes of conduct … [and] all forms of authority and tyranny' (p. 29). We achieve self-realisation by freeing ourselves from all such power structures and entering into our ground-state. By attending to anxieties generated by dualistic systems and forces, we become reflexively aware of them, step back from their enveloping power, and learn to observe them in a manner devoid of negative thoughts, judgements and emotions. This dialectical process of inaction is designed to expand awareness of our situation, bring greater clarity to it, and thereby clear a space from which we can access our ground-state and achieve enlightenment. Meditative witnessing and attentive watchfulness enables us to dis-identify ourselves from the dual world and shed its hegemony, thereby freeing us to embody the presence of the ground-state. In moments of self-realisation the embodied personality in the dual world becomes one with her transcendentally real self in the non-dual world. This dialectic of inaction opens up the possibility of a dialect of action concerned to transform the dual world. In accessing our ground state we remain in the dual world, so that a realised being is not necessarily in its ground-state, but merely in a state consistent with its ground-state. Though a non-dual being will not always be in a state of transcendental consciousness, her ground-state will still be ingredient in everything she does and embed her activity within the totality of the cosmic envelope.

The fundamental problem of the human condition is that we inhabit a dualistic world in which we are alienated from ourselves, from other people, from the natural world and from our ground-state. Despite this, we are already engaged in non-dual states that ground and support our alienated dual states. We exist in the dual world as emergent embodied personalities possessing egos that separate us from our true selves and other true selves. However, our egos are in reality no more than causally efficacious illusions: our true, transcendentally real selves inhabit the world of non-duality in which everything is united in the cosmic envelop. Since our illusory egos are dependent on our transcendentally real selves, we can learn to access our true selves by expanding the holistic field of our consciousness, awareness and perception. In doing so, we enter into our ground-state and transform our embodied personalities, thereby dissolving the ego and terminating our self-alienation. In overcoming alienation we achieve enlightenment and self-realisation, and thereby contribute to the evolution of the cosmos towards universal self-realisation.

In the dual world we employ our discursive intellects to reify and objectify reality, thereby generating forms of alienation and contributing to the breakdown of personal relationships and social systems. In the non-dual world we act spontaneously, intuitively and creatively. Creativity is essentially the production of something new that previously existed only as a potential, the making present of that which was previously absent. Hence the act of creation entails the irruption of something new from the non-dual world into the dual world. Because the fruits of creativity are already implicit in being as potentialities awaiting realisation, we already know implicitly that which we will create. However, in the dual world such implicit knowledge remains unmanifest and forgotten, and consequently needs to be recovered by an intuitive process of anamnesis or recollection. Learning via anamnesis is driven by an initial impulse or primal urge in which the role of the discursive intellect is circumscribed. We do not learn to be creative through the discursive intellect, but by suspending the discursive intellect and trusting in those non-discursive intuitions that emanate and flow from the non-dual world. Where religious mystics follow the creative path of spiritual intuition, religious institutions undermine creativity by overlaying it with discursive doctrines. The creativity intrinsic to the natural sciences is grounded in the priority of intuition over discursive thought, so that we can identify the scientist as a practical mystic. Transformative praxis requires creative acts that draw their primal source and energy from intuitive engagements with the non-dual world.

Love in the non-dual world underpins all emotions in the dual world. Negative emotions in the dual world, such as fear and pride, are entirely dependent on the absence or incompleteness of love, and exist only insofar as they are sustained by, and derive their energy from, the ground-state quality of love in the non-dual world. When we are in our ground-state we absent all negative emotions and experience only pure unconditional love. Such love 'is the totalising, binding, unifying, healing force in the universe' (p. 175). In the dual world pure love is contaminated by negative emotions and becomes conditional, so that the relationship between lovers becomes bound to economies of attachment and manipulation. Pure unconditional love absents negative emotions and the hegemonic forces they generate, so that the relationships between lovers serve to sustain and enhance their mutual self-realisation and fulfilment. As we access out higher transcendentally real selves, so we access unconditional love, and in doing so we learn 'to act more spontaneously, rightly, creatively, compassionately, tolerantly than we normally do' (p. 174). Emotions are not in themselves actions, but rather the grounds for actions. It is not always transparent what the motivations underlying any given action are: the giving of a gift may be an act of patronisation driven by feelings of superiority, or an act of altruism driven by feelings of love. This means that if we wish to act out of unconditional love we must first access our ground-state, clear away all passions, and follow through our intentions without hesitation, since the act of hesitation implies the occlusion of pure unconditional love by negative emotions.

In unpacking his understanding of action, Bhaskar invokes the principal of self-referentiality: we can only ever act as individual agents, but in doing so we necessarily impact upon the whole of reality. This means that ultimately the better the person, the more in tune they are with their ground-state, then the more efficient their actions and the more positive their impact upon reality. Successful right action requires us to be in our ground-state, in which we act effortlessly, spontaneously, unconditionally and joyfully. However, this needs qualifying, since acting in harmony with our ground-state cannot guarantee the success of our actions in the short term. This is because in the current stage of cosmic evolution we necessarily act within a dual world in which other, negative, causal mechanisms are at work. It follows that the best actions are not necessarily the most effective ones, and that non-dual actions grounded in unconditional love cannot always avoid, in the short term at least, dualistic and alienating consequences. Despite this, it is good to act in harmony with one's ground-state because it ultimately allows for more efficacious, creative and successful activity, constitutes the best way of ensuring lasting success in the dual world, hastens the cosmic evolution of the non-dual world, and constitutes the only means of saving the planet and ensuring the survival of the human species. All actions in the dual world, including actions that are conditional, instrumental and oppressive, are ultimately sustained by spontaneous unconditional acts in the non-dual world. Ultimately, the only way of absenting negative actions is to persist in carrying our positive actions in harmony with our ground-state and grounded in unconditional love.

Bhaskar understands perception as the immediate recognition of the world as intrinsically meaningful and valuable. Modernity oversaw the occlusion of the semi-enchanted world of those classical philosophical discourses that recognised the reality of God, and its replacement by a radically disenchanted world marked by multiple forms of alienation that threaten to destroy the material conditions necessary for the survival of the human species. Where reflective philosophies in the dual world serve to disenchant reality, the non-reflective philosophy of meta-Reality in the non-dual world functions to re-enchant being. Re-enchantment entails the collapse of any subject–object duality that distinguishes between fact and value and consigns the realm of value to the level of a subjective classification devoid of any ontological purchase on reality. This brings with it the collapse of the distinction between the sacred and profane, so that the whole of life and totality of reality is intrinsically and implicitly sacred, and embarked on a course that will eventually lead to the subsuming of the dual world within its non-dual source. As we perceive the world to be essentially sacred and enchanted, so such perception impacts positively upon our identities and actions.

The perception of the world as essentially sacred and enchanted establishes the conditions for self-realisation, the triumph of freedom, the victory of unconditional love, and the establishment of a truly eudemonistic society. Since we already possess within ourselves, and are transcendentally identified with, the totality of being, our only task is to awaken in our minds and realise in our actions that ontological fact. Here Bhaskar returns, albeit briefly, to the theological language employed in *From East to West*: self-realisation is God-realisation. God, the cosmic envelope, is categorically ingrained in reality and dispositionally oriented towards the self-realisation of the totality of being, so that 'an amoeba can in principle and eventually fully realise the essence of Mozart or Leonardo, of Socrates, Jesus or Buddha' (p. 325).

> At this point we have become one with the totality and at this point we can talk of god-realisation. This unification with god is presumably what Jesus meant when he said, 'to him that overcometh will I grant to sit with me in my throne, even as I also overcome, and am set down with my father in his throne' (Revelation 3:21). Now if any one thing can become one with all other beings, then by the same token all beings can become one with all other beings, that is with every being. At this point we have universal god-realisation.
>
> (Bhaskar 2000, p. 324)

3 Christianity and critical realism

While Bhaskar was working on *From East to West* three pre-eminent critical realists, Margaret Archer, Andrew Collier and Douglas Porpora, were already beginning to tease out the implications of critical realism for their own Christian faith. Where Bhaskar adopted a position that sought to be inclusive of a range of different religious traditions, Archer, Collier and Porpora were committed to the specific tradition of Christianity: Archer and Porpora as Roman Catholics with particular interests in, respectively, Catholic mysticism and process theology; Collier as an Anglican with a particular interest in theological epistemology and political theology. Their efforts resulted in a flurry of publications that, in effect, constituted a second wave of critical realism's spiritual turn. One of the results of their work was the identification of a tradition of Christian critical realism that first emerged in the late 1950s, and which until recently operated almost entirely independently of the tradition of critical realism associated with Bhaskar. In the following sections we will first outline the work of Archer, Collier and Porpora, and then introduce the independent tradition of theological critical realism as it sought to explore (1) the interface between Christian theology and natural science, (2) the epistemic status of Christian doctrine, and (3) the relationship between historiography and biblical interpretation.

Transcendence: critical realism and God

Archer's co-authored *Work and Human Fulfilment* – based on seminars held at the Pontifical Academy of Social Sciences, founded by Pope John Paul II in 1994 to encourage dialogue between the social sciences and the Church's social teaching – defended the Roman Catholic Church's long-standing affirmation of the primacy of labour over capital (Malinvaud and Archer 2003). Porpora's *Landscapes of the Soul* presented empirical evidence of 'a culturally pervasive lack of orientation in metaphysical space, an inability to place ourselves meaningfully in the cosmos … [resulting in] an equally pervasive void in our own sense of self' (Porpora 2001, p. 152). He suggested that since beliefs are causal mechanisms, sociologists seeking to explain the place of religion in society cannot legitimately avoid questions regarding the truth or otherwise of religious beliefs. Since religious and theological beliefs have a significant impact on society, since atheism cannot be appropriated as an indisputable default position, and given the fact that within a critically realistic framework there are no grounds for questioning the status of theology as a publicly accessible rational discourse, sociology can no longer continue 'to keep religion exclusively as an object of study and not an intellectual partner' (p. 8). Collier's *Being and Worth* defended a realistic approach to morality, and argued for the intrinsic value and worth of the realms of both nature and human culture (Collier 1999). This naturally raised the question of the source and ground of such worth, a question Collier sought to pursue in two further books. *Christianity and Marxism* set out to contribute to the philosophical reconciliation of these two apparently diametrically opposed traditions (Collier 2001). *On Christian Belief* defended a realistic theological epistemology, arguing that Christian truth claims, insofar as they are based on reliable testimony, subject to critical analysis and tested by experience, are little different to claims to knowledge in other epistemic spheres (Collier 2003).

In 1999 Bhaskar met with Archer, Collier and Porpora to plan a joint book on God, transcendence and critical realism. All four interlocutors 'found a high degree of convergence, but not identity' in their respective positions (Bhaskar 2002a, p. 148). However, the failure to reach full agreement proved decisive: despite agreement on the possibility and importance of addressing questions of God and transcendence, Bhaskar elected to withdraw from the project. Both parties agree that the critical issue dividing them was that of God's immanence in the world, and in particular Bhaskar's panentheistic conceptualisation of the immanence of God within, yet teleologically transcendent of, the world's categorical structure (Bhaskar 2002a, pp. 145ff.; Archer *et al.* 2004, p. 1).

As we have seen, Bhaskar contends that God is ontologically immanent within the world as its ultimate ingredient and categorical structure: if this were not so it would be impossible for human beings to experience Him. Yet, at the same time, God is also ontologically transcendent, since he can neither be saturated by, nor exhaustive of, the world: if this were not so, non-experience of God would be impossible. Consequently, the existence of a God of infinite love is 'consistent with error, evil, illusion and "structural sin"' (Bhaskar 2002a, p. 146). Whereas God provides the conditions for human activity, it is ultimately human beings themselves who bear the burden of the future flourishing of the cosmos. As we have already noted, Bhaskar's theology is consistent with panentheism: the doctrine that God *both* interpenetrates the world *and* transcends it.

Though it is possible to discern slightly differing understandings of God in the work of Archer, Collier and Porpora – the latter's commitment to process theology placing him closer to Bhaskar than the other two – they all subscribe to a Christian theism that, despite affirming the simultaneous transcendence and immanence of God, is ultimately incompatible with Bhaskar's panentheism. The Trinitarian God is absolutely transcendent insofar as he creates *ex nihilo* (out of absolutely nothing) an ordered-yet-contingent creation that relies entirely on the creative and sustaining agency of God for its being and continuing existence. As such, and contra Bhaskar, the created order is not in any sense divine. The

Christian God is indeed omnipresent within creation, but not as the world's ultimate ingredient or categorical structure. Rather, God is immanently present within his creation in the person of the Holy Spirit, the active divine agent responsible for bringing to completion the divine project of reconciling the Creator and his creation, a project that came to a head in the incarnation of God in the person of Jesus Christ. For Christian theology there can be no panentheistic confusion between God and his creation; to affirm otherwise opens the door to pagan idolatry and the worship and service of created things rather than the Creator (Romans 1:25). The immanence of God within his creation as an all-loving and graciously active agent reflects the fact that salvation is not dependent on humankind's ability to transform its spiritual nature and/or establishing a set of social structures conducive to human flourishing. The achievement of salvation, enlightenment and human flourishing is beyond the scope of human agency, and is instead dependent on the unconditional gift of grace offered by the all-loving Triune God.

In *Transcendence: Critical Realism and God* Archer, Collier and Porpora set out to explore the significance of critical realism for theology in general and Christian theology in particular (Archer *et al.* 2004). Their basic thesis is that the question of the existence and nature of God (ontological realism) is subject to a range of different responses (epistemic relativism) that are open to critical assessment (judgemental rationality). '*Ontological realism about God in the intransitive dimension is consistent with epistemic or experiential relativism in the transitive dimension*' (Archer *et al.* 2004, p. 5; emphasis in original). By employing judgemental rationality 'we can publicly discuss our claims about reality, as we think it is, and marshal better or worse arguments on behalf of those claims' (p. 2). This runs contrary to two common secular assumptions: (1) that some form of naturalism – in which the natural order-of-things is deemed self-generating, self-sustaining and self-rationalising – constitutes a secure and self-evident default position that does not require rational justification; and (2) that religious beliefs do not constitute cognitive truth claims open to rational debate. Critical realism offers the possibility of dialogue across different religious and secular traditions, thereby undermining the twin monologues of worldview exclusivism and thoroughgoing worldview relativism, both of which generate premature epistemic closure.

Despite the cogency of their arguments, the authors are clearly aware of the book's limitations. They are self-confessed 'auto-didacts in "unsystematic" theology', under-labouring for theology as sociologists and critically realistic philosophers, rather than as specialist theologians (p. 24). As such, they do not pretend to establish definitive positions, but instead seek to stimulate further debate, both within the community of critical realists and with those Christian theologians who have advanced forms of theological critical realism independently of the work of Bhaskar. With regard to ontology, they work with a 'minimal consensus' regarding the nature of God as 'the alethic truth of the world, the source from which the world originates and its ultimate meaning' (pp. 24ff.). With regard to epistemology, they seek to ground knowledge of God primarily in spiritual experience rather than philosophical argument. In doing so, they are clearly aware of the Christian commitment to revelation as a basic epistemic category, and of theological critiques of appeals to spiritual experience as potentially subjective and devoid of any cognitive purchase on reality. Though *Transcendence* is broadly successful in establishing its basic thesis, it inevitably leaves fundamental questions regarding Christian ontology and epistemology tantalisingly open.

In 2000 Brad Shipway drew attention to the existence of a parallel, and largely independent, tradition of Christian critical realism operating at the interface of theology and the natural sciences, with a particular interest in the congruence between scientific and theological epistemology (Shipway 2000, 2011, pp. 11ff.). His call for dialogue between the two traditions was subsequently repeated by Alister McGrath: 'critical realism has engaged with Christian theology in a manner which is sporadic and eclectic. ... Christian theologians have mounted an equally unsatisfactory engagement with critical realism.... More and deeper conversations need to happen' (McGrath 2012, p. 203). Alongside this tradition run two further and parallel streams of critically realist Christian theology upon which Shipway touches only tangentially: the first is concerned primarily with the epistemic status and ontological substance of Christian doctrinal truth claims; the second with the relationship between historiography and Christian Scripture.

Theology and natural science

The ongoing debate between science and religion took a fresh turn with the publication of Ian Barbour's *Issues in Science and Religion* in 1966 (Barbour 1966; cf. Barbour 1968, 1974, 1990, 1998, 2000, 2002; Russell 2004). His basic thesis was that the potentially creative interface between the two spheres of knowledge had become moribund as a result of the all-pervasive influence of forms of empiricism and idealism. Barbour's *a posteriori* reflection on the epistemic and ontological presuppositions of both natural science and Christian theology led him to conclude that they shared a common – if rarely articulated – critically realistic framework, and to propose that this provides a basis for a renewed exploration of the interface between the two disciplines. Barbour effectively established a tradition of critically realistic scholarship in the field that continues to flourish today. Of the many protagonists in the debate, three are particularly worthy of note: Arthur Peacock, John Polkinghorne and Alistair McGrath (Peacock 1971, 1979, 1984, 1986, 1990, 2001, 2004; cf. Clayton 2007; Polkinghorne 1986, 1988, 1994, 1996, 1998; McGrath 2001, 2002, 2003, 2004a, 2004b, 2006, 2008, 2009a; McGrath and Collicutt 2007). Together they form a distinctive triumvirate of scientist-theologians: all three have doctorates in natural science (biochemistry, mathematical physics and molecular biophysics, respectively), all are ordained Anglican priests, all have held Oxbridge chairs, all have delivered the prestigious Gifford Lectures on Natural Theology, and two have received the esteemed Templeton Prize for Progress in Religion. Significantly, of the three only McGrath has developed his understanding of critical realism by engaging with the work of Bhaskar (McGrath 2002, pp. 195ff., *passim*). Within the limits imposed by the constraints of a brief overview, and conscious of the dangers of conflating distinctive voices within a singular description, I will outline this particular tradition of scholarship in terms of (1) its critique of the understanding of the relationship between science and theology generated by the Enlightenment; (2) its development of a common critically realistic epistemology, and (3) its explorations of the substantial ontological issues raised by the dialogue between science and religion.

(1) Natural science, theology and the Enlightenment. Barbour presents his turn to critical realism as a response to two flawed approaches to the relationship between science and religion generated by the Enlightenment.

The first is the naïve assumption that science and theology share a common agenda in seeking to explain the origins and workings of the natural order-of-things, which they follow through in two distinct and mutually incompatible ways. On this view theology, in attempting to make sense of the universe, employs an archaic pre-modern methodology that, given the greater explanatory power of the natural sciences, has been exposed as little more than a primitive pseudo-science. Medieval scholastic Christianity identified two sources of knowledge of God: revealed and natural theology. Revealed theology, grounded in the Hebraic religious tradition, focused on God's self-revelation as a divine agent in a historical process culminating in the incarnation, as attested to in Christian Scripture; natural theology, grounded in the Greek philosophical tradition, sought evidence of God's existence and nature through philosophical reflection on the natural order-of-things. Partly due to the Protestant insistence on the primacy of revealed theology, and partly due to the Enlightenment's conviction that accidental truths of history can never become necessary truths of reason, the modern debate between science and religion came to focus almost exclusively on natural theology, and took as its starting point the deistic God of philosophical reflection rather than the Trinitarian God of revealed faith. The fact that the emergent natural sciences were driven by empirical observation rather than philosophical argument effectively placed natural theology on the back foot. As attempts to demonstrate the existence of God as the first cause and source of order in the universe became increasingly peripheral to the actual process of scientific discovery, so the 'god-hypothesis' appeared increasingly redundant. Attempts to assert a place for theology within natural science, whether by invoking a 'god-of-the-gaps' in scientific knowledge, appealing to evidence of 'intelligent design' in the universe, or resorting to literalistic readings of the Bible in an attempt to establish scientific knowledge simply served to confirm the fact that theology had no positive contribution to make to the ongoing progress of scientific discovery, and drew the conclusion that it must consequently be rejected as a legitimate academic enterprise.

The second approach was the equally naïve assumption of a rigid demarcation between the tasks of science and religion. Faced with the apparent failure of theology to engage productively with modern science, and the concomitant assertion of its redundancy as an intellectual enterprise, attempts were made to preserve the integrity of theology by insisting on a fundamental divergence between science and religion. In neo-orthodox Christian circles this was marked by a recovery of revealed theology and rejection of natural theology. God, on this view, can only be known insofar as he chooses to reveal himself: through the Word of God incarnate in Jesus of Nazareth, written in Scripture and proclaimed by the Church. Since such knowledge has nothing *directly* to do with the natural world explored by scientists, there can be no conflict between science and religion. Theology and science are discrete activities, each with its own specific object of knowledge. Consequently there can be no useful engagement, positive or negative, between the two. With respect to the natural world, neo-orthodox theologians were content to do no more than affirm, on the basis of biblical revelation, divine sovereignty over an ordered, meaningful and essentially good creation. A demarcation of science and theology was also asserted, albeit in a different way, by liberal Christian theologians. They drew a distinction between the objective gathering of essential facts and the subjective exploration of existential meaning: scientists tell us how the universe works, theologians help us to live fruitful spiritual lives. This fact–value divide tended to loosen the link between theological statements and reality, so that expressions of spiritual experience increasingly appeared to possess little cognitive or propositional value, and therefore to be devoid of any substantial epistemic purchase on the actual ontological order-of-things. This was reinforced by the rise of linguistic philosophy, which opened the door to the identification of two autonomous and utterly unrelated language games, that of science and religion: the one functioning realistically to explain nature, the other functioning non-realistically to enrich humankind's spiritual life.

If to conflate science and religion was to expose theology as a pseudo-science, then to insist on their rigid separation was to preserve theology – but only at the cost of weakening, or even severing, the already apparently tenuous link between theological statements and reality. Barbour refused both options, arguing that the problem lay not in science or theology per se, but in the dominance of science by a naïve empiricism and theology by an equally naïve idealism, the combination of which served to reinforce a fundamental dualism between objective empirical fact and subjective idealised value. Attempts to overcome these binary opposites led Christian theologians to generate a critically realistic epistemology and explore its ontological significance.

(2) Towards a critically realistic epistemology. At the epistemic level, critical realism offers a more powerful account of the methods of theology and natural science than both empiricism and idealism, and thereby opens up the possibility of a renewed dialogue between science and religion. If accounts of critical realism within the theological tradition have not always achieved systematic precision, it is nevertheless possible to identify a set of recurrent themes that bear remarkable resemblances to and anticipations of Bhaskar's work. (i) Both the natural and theological sciences, as interpreted within a critically realistic framework, embrace a set of *a posteriori* assumptions about the necessary conditions

for knowledge: given our established knowledge relationships, it is reasonable, and indeed indispensable, to suppose that reality must be ordered and structured if it is to be intelligible, and at the same time open and contingent if it is to be susceptible to empirical observation. (ii) Both reject idealism: knowledge of reality is dependent not on *a priori* principles but on *a posteriori* engagement with the external world. (iii) Both reject naïve realism and literalism: reality cannot be reduced to its surface appearances, and descriptions of that which is not immediately present to the senses cannot avoid metaphor: 'the metaphors of theological models that explicate religious experience can refer to and can depict reality without at the same time being naively and unrevisably descriptive, and they share this characteristic with scientific models of the natural world' (Peacocke 1990, p. 15). (iv) Both acknowledge the priority of ontology over epistemology: despite operating within a common critically realist framework, both natural science and theology must adapt their epistemic tools in the light of the distinctive nature and demands of their particular objects of investigation. (v) Both assert that the search for universal meaning cannot avoid epistemic relativism: in both science and theology 'concepts and models should be regarded as partial and inadequate, but necessary and, indeed, the only ways of referring to the reality which is named' (p. 14). (vi) Both insist that their primary task is to establish and progressively refine the best possible explanatory hypothesis regarding their particular objects of investigation. (vii) Both accept that, since knowledge is constituted by the relationship between the knower and the object of knowledge, the observer plays a fundamental role in the epistemic process, as explicated by Polanyi's insistence on the personal coefficient of knowledge (Polanyi 1958). (viii) Both recognise that the attempt to make sense of reality is a communal activity, 'so that there is also a continuous community and interpretative tradition, in comparison and in contrast with which one's own experiences can be both enriched and checked' (Peacocke 1990, pp. 15ff.). (ix) Both affirm that a mark of a robust tradition is its openness to both external and internal criticism, which enables knowledge to be progressive even though the line of progression can never be assured: the claim that religions 'are a consequence of successive generations testing, correcting, confirming, extending, changing, the accumulated wisdom of experience' can be applied equally to scientific traditions (p. 18). (x) Both allow that the success of modern science does not necessarily support a naturalistic metaphysic: it remains open whether naturalism or theism constitutes the more powerful, least impoverished, retroductive explanation of reality.

(3) The significance of ontology. As one would expect, given the affirmation of epistemic relativism, at the ontological level the application of critical realism to the debate between science and religion does not share the broad epistemic consensus outlined above. It is possible to identify two broad trajectories that – significantly in the current context – broadly reflect differences between Bhaskar's meta-Reality and Christian theism.

On the one hand, the work of Barbour and Peacocke appears close to Bhaskar's meta-Reality. One of Barbour's primary concerns was to draw science and theology into dialogue without slipping into the danger of presenting theology as a legitimate alternative to scientific investigation of the natural order. He achieves this by invoking a panentheistic 'process theology' in which God is simultaneously ingrained within the natural order yet transcendent of it, and the being-in-becoming of the universe coincides with the being-in-becoming of God. This position is picked up by Peacocke, who insists that evolution constitutes the working out of divine providence, so that scientific descriptions of the emergent universe converge with theological descriptions of God's providential actions. The course of nature thus both reveals and constitutes the will of God, as it draws creation towards the future through the interplay of chance and necessity. With the emergence of humanity, God is joined by a set of co-workers engaged in the task of guiding creation to its final consummation. According to Barbour and Peacocke, such panentheism enables theology and science to offer equally valid descriptions of the same reality, and thereby undermines any notion of an irreconcilable dispute between the two disciplines.

On the other hand, Polkinghorne and McGrath offer an alternative account of the relationship between science and theology, one that rejects panentheism in favour of classical Trinitarian theology. The basic point of contention revolves around the different theological models adopted by each side in the dispute, and the status attributed to them. McGrath suggests that Barbour and Peacocke work 'on the basis of significantly weakened variants of the classical statements of Christian orthodoxy' (McGrath 2001, pp. 37ff.). Under the influence of liberal theologians concerned to defend Christian theology by bringing it into conformity with the legacy of the Enlightenment, Peacocke replaces the doctrine of incarnation with an account of Christ as the model human being, rejects the doctrine of original sin in favour of a notion of a failure of humanity to fulfil its potential, and supplements the doctrine of creation *ex nihilo* with a notion of emanation that opens up an understanding of God and God-in-nature operating together as co-creators. Viewed from the vantage point of classical Christian theology, God ceases to be the sovereign creator and sustainer of a created order other than himself, and becomes instead a co-regent, sharing power with, and forced to work within, the limits and constraints of the natural order.

McGrath responds to Peacocke's revisionism by exploring the implications of retaining the classical Trinitarian theological model and seeking to understand it, and its relationship to natural science, within a critically realistic framework. He seeks to achieve this by approaching the natural world from the *a posteriori* retroductive claims of the classical Trinitarian doctrine of creation. According to this doctrine, the universe was created by, and is continuously sustained by, a creator God who created *ex nihilo* and allows creation the freedom to exist apart from himself, yet without ceasing to engage creatively with it. This immediately rules out panentheism: according to Trinitarian theology, God cannot be identified even partially with nature, so that nature is in no sense divine. That being the case, natural theology's attempts to infer the existence and nature of God from the natural order-of-things without reference to God's self-revelation in historical events can, at best, only arrive at a partial and contracted understanding of God. God may well be partially visible in the structures of his creation, just as something of the personality of an architect may be visible in the buildings she designs; but that is no substitute for the knowledge of God revealed in the life, death and resurrection of Jesus Christ and the ongoing creative work of the Holy Spirit, just as viewing a building is no substitute for lunch with its architect (given that one's basic interest is in the Creator rather than his creation, and the architect rather than her designs). Once we know the architect, so we can approach the buildings she designs in a new, deeper and more profound light. In the same way, once we know the nature of the Creator as revealed in the incarnate Christ, so we can approach his creation with new understanding and insight. For this reason McGrath rejects any notion of natural theology providing a basic, alternative or supplementary foundation for knowledge of God apart from divine revelation. Instead of natural theology, he advocates a 'theology of nature', in which the natural order, as described and explained by natural science, is explored from the perspective of the retroductive explanatory model provided by Trinitarian theology.

> We argue that if nature is to disclose the transcendent, it must be 'seen' or 'read' in certain specific ways – ways that are not themselves necessarily mandated by nature itself. It is argued that Christian theology provides an interpretative framework by which nature may be 'seen' in a way that connects with the transcendent. The enterprise of natural theology is thus one of discernment, of seeing nature in a certain way, of viewing it through a particular and specific set of spectacles.
>
> (McGrath 2008, p. 3)

Since God and nature – at least according to classical Christian doctrine – do not occupy the same strata of reality, theology is no more equipped to answer scientific questions, or natural scientists theological questions than, say, sociologists are equipped to answer biological questions and vice versa. The stratified nature of reality rules out the reduction of explanations at one level of stratification to explanations at another level: this being the case, natural science and theological science are discrete activities concerned with particular strata of a singular reality. The natural scientist strives to tell us about the natural world, while the theological scientist strives to tell us about God. On this reading theology and natural science each have their proper objects of knowledge and designated strata to explore, and as such cannot be expected or required to pontificate on strata of reality other than their own – and, indeed, it is illegitimate for them to attempt to do so. That said, the fact that different strata constitute a single, albeit differentiated, reality does not rule out the possibility of communication across strata: we are not dealing here with incommensurable paradigms. The criterion of congruence holds that a truth claim at one level of stratification is more likely to be true if it is congruent with truth claims in other levels of stratification. It follows that cross-disciplinary exploration may legitimately compare and contrast explanatory accounts offered at different levels of stratification in order to test for commensurability and guide the refinement of explanatory models. Similarly, the criterion of fertility suggests that an explanatory hypothesis is more likely to be true if it is able 'to permit the acceptance of alien beliefs when its holders meet with good reasons to hold them true' in a manner that does not compromise the integrity of the original hypothesis (Marshall 2000, p. 147).

Polkinghorne and McGrath seek to affirm both the congruence of natural science and theology and the fertility of theology with respect to natural science. Both disciplines affirm that the natural world is simultaneously ordered and structured, and open and contingent. Beyond that, theology recognises both that it has nothing of primary substance to offer to a scientific understanding of the natural world, and that natural science has nothing of primary substance to offer to an understanding of God: if the theologian *qua* theologian cannot legitimately pontificate about the results of natural science, so the natural scientist *qua* natural scientist cannot legitimately pontificate about the results of theological science. Despite the initial negative response of some Christians to the insights of Galileo and Darwin, Christian theology has proved remarkably fertile in its ability to embrace the results of natural science without damaging the integrity of orthodox Christian theology. Indeed, the ongoing Christian engagement with natural science raises a range of issues that warrant further investigation: empirical evidence of widespread experience of, and apparent inbuilt psychological drive towards, the transcendent; the evident anthropic fine-tuning of the universe; the inherent intelligibility of reality; and the apparent contingency of the universe, which generates the ultimate question 'Why is there something rather than nothing?' and raises the further question as to whether the hypothesis of a self-generating, self-sustaining and self-rationalising universe constitutes a more powerful explanation than the alternative hypothesis of God as the ultimate self-generating, self-sustaining and inherently rational being.

Theology and epistemology

Thomas Torrance's *Theological Science*, based on lectures delivered in 1959, was the first major exploration of the interface between Christianity and critical realism (Torrance 1969; cf. McGrath 1999). Though some of the book explores the relationship between theology and the natural sciences, its primary concern was for the nature and status of Christian doctrine rather than the interface of science and religion. Torrance's core motivation was to under-labour for the Christian community by reaffirming the inherent intelligibility of its classical Trinitarian doctrines and thereby recalling it to its proper object of worship (T.F. Torrance 1993, 1996). Just as Alan Norrie identified Karl Marx as a proto-critical realist, so Torrance identifies a theological lineage of proto-critical realists, which he traces back from the twentieth-century theologian Karl Barth, through the Protestant Reformer John Calvin, to the Greek Patristic theologians of the early Church primarily responsible for the retroductive formulation of the classical doctrines of Incarnation and Trinity – in particular Athanasius and the three Cappadocian Fathers, Basil of Caesarea, Gregory of Nyssa and Gregory of

Nazianzus (Torrance 1988, 1990, 1995, 2000). Torrance argues that from the outset the Christian Church accepted the ontological reality of God, acknowledged its own epistemic fallibility, and affirmed the possibility of deeper levels of theological knowledge and spiritual discernment.

Torrance's theology was, in part, a response to the crisis in theological method provoked by the Enlightenment. The claim that God is most fully revealed in the life of Jesus of Nazareth drew the retort that the accidental truths of history can never become necessary truths of reason. Consequently, Enlightenment philosophers sought God not in contingent historical agency but in the universal structures of the natural world. As a result, their debates assumed an impersonal deistic God disassociated from the world apart from the initial act of creation, rather than the personal Triune God intimately concerned for the well-being of humanity. From the Christian perspective deism fell far short of Trinitarian orthodoxy: not only was the deistic God unworthy of worship, but knowledge of its existence was dependent on increasingly tenuous philosophical arguments. Romanticism appeared to offer Christianity a way out of this impasse. Just as the romantic philosophers had posited reflexive moral and aesthetic experience as a path to universal knowledge, so romantic theologians identified intuitive spiritual experience as a path to knowledge of God. Thus Schleiermacher sought to ground theology in the numinous experience of 'absolute dependence' (Schleiermacher 1958, 1976). In doing so, he rejected the traditional cognitive-propositional model of theological statements in favour of an experiential-expressive one: Christian doctrines are primarily expressions of subjective spiritual experience rather than descriptions of objective realities. The effect of this was to ease the purchase of theological language upon reality and open the door to increasingly non-realistic accounts of Christian doctrine. Liberal Christian theologians increasingly viewed the doctrine of incarnation pragmatically, as a spiritually useful myth, rather than as a cognitive account of a substantial ontological event (Bultmann 1958). In time, forms of post-Christian theology emerged that viewed Christianity as one among many equally valid and culturally relative expressions of a universal accessible spiritual experience, and thereby denied Christian doctrine any substantial epistemic purchase on ontological reality (Hick 1989). This progressive dislocation of Christian doctrine from reality came to a head in the thoroughgoing anti-realist theology of Don Cupitt, who argued that the projection of subjective experience onto a putative-yet-hegemonic objective reality was necessarily detrimental to the spiritual autonomy and progress of humanity (Cupitt 1980).

Karl Barth's rejection of this subjective turn in theology proved decisive for the development of classical Trinitarian theology in the twentieth century (Barth 1936–1952; cf. McCormack 1997). Barth argued that a liberal theology grounded in spiritual intuition constituted – like its mirror image, rationalistic natural theology – nothing more than a human attempt to ascend to God. Classical Trinitarian theology, on the other hand, tells the story of God's descent to humanity in the incarnation. If the proper object of theology is the ontological being of God, rather than merely our epistemic experiences of God, then theology must engage, and can only engage, with God as he chooses to be known: in the incarnation, as witnessed by Christian Scripture and proclaimed by the Church. Barth's insistence on doing theology within a Trinitarian framework, and denial that knowledge of the self-revealing God can be grounded, buttressed or justified by appealing to rational argument or spiritual intuition, led to charges of fideism: if Barth had recovered classical Christian theology for the Church, he did so, according to his liberal critics, at the unacceptable cost of requiring of Christians an irrational leap of blind faith.

Torrance follows Barth in affirming the centrality of classical Trinitarian doctrine:

> It is ... the Lord Jesus, the very Word and Mind of God incarnate in our humanity ... [who] prescribes for us in Christian theology both its proper matter and form.... While the Lord Jesus Christ constitutes the pivotal centre of our knowledge of God, God's distinctive self-revelation as Holy Trinity, One Being, Three Persons, creates the overall framework within which all Christian theology is to be formulated.
>
> (T.F. Torrance 1996, pp. 1ff.)

However, he departs from Barth by seeking to tease out the cognitive basis of such a claim, not primarily for apologetic purposes, but in order to enable the Church to recover an understanding of the intrinsic intelligibility of Christian doctrine.

Torrance argues that theological science and natural science share a common critically realistic framework. Both are concerned with objective realities existing independently of the perceiving subject. The possibility and actuality of knowledge of these objects is determined by their ontological actuality and intrinsic intelligibility, rather than by the cognitive structures of the mind. Contra Kant, we discover knowledge not by imposing our own frames of meaning on objects of knowledge, but by progressively adapting our frames of meaning so that they increasingly conform to the ontological nature of such objects, thereby allowing them to reveal themselves to us. Theological science and natural science are thus both *a posteriori* activities. They differ only to the extent that the nature of their respective objects differs: nature is impersonal and reveals itself in the contingent structures of the natural order, whereas God is personal and reveals himself primarily through historical agency. Both theological and scientific knowledge are grounded in faith, insofar as an initial encounter with any object of knowledge takes the form of a tacit and intuitive anticipation of its potential intelligibility. This intuition takes cognitive form as the learning community develops retroductive explanatory models that, insofar as they engage truthfully with the intrinsic intelligibility of the object, function as disclosure models. Such models can become fiduciary frameworks that inform the worldview of those who accept them, and which are abandoned only when a more powerful worldview becomes available. All sciences have proper limits, demarcated by the nature of their objects of study: thus natural science is concerned to explain the contingent order of the natural world, while theological science is concerned to explain divine nature, both as it is-in-itself and as it engages with the created order. Both are concerned with different aspects of one reality: natural science with the natural order-of-things, theological science with the Triune creator. Further, both recognise that their retroductive models are fallible and contingent, and as such must be subject to ongoing iterative refinement and development, on the basis of a judgemental rationality grounded not in any pre-established epistemic criteria but in the intrinsic intelligibility of the object of investigation as it progressively reveals more of its nature to us as a result of our ongoing attentive engagement and scrutiny.

Theology and historiography

Two New Testament scholars, Ben Meyer and Tom Wright, have pioneered the application of critical realism to biblical interpretation. Both apparently developed their theoretical frameworks independently of one another and neither engages with the work of Bhaskar, though Wright does refer – albeit only tangentially – to the work of the Christian critical realists discussed in the previous sections.

Ben Meyer's pioneering application of critical realism to New Testament exegesis is grounded in the work of the Jesuit theologian, philosopher and economist Bernard Lonergan. In two seminal works, *Insight: A Study of Human Understanding* and *Method in Theology*, Lonergan developed a critically realistic 'generalized empirical method' applicable across all academic disciplines, including the natural and human sciences, history, philosophy and theology (Lonergan 1957, 1973). Following Aristotle and Aquinas, and influenced by Kant, Hegel and Marx, Lonergan argues that the mind achieves realistic knowledge via a critical progression from the twin data of sense experience and consciousness, through inferential hypothesis, to verification. (1) Lonergan's *cognition theory* identifies two modes of knowledge, commonsense and theoretical, and argues that in both modes the attainment of knowledge passes through four stages of consciousness: attentiveness to the data of sense experience and consciousness; intelligent sifting, analysis and modelling of the data; reasonable judgement with regard to any emergent understanding of the data; and responsible action in the light of such emergent understanding. The authenticity of a person's engagement with reality is thus rooted in their ability to follow four transcendental precepts: attentiveness, intelligence, reasonableness and responsibility. The attainment of knowledge constitutes a process of self-transcendence, in which the solitary self engages, in wonder and care, with other selves, the natural order and the transcendent realm. (2) Lonergan's *epistemology* views the correlation between subjective cognition and objective reality as an ideal to be striven for via the development of a progressively more attentive, intelligent, reasonable and responsible worldview. The pursuit of knowledge is a communal process grounded in received tradition and driven by the composite of experience, understanding and judgement: experience is grounded in the objective actuality of data; understanding is grounded in the norms of subjective reasoning; while judgement looks beyond experience and understanding to the deep ontological structures of reality itself. Lonergan follows Aristotle in identifying true judgement as the correspondence between reality and our beliefs about reality. (3) Lonergan's *metaphysics* seeks to penetrate beyond objective data and subjective cognition to reveal the basic structures of reality itself. Since the subject indwells reality, any metaphysics must take into account the dynamic relationship between the known and the knower, between the deep mystery of reality and the profundity of human wonder. Metaphysics seeks to explain things, both as they currently function and as their functions develop over time, by establishing both direct insights into a single developing function and inverse insights into the complex dialectical web of interactions between different developing functions. Because reality is dynamic and open, the intelligibility of a process may only become apparent in the future. Human action is radically unintelligible whenever an individual acts against her better judgement intentionally in the absence of coercion, social conditioning or ignorance. Lonergan claims that insight into cognition, epistemology and metaphysics makes possible the development of a generic, critically realist methodological framework adaptable to the particular requirements of specific academic disciplines.

Meyer applies Lonergan's methodology to New Testament exegesis (Meyer 1989, 1995; McEvenue and Meyer 1992). Biblical interpretation cannot bypass the subjective interiority and conscious intentionality of the historian. The fact that human beings are capable of self-transcendence, of looking from the present to the future and from the finite to the infinite, means both that the historian's understanding of the past is limited, and that wonder, questioning, bafflement and the desire for discovery drive the pursuit of deeper historical understanding. Since true understanding correlates with reality, the subjective historian must strive for objective, realistic, knowledge. 'As the scholastic adage has it, *ens per verum innotescit*: reality becomes known through the (act of finding out what is) true' (Meyer 1989, p. xi). The biblical scholar is concerned with concrete historical events that are imbibed with divine and human meaning, and cannot, contrary to the claims of positivistic historiography, be reduced to a mere chronology of empirical facts. The meaning of historical texts lies in their ontological purchase on unique and unrepeatable historical events, and as such cannot, contra idealistic historiography, be reduced to the level of abstract universal generalisations. Consequently, it is a false move to attempt to emplot particular historical narratives within any pre-existing idealistic meta-narrative and thereby approach historical events as merely exemplars of abstract universal values. In interpreting historical texts the historian must recognise that the intended sense of the text has a *prima facie* claim for attention that is constitutive of historical discourse. Historical texts constitute particular utterances (*parole*) rather than language in general (*langue*). Contra post-structural linguistic idealism, language-in-itself does not generate meaning; rather, meaning is generated by the fit between historical text and historical event. Hence the task of the historian is to employ judgemental rationality in attempting to reconstruct and explain past events on the basis of historical evidence and testimony. Historical understanding is impossible without critical judgement, and such judgement is intrinsically hypothetical, requiring the construction of explanatory hypotheses designed to expose the deep meaning of events. Historical hypotheses must be tested against historical reality as mediated by historical texts; though they can never be certain, some may be shown to be more intelligent, reasonable, responsible and attentive to the evidence than others. The indispensability of historical judgement rules out any immediate, unmediated, quasi-positivistic grasp of historical meaning obtainable by naïvely reading off the surface of the historical text. A critical realist historiography operates with a hermeneutical triangle of reader, historian and referent, in which the dialectical interplay between the three seeks to bring the subjectivity of both historian and reader into an increasingly truthful relationship with objective historical events as these are mediated by historical evidence. Both primary and secondary historical texts generate traditions of interpretation.

> The tradition may be authentic, representing remarkable reach and a steady accumulation of insightful adjustments and reinterpretations. On the other hand, it may be inauthentic, representing a falling-off, a watering down, a tailoring of the text to the mediocrity of its readers.
>
> (Meyer 1989, p. xiii)

To the extent to which a historical text and its tradition of interpretation is truthful to past events properly understood in their ontological depth, then the subjectivity of the interpreter will itself be vulnerable to interpretation and measurement by the objectivity of the text. The past, that is to say, may reveal a more powerful way of understanding reality and thereby enable us to understand ourselves and the world we inhabit in deeper and better ways. This has tantalising implications for self-transcendence and human emancipation. 'When the literature to be interpreted is great, it may well call for an understanding of the world and a self-understanding on the part of the interpreter that at the moment are simply beyond him' (p. xiii). Here Meyer quotes Lonergan: some historical texts may make claims about the past and about the ultimate structure of reality that 'are beyond the initial horizon of their interpreters' and may consequently 'demand an intellectual, moral, religious conversion' (p. xiii). This means that any interpretation of Christian Scripture must take into account the theological claims made within the text and respond to them attentively, intelligently, reasonably and responsibly. To set aside the text's theological claims as irrelevant to the task of exegesis is to make an *a priori* decision, either about ontology (there is no God) or epistemology (talk of God has no epistemic purchase on reality), that can only do violence to the integrity of the text, regardless of the results of any ensuing assessment of its theological claims.

Tom Wright, Anglican theologian and former Bishop of Durham, outlines his critically realist approach to biblical exegesis in the first five chapters of the opening volume of *Christian Origins and the Question of God*, his ongoing systematic presentation of New Testament theology (Wright 1992, pp. 3ff.; cf. Wright 1996, 2003). Though he identifies critical realism as a term 'used quite widely in various disciplines', the immediate sources of his particular understanding of critical realism are something of an enigma (Wright 1992, p. 32, n. 4). He makes no reference to the philosophical tradition associated with Bhaskar, and engages with the tradition of Christian critical realism only tangentially. Though he acknowledges a broad similarity between his own work and that of four advocates of classical Trinitarian Christianity – Torrance, Louth, Gunton and Thiselton – he does not engage with them directly in any depth (p. 32, n. 3; cf. Torrance 1976; Louth 1983; Gunton 1985; Thiselton 1992). While acknowledging that in Meyer's *Critical Realism and the New Testament* 'a good deal of what I was trying to say is spelled out, argued for, and given (to my mind) solid foundations', he notes that he only read Meyer's work after completing the first draft of his own argument (Wright 1992, p. 32).

Wright argues that a theology of the New Testament cannot be reduced to a history of the origins of the Christian faith: because the first Christians made substantial claims about the nature and actions of God, an account of the New Testament cannot avoid addressing the question of the truth or falsehood of these claims. Following the Enlightenment a naïve pre-critical reading of the New Testament is no longer possible. A critical reading must embrace four interlinked and complementary approaches to the text, and recognise that to privilege any one over the other three can only lead to distorted readings (Wright 1992, pp. 7ff.). (1) It is important not to ignore pre-modern insights into the issue of biblical authority: texts in all domains of knowledge have the potential to become authoritative, insofar as they provide truthful and authentic explanations of the object under investigation; this is potentially so for the Bible, though as with all texts such authority must be internal to the text itself, rather than imposed by some external power. (2) The New Testament text must be subject to rigorous historical scrutiny. (3) The text must also be subject to equally rigorous theological analysis. (4) Biblical interpretation must embrace postmodern literary criticism insofar as it focuses attention on the act of reading itself. These four approaches, when located within a critical realist frame, have the potential to overcome the failings of two alternative approaches that have dominated investigation of the New Testament since the Enlightenment: the naïve positivistic affirmation of objective fact, and the equally naïve phenomenalistic celebration of subjective impression (pp. 32ff.). Critical realism offers a path between and beyond these extremes:

> [Critical realism] is a way of describing the process of 'knowing' that acknowledges the *reality of the thing known, as something other than the knower* (hence 'realism'), while also fully acknowledging that the only

> access we have to this reality lies along the spiralling path of *appropriate dialogue or conversation between the knower and the thing known* (hence 'critical'). This path leads to critical reflection on the products of our enquiry into 'reality', so that our assertions about 'reality' acknowledge their own provisionality.
>
> (Wright 1992, p. 35; emphases in original)

The knowledge relationship is such that, though historical realities exist (ontologically) independently of the knower, they can never be known (epistemically) independently of the knower.

Part II

Epistemic relativism

The ambiguity of Christianity

4 The identity of Christianity

In Part II, we consider Christianity from the perspective of epistemic relativism. To be epistemically relative about Christianity within a critically realistic framework is to do more than merely acknowledge that Christianity is a controversial and disputed phenomenon. Specifically, it is necessary to recognise: (1) that Christianity makes alethic truth claims about the ultimate ontological ground, nature and structure of reality; (2) that these truth claims are contested, both within and beyond the Christian community; and (3) that it is nevertheless possible to make informed judgements regarding the veracity of Christian truth claims, on the understanding that such judgements are themselves open to further evaluation. These criteria apply to Christians themselves, as well as to those outside the Christian community. It is certainly true that there are many Christians and non-Christians who reject one or other of them: for example, fundamentalist Christians who insist that their truth claims are self-evident and beyond contestation, and secularists who deny that Christian truth claims are amenable to critical evaluation. A fundamental question, in the present context, is whether those who deny that Christianity is epistemically relative have a legitimate warrant to do so. In Chapter 5 we will consider the extent to which Christianity is capable of embracing epistemic relativism, and in Chapter 6 we will ask whether Christian claims to possess exclusive access to ultimate truth are compatible with epistemic relativism. The present chapter sets out to prepare the ground for those discussions by addressing the question of the identity of Christianity. 'Christianity' is a significantly underdetermined umbrella label that covers a multitude of different spiritual experiences, communal traditions and ultimate truth claims. Unless we first identify the particular version of 'Christianity' we wish to discuss, we run the risk of embarking on an over-generalised exchange, whose conclusions will be vulnerable to a plethora of qualifications. In addressing the question of the identity of Christianity we will call upon the under-labouring services of critical realism. I will argue that accounts of Christian identity that function within essentialist and nominalist frameworks are not merely insufficiently robust to illuminate the issue of epistemic relativism, but – and more importantly – contain within themselves an inbuilt drive to epistemic closure. Critical realism, on the other hand, provides an interpretive framework sufficiently powerful to shed light on the question of Christian identity in a manner that foregrounds, and does full justice to, the issue of epistemic relativism. In addition, a critically realistic exploration of question of Christian identity lends strong support to our decision to focus primarily on the classical tradition of orthodox Trinitarian Christianity in the rest of this study.

Essential Christianity

In this section we are concerned with predominantly modern attempts to identify Christianity with some idealised abstract 'essence' above and beyond the contingencies of history. Classical idealism sought to establish epistemically secure descriptions of those stable essences or unchanging universals held to transcend the ephemeral realm of fleeting appearances. According to Plato, the bedrock of reality is to be found in the eternal Forms that transcend the shifting and unstable realm of ordinary experience. Hegel took up this concern in the modern era, arguing, in the *Phenomenology of Spirit*, that behind the rich diversity of religious phenomena stands an underlying essence (*Wesen*) or unity of Spirit (*Geist*) (Hegel 1977, pp. 410ff.). 'Hegel's ultimate objective was that of discerning unity behind diversity, of reaching an understanding of the one essence of religion behind its many manifestations' (Sharpe 1986, p. 221).

In insisting that it is grounded in contingent historical events surrounding the life of Jesus of Nazareth, and that the essential being of God cannot be separated from his historical acts, classical Trinitarian theology rejects any attempt to idealise Christianity by seeking to identify some timeless essence. The Christian commitment to the particularity of God's active engagement in history, and insistence that the immanent created world is real-in-itself rather than a mere shadow or cipher of some greater transcendent reality, requires the rejection of Gotthold Lessing's dictum that '*accidental truths of history can never become the proof of necessary truths of reason*' (Lessing 1957, p. 53; emphasis in original). Despite this, Augustine's influential Platonic reading of Christian doctrine undoubtedly helped to prepare the ground for modern attempts to generate idealised accounts of the essence of Christianity.

The theological legacy of Augustine is currently the subject of intense scrutiny among Christian theologians (Gunton 1990; Hanby 2003). Colin Gunton argues that a major weakness of Augustinianism was its failure, under the influence of Platonic idealism, to attribute full reality to particular embodied objects and events, in direct opposition to the orthodox Christian affirmation of the essential goodness and intrinsic rationality of the created order (Gunton 1993, pp. 46ff., passim). This generated a negative attitude towards creation that led directly to the denunciation of the material world, the equation of sexuality with sin, a deeply rooted suspicion of political activism, and the emergence of instrumental attitudes towards nature (Hanby 2003, p. 6). On this reading, Augustinian Christianity offers a means of spiritual escape from the demi-reality of the created order, rather than a path to its material regeneration and restoration. This is reflected in the Anglican *Book of Common Prayer*, which affirms an Augustinian understanding of Christian sacraments as 'outward and visible signs of an inward and invisible grace'. This Platonised sacramental theology effectively reduces the material world to a shadow-reality offering signs and intimations of a greater transcendent reality wherein true salvation lies, and stands over against an alternative understanding of sacraments as material objects and events in which God is uniquely active and present in the world conferring grace upon his creatures. In identifying the immaterial soul as the essence of human nature, and the material body as an accidental and disposable accretion, Augustinianism engineers a breach between appearance and reality. Though I may *appear* to be the person whose identity is grounded in my mutually constitutive relationships with other persons as mediated through the material world, in *reality* my essential identity bears no necessary connection to the embodied particularity of these relationships (Gunton 1993, p. 49). The orthodox Christian doctrine of creation rejects any such divide between appearance and reality, and – contra Bhaskar – between illusory egos and transcendentally real selves. This is why Christianity insists on a bodily resurrection and rejects any notion of reincarnation or the transmigration of disembodied souls (p. 49). Ultimately Augustinian idealism is rooted in the Parmenidean affirmation of the priority of the One over the Many and the concomitant rejection of the Heraclitean assertion of the priority of the Many over the One; as such, it suppresses the embodied particularity of a diversity of immanent created objects as valorised in orthodox Christian teaching, and views them instead as mere reflections or ciphers of a singular transcendent and disembodied divine order (Gunton 1993, pp. 16ff.; 1998, pp. 25ff.).

Feuerbach's account of the *Essence of Christianity* argues that Christianity involuntarily projects the essential attributes of humanity onto a putative deity, and in doing so alienates human beings from their true nature (Feuerbach 1957). On this reading Christianity is a religious system that correctly identifies the intrinsic worth of humanity, but incorrectly attributes the highest human qualities to God rather than to humanity. Love is an essential human quality; however, in identifying God as love and insisting on the fallen nature of humanity, Christianity effectively dislocates human beings from a quality essential to their nature. Feuerbach's rejection of Hegelian idealism and affirmation of materialism, which is replicated in his rejection of the other-worldly orientation of Christianity in its heterodox Platonic form (which he mistook as normative of Christianity as a whole), undercuts any thoroughgoing idealistic interpretation of his argument. Despite this, the essential human qualities of love, wisdom, justice, etc. retain for Feuerbach the status of universal attributes capable of being realised under material conditions. The essence of Christianity, according to Feuerbach, lies not in the historical particularities of Christian belief and culture, but in its identification of essential attributes of human nature that it has inappropriately projected onto a putative transcendent God.

Adolf von Harnack's *The Idea of Christianity* followed Feuerbach's lead in attempting to identify a timeless essence of Christianity (Harnack 1904). Harnack was a leading representative of the tradition of Liberal Protestantism that sought to reconcile Christianity with the culture of modernity. The essence of Christianity, he argued, is to be found in Jesus' proclamation of a universal ethic of altruistic love. The Hellenisation of Christianity through its encounter with Graeco-Roman philosophy generated a doctrinal system that focused on metaphysical and onto-theological questions rather than ethical ones. As a result, Jesus' original ethical gospel was replaced by a derivative theological gospel that focused on the person of Jesus rather than on his ethical teaching. By stripping away this extraneous layer of myth and dogma it is possible to recover the original moral essence of Christianity, and thereby empower the Church to claim its rightful place within the culture of modernity. Harnack interprets Jesus' ethical teaching in terms of a quasi-Hegelian understanding of humankind's inevitable moral progress through history. The Church is a worldwide community that, provided it sets aside its myth and dogma, has the potential to nurture and sustain Jesus' ethic of altruistic love and universal fraternity. Once it realises this potential it will be in a position to guide and support the progressive emergence of an ethically grounded society – one that, according to Harnack's critics, bears an uncanny resemblance to the society espoused by nineteenth-century European bourgeois culture.

John Hick's universal theology constitutes a further attempt to identify a timeless essence of Christianity. Affirming the 'modern awareness of religious plurality and conceptual relativity', Hick identifies Christianity as a particular instance of generic 'religious thought and experience' located within 'a global continuum containing an immense variety of forms in a history moving from archaic beginnings to the present still-evolving state of the great world traditions' (Hick 1989, p. 9).

> [T]his vast and multifarious field of human faith is ... not wholly projection and illusion – even though there is much projection and illusion within it – but constitutes our various transparent and opaque interface with a mysterious transcendent reality.

Christianity is a culturally bound manifestation of the genus 'religion', a historically rooted vehicle that, like other historically rooted religions, is capable of facilitating humankind's experiential encounter with transcendent reality. Participation in the Christian community, or any religious community, may potentially enable individuals to experience 'the Transcendent, the Ultimate, Ultimate Reality, the Supreme Principle, the Divine, the One, the Eternal, the Eternal One, the Real' (p. 10). The distinctive Christian doctrine of the incarnation is consigned to the level of instrumental myth rather than cognitive proposition, whose value lies in its potential ability to stimulate experiential awareness of the 'Transcendent' rather than provide a retroductive explanatory account of the ultimate order-of-things. The essence of Christianity is not to be found in its existence as a contingent historical phenomenon, but in its *function* as an immanent time-bound vehicle through which its adherents may access a transcendent timeless encounter with ultimate reality. There are clear affinities here with Bhaskar's philosophy of meta-Reality, the fundamental difference being that he replaces Hick's broadly dualistic distinction between transcendence and immanence with a panentheistic account of ultimate reality ingrained within the demi-reality of the world of everyday experience.

Though Feuerbach, Harnack and Hick offer different accounts of the timeless essence of Christianity, neither provides *in themselves* an appropriate basis for a reading of Christian alethic truth claims within the framework of epistemic relativism. This is because all three effectively bypass the historical particularities of orthodox Christian belief by superimposing an essentialised, idealised and ahistorical reinterpretation of Christianity onto the received Christian tradition. To be genuinely epistemically relative about Christianity, it would be necessary to: (1) identify the particular historically rooted truth claims proclaimed by orthodox Christianity; (2) recognise the fundamental differences between these truth claims and the truth claims embedded in the heterodox reinterpretations of Christianity offered by Feuerbach, Harnack, Hick and others; (3) acknowledge that no party can lay claim to epistemic closure; and (4) employ judgemental rationality in an attempt to identify which (if any) of the conflicting readings of Christianity offers the most comprehensive and powerful expression of alethic truth. This is *not* to argue that the reinterpretations of Feuerbach, Harnack, Hick and others lack alethic truth simply because, from the perspective of orthodox Christianity, they are heterodox. But it is to suggest that to accept the essential truth of one or other without first engaging in appropriately informed debate with the Christian orthodoxy from which they diverge is to bypass the demands of both epistemic relativism and judgemental rationality.

Feuerbach's identification of the essential attributes of human nature, Harnack's identification of a universal altruistic ethic, and Hick's identification of experience of the Transcendental Real all tend to follow the idealist tradition in assuming the intrinsic coherence and self-evidential truth of these ideals. Both the human values of love and justice, and the reality of the 'Transcendent', are known directly and intuitively via unmediated, pre-linguistic and pre-discursive experience. Insofar as Bhaskar's 'spiritual turn' may be interpreted idealistically (and such a reading remains contentious), then its failure to attend to the particularities of Christian doctrine, and concomitant tendency to overlay an idealistic account of meta-Reality onto the received Christian tradition, is vulnerable to the charge of assuming, rather than defending, the greater explanatory power of the philosophy of meta-Reality vis-à-vis orthodox Christian theology. It may well be alethically true that by demythologising orthodox Christianity and stripping away its doctrinal accretions we can arrive at some timeless essence behind the historically bound Christian tradition; but it may equally be true that the alethic truth of Christianity lies in the historical particularity of its claims regarding God's concrete actions within history. This dispute cannot be resolved by appeal to supposedly self-evident idealistic truths, but only through informed debate guided by the critically realist principles of epistemic relativism and judgemental rationality. Insofar as essentialist accounts tend to present the purported timeless essence of Christianity as self-evidently true, they bypass alternative renditions of Christianity and thereby occlude judgemental rationality and reveal an inherent drive towards premature epistemic closure.

Nominal Christianity

We turn now to predominantly empiricist and postmodern attempts to eschew idealism and interpret Christianity in the light of its historical particularity. Our discussion takes as its point of departure the work of Robert Jackson, an eminent professor of religious education of international repute, as he addresses the question of how religious traditions are best represented in religious education classrooms. Jackson correctly rejects idealistic approaches to the pedagogic representation of religions 'which posit universal "essences" or entrap insiders within schematic formulations of key beliefs and concepts' (Jackson 1997, p. 69). Ethnographic research reveals significant levels of disparity between the perspectives of 'ordinary' adherents of religious traditions and normative representations of them generated by community gatekeepers and academic scholars. The notion that religions constitute stable and structured belief systems is a product of the European Enlightenment in general, and attempts by Christian theologians to interpret non-Christian religions within Christian doctrinal schemata in particular (pp. 49ff.). Consequently, there can be no place for any representation of religions in which 'the fuzzy edges of real life are trimmed off, and the personal syntheses and multiple allegiances revealed by ethnographic study are interpreted as deviations from doctrinally pristine religious narratives' (Jackson 2004, pp. 81ff.).

In place of idealist representations of religious traditions, Jackson advocates a form of 'contextual religious education' that focuses on concrete manifestations of religious belief and praxis rather than abstract generalisations. Representations of religious traditions should respect their internal diversity and open borders by focusing on the life-worlds of ordinary adherents, thereby creating space for 'more personal accounts which link individual experience to social experience' (Jackson 1997, p. 69). Methodologically this means that the representation of religious traditions in the classroom must be grounded in ethnographical studies of individuals, families and local communities. The basis of a religious tradition is 'the individual's language, experience, feelings and attitudes in a religious context' (p. 62). Jackson accepts that any interpretation of religious life must proceed by 'examining the relationship between individuals, relevant groups to which the individuals belong, and the wider religious tradition' (p. 66). However, his weighting of the hermeneutical circle, according to which parts and the whole reciprocally illuminate one another, clearly gives precedence to the individual over the communal: the teacher must be careful 'not to identify individuals with the key concepts and practices of a "constructed" religion', and 'be wary of giving a prescriptive account of a tradition or of locking individual persons into stereotypical group identities' (p. 66).

I have argued elsewhere that this account of religious traditions is basically nominalist. On Jackson's view, religious traditions 'do not enjoy any substantial formal or structural identity: as merely the accidental sum of individual spiritualities, they are human constructs existing "in name only"' (Wright 2008, p. 5; cf. Jackson 2008). Jackson is a realist insofar as he recognises that religious phenomena exist independently of human perception; however, his reading of the phenomena is basically nominalist insofar as he views any holistic account of a religious tradition to constitute a human construction that, though it may possess pragmatic or heuristic value, cannot legitimately claim any epistemic purchase on ontological reality. This resonates strongly with Humean empiricism: we observe a series of atomistic facts, or experience individual packets of sense data, and since we are unable to discern any underlying causal mechanisms capable of explaining the relationship between them, we have no option other than to resort to two elementary criteria for the construction of holistic accounts of our observations and experiences: (1) the regular conjunction of events, and (2) social convention.

(1) According to David Hume, empiricists can make sense of the world by identifying the 'constant and regular conjunction' of patterns of objects and events, or experiences of objects and events, without the need to invoke any underlying, and empirically inaccessible, notion of causality (Hume 1902, p. 111). Gavin Flood defines religions as '*value-laden narratives and behaviours that bind people to their objectives, to each other, and to non-empirical claims and beings*' (Flood 1999, p. 47; emphases in original). There are, I suggest, significant merits in this definition. However, in seeking to clarify the identity of specific religious traditions within this generic definition he resorts to a notion of 'prototypical properties'. Thus he suggests that the 'beliefs and practices of a high-caste devotee of the Hindu god Viṣṇu, living in Tamilnadu in south India, fall clearly within the category of "Hindu" and are prototypical of that category', whereas the

> beliefs and practices of a Radhasaomi devotee in the Punjab, who worships a God without attributes, who does not accept the Veda as revelation and even rejects many Hindu teachings, are not prototypically Hindu, yet are still within the sphere, and category, of Hinduism.
>
> (Flood 1996, p. 7)

Flood's 'prototypical categories' resonate with Hume's 'constant and regular conjunction' of events, so that the more regularly an aspect of a particular religion manifests itself, the more prototypical it is considered to be. However, such a reading fails to penetrate beneath surface appearances and provide causal explanations of the emergence and development of specific aspects of particular religious traditions. The empirical observations that prototypically swans are white and initiation into the Christian community is through infant baptism may well be able to accommodate within their orbit the recognition that some swans are black and that Baptists practise adult baptism, but it does nothing to *explain* why swans are the colour they are, or *explain* the theological significance (putative or real) of the sacrament of baptism as a rite of initiation into an ontological participation in the 'Body of Christ'.

(2) With regard to the criterion of social convention, Judith Everington follows Jackson in suggesting that disparities between accounts of religious traditions constructed by different insiders and outsiders should be resolved by negotiation

between individual religious adherents, community gatekeepers, teachers and academic scholars. They are given the collective responsibility of establishing a consensus regarding the way in which accounts of religions should be approved prior to their representation in school classrooms (Everington 1996, p. 72). This resonates with the Humean appeal to social convention as a means of organising atomistic sense data. There is no recognition of the critically realist insistence that the truth value of constructed accounts of religious traditions lies in their ability to penetrate beneath surface appearances and provide more comprehensive, powerful and truthful causal explanations of religious phenomena. If the best representations of religions are simply those that achieve maximal consensus, then they are reduced to social constructions generated by economies of power. Jackson's democratic solution to the problem of representing religions in school classrooms is to seek to dissipate the hegemony of community gatekeepers and academic scholars by shifting the balance of power to accommodate religious adherents themselves; the critically realistic alternative would be to offer students access to the most powerful and comprehensive retroductive explanations currently available.

There is no disputing the fact that at the socio-cultural level Christianity is an extraordinary diverse phenomenon, to the extent that, for example, some Amish Mennonites in Pennsylvania might struggle to identify members of the Ethiopian Orthodox Church as fellow Christians, and vice versa. However, this does not warrant the nominalist conclusion that 'Christianity is not a single thing, but a kaleidoscope of different lived interpretations of the meaning of faith, some good, some bad, some (it may be thought) more normative, others less so' (Smart 1979, p. 7). If part of what it means to be epistemically relative about Christianity is to recognise that Christians affirm truth claims about the ultimate nature of reality, any realistic account of Christianity must penetrate beneath the surface level of socio-cultural appearances and address ontological questions of the underlying causal mechanisms (natural, human and possibly divine) that generate and sustain Christian culture. If Christian communities are reduced to nominal entities, then epistemic priority will inevitably be given to the experiences of individual Christian adherents. In this case there will be as many versions of Christianity and Christian ontology as there are Christian believers – indeed more so, if we accept that the beliefs of individual Christians may change over time. Nominalist readings fail to provide adequate account of the identity of Christianity because they deny: (1) the possibility of penetrating beneath the surface of Christian culture and revealing the various causal mechanisms that constitute and order the cultural phenomenon of Christianity; and (2) that constructed accounts of Christian beliefs possess epistemic purchase upon ontological reality. This rules out the exercise of judgemental rationality and leads directly to premature epistemic closure.

Realistic Christianity

I have suggested that to be epistemically relative about Christianity is to recognise that Christianity makes contested transcendent truth claims that are open to critical evaluation. Though essentialist readings of Christianity recognise that Christianity makes transcendent truth claims, these tend to be affirmed as self-evident and imposed upon Christianity from outside, in a manner dismissive of both the historical claims of Christianity and the historical tradition itself. Nominalist readings, on the other hand, reject the notion that Christianity per se proffers any substantial transcendent truth claims, and in doing so is equally dismissive of the transcendent truth claims embedded in the historical continuity of the Christian tradition. This leaves little space for any recognition of the relativity of their own readings: essentialism offers readings that are assumed to be self-evident, while nominalism, despite a veneer of openness to the diversity of different interpretations of Christianity affirmed by its various adherents, effectively adopts a singular non-realist understanding of their cognitive purchase on reality. In both cases there is little, if any, room for the employment of judgemental rationality: the presuppositions of both lead ultimately to premature epistemic closure – essentialism by affirming a singular self-evident transcendent truth, the latter by denying any substantial access to transcendent truth. Whereas essentialism identifies Christianity as a vehicle of transcendent truth dislocated from its historical contingency, nominalism affirms Christianity as a historically contingent entity devoid of any contact with transcendent truth. Might the interpretive framework provided by critical realism offer a way beyond this impasse?

Bhaskar distinguishes between the domains of the 'empirical', the 'actual' and the 'real' (Bhaskar 1997, *passim*). The *empirical domain* is the sphere of our personal experiences of the world. Such experiences are epistemically limited, since we are capable of error and self-deception, and direct unmediated apprehension of an object or event is no guarantee of truthful discernment. The *actual domain* consists of the totality of objects and events in the world (including, but not restricted to, our own personal experiences). Our epistemic access to this domain is similarly restricted, since we cannot possibly experience and fully comprehend the totality of actual objects and events in the world, past, present and future. The *real domain* is the sphere of the network of causal mechanisms that generate and sustain different configurations of objects and events in the actual domain, and make possible particular experiences in the empirical domain. The paradigmatic form of the epistemic fallacy is the assumption that the empirical domain is ontologically basic, and that the domains of the actual and real must conform to our empirical experiences and intuitions. Critical realism, in rejecting the epistemic fallacy, identifies the domain of the real as ontologically basic, and recognises that without underlying causal mechanisms there can be no objects or events in the world, and hence no empirical experience of them. Access to the domain of the real is achieved not by expressing experience in the empirical domain, nor by producing surface descriptions of phenomena in the actual domain, but by generating and iteratively testing retroductive models of reality in the real domain. Such retroductive models seek to identify the causal mechanisms that actualise and configure objects and events in the actual domain and make possible experience of them in the empirical domain. Both idealised expressions of personal experience in the empirical domain and nominal surface descriptions of events in the actual domain fall short of retroductive causal explanation.

There are close affinities between Bhaskar's account of these three domains and George Lindbeck's identification of three contrasting models of the nature of Christian doctrine (Lindbeck 1984). The *experiential-expressive* model 'interprets doctrines as noninformative and nondiscursive symbols of inner feelings, attitudes, or existential orientations', so that Christian doctrines constitute linguistic expressions of an individual's pre-linguistic intuitive experience. (p. 16). The *cultural-linguistic* model – championed by Lindbeck himself, and influenced by Clifford Geertz's social anthropology and Wittgenstein's account of meaning-generating communal practices and language games – understands Christian doctrines as 'communally authoritative rules of discourse, attitude, and action' that regulate the life of the Christian community, regardless of whether the rules enjoy any purchase on transcendent reality (p. 18). The *cognitive-propositional* model understands Christian doctrines as cognitively grounded 'informative propositions or truth claims about objective realities' (p. 16). There are, I suggest, significant similarities between: (1) the empirical domain of personal experience, the experiential-expressive model of religious language and the essentialist claim to self-evident pre-linguistic access to transcendent realty; (2) the actual domain of the sum total of existing objects and events in the world, the cultural-linguistic model of religious language and nominalist accounts of religious traditions; and (3) the real domain of the causal mechanisms underlying reality, the cognitive-propositional model of religious language and the critical realist commitment to retroductive explanatory modelling.

(1) In the *empirical domain* we enjoy particular experiences. As I write, I can look through my study window and observe snow falling in East Sussex. In the same way, different individuals may encounter experiences of transcendence triggered by particular circumstances: the contemplation of nature while walking the dog, the ingestion of Ecstasy (MDMA) while out clubbing, participation in an emotionally charged act of charismatic worship, etc. According to the experiential-expressive model, Christian doctrine functions instrumentally to express and further stimulate intuitive pre-linguistic experiences of transcendence in the context, for example, of communal acts of worship and private acts of prayer and meditation. However, it is entirely possible that identification of a particular experience as an encounter with a transcendental object may be mistaken: experience of God in an emotionally charged act of worship *may possibly* be delusional, and best explained in terms of emotional immaturity and intellectually gullibility rather than an encounter with transcendence. Contra the Cartesian-inspired Romantic insistence on the self-authenticating nature of intuitive unmediated experience, our understanding of such experiences can be misplaced and out of phase with reality. It may well be true that there is a constant and regular conjunction between the ingestion of Ecstasy and experiences of euphoria, but it does not necessarily follow that such euphoria is best explained as an objective encounter with transcendent reality.

Personal experience in the empirical domain is inextricably linked with communal experience in the actual domain. With reference to the game of chess, John Searle suggests that it:

> is not the case that there were a lot of people pushing bits of wood around on boards, and in order to prevent them from bumping into each other all the time and creating traffic jams, we had to regulate the activity.
>
> (Searle 1996, pp. 27ff.)

In order for an individual to play chess it is necessary to participate in an antecedently existing communal practice in the actual domain: 'the rules of chess create the very possibility of playing chess' (p. 28). By analogy, it makes little sense to suggest that acts of Christian worship are the result of isolated individuals coming together to collectively stimulate their private experiences of transcendence, and that in order to regulate such gatherings theologians construct the doctrinal system of Christianity. On the contrary, the evidence is incontrovertible: Christianity arose in direct response to the first disciples' encounter with the life, work and teaching of Jesus of Nazareth, rather than as a club for those seeking to express their private, intuitive, pre-linguistic experiences of transcendence. By the same token, Trinitarian theology was generated by the desire to explain the theological significance of Jesus of Nazareth as experienced within the collective life of the Christian community, rather than by the desire to articulate and foster private transcendental experience in a communal setting. The intentional stimulation of pre-linguistic spiritual experience, though apparently common among Gnostic sects and the various esoteric Mystery Religions, was looked on with suspicion by the early Christians. Paul, for example, insists that all such spiritual experience, together with all expressions of such experience, must be carefully assessed against the three-fold criteria of normative Christian belief regarding Jesus Christ and the Holy Spirit, the flourishing of the Christian community, and the supremacy of the Christian ethic of love:

> Now concerning spiritual persons.... You know that when you were pagans, you were enticed and led astray by idols that could not speak. Therefore I want you to understand that no one speaking by the Spirit of God ... can say 'Jesus is Lord' except by the Holy Spirit.... To each is given the manifestation of the Spirit for the common good.... If I speak with the tongues of mortals and of angels, but do not have love, I am a noisy gong or a clanging cymbal. And if I have prophetic powers, and understand all mysteries and all knowledge, and if I have faith, so as to remove mountains, but do not have love, I am nothing.
>
> (1 Corinthians 12:1ff.)

In insisting on the cognitive-propositional nature of Christian doctrine, the first Christian theologians established an explicit and particular understanding of the ultimate nature of reality that questioned the value of esoteric spiritual experience and brought the Christian community into direct conflict with the various Gnostic sects.

(2) In the *actual domain* the totality of objects events in the world exceeds our ability to experience them. Though I can look through my study window and observe that it is still snowing outside, I am aware that other human beings, in different geographical contexts, are currently experiencing very different weather conditions. In the same way, I am aware that diverse individuals and communities inhabit worldviews and subscribe to a range of religious and secular belief systems very different from my own. Further, I am also aware that my own Christian beliefs and practices are not the result of my first experiencing and then seeking to articulate particular encounters with transcendent reality. Were this the case, Christianity would consist of an amalgam of private and hermeneutically incommensurable language games, and as such would be vulnerable to Wittgenstein's critique of private language (Wittgenstein 1968). Rather, my Christian beliefs and practices are the result of my encounter with, and conscious decision to participate in, and subscribe to the teachings of, the Christian Church, which existed antecedently to any personal experience of transcendence I may or may not have had. According to Lindbeck's cultural-linguistic model, Christian doctrines are regulative of Christian communal beliefs and practices, and as such are precursors of, and enjoy priority over, the particular transcendent experiences of individual Christians.

This coheres with Searle's suggestion that the collective intentionality of a community is greater than the sum total of individual intentions. If, as Searle claims, it is impossible to play chess without a prior collective agreement regarding the rules of chess, so it is equally impossible to participate *meaningfully* in the Christian tradition without a prior collective agreement regarding the regulative rules and doctrines of Christianity. Both the 'collective intentionality of American citizens to treat certain pieces of paper as twenty-dollar bills', and the collective intentionality of 'Roman Catholics to treat baptism as the means of entry into the Christian community', constitute 'social facts that cannot be altered by any *individual* act of will' (Wright 2007, p. 154; emphasis in original). Social facts are normally rule driven, though the degree to which such rules are explicit (e.g. laws passed by the UK Parliament) or implicit (e.g. the widespread, though unlegislated, advocacy in the UK of the twin liberal principles of autonomy and tolerance) varies. Constitutive rules establish the very possibility of certain actions (e.g. without the rules of cricket it would be impossible to play cricket), while regulative rules order antecedently existing activities (e.g. the rule that we drive on the left in Britain and on the right in continental Europe regulates the antecedently existing practice of driving). In the case of both constitutive and regulatory rules, the formula 'you intend + I intend ∴ we intend' does not apply; because collective intentionality is dependent on a shared agreement (whether explicit or implicit) to act together, the proper formula is 'we intend ∴ you intend + I intend'. It is impossible for an orchestra to play a Mahler symphony without the collective intention to do so. This suggests that Christian doctrines are: (1) *constitutive* of Christian faith and praxis, so that Christianity could not exist apart from specific understandings of ultimate reality articulated by Christian doctrines; and (2) *regulative* of Christian faith and praxis, insofar as they function to establish order in particular Christian communities. If, in philosophical terms, the actual domain of the totality of objects and events is ontologically antecedent to the empirical domain of the personal experiences of individuals, so in theological terms Christian doctrines, in both constituting and regulating Christian communities, enjoy priority over the particular transcendent experiences of individual Christians.

(1) In the *real domain* a multiplicity of causal mechanisms generate and sustain different configurations of events in the actual domain, and make possible particular experiences in the empirical domain. As I write, I can still see snow falling outside my study window, and I remain aware that other people continue to experience very different weather conditions; however, to *explain* why this is the case I must look beyond (essentialist) expressions of my own experience and (nominalist) surface descriptions of the experiences of others, and seek to account (both critically and realistically) for the various causal mechanisms that generate and sustain actual weather systems and make possible particular empirical experiences of the weather. In other words, I must engage with the science of meteorology. According to the cognitive-propositional model, Christian doctrines are the products of theological science that, on the basis of an abductive hypothesis concerning the universal theological significance of Jesus of Nazareth, seeks to generate retroductive explanatory models of the ultimate nature of reality, and to subject them to iterative testing and revision. In doing so Christian theology is predicated primarily on the creative, reconciliatory and redemptive actions of the Trinitarian God that come to a head in the incarnation, and secondarily on the various responses of human beings to these divine actions. The cognitive-propositional model thus assumes that Christian doctrine possesses an epistemic purchase on deep ontological realities. It is here, I propose, that we gain access to Christian truth claims, rather than in the spiritual experiences of individual Christians or the cultural phenomena of Christianity.

On this realistic reading, the constitutive function of Christian doctrine envisaged by the cultural linguistic model of Christian doctrine draws its ontological legitimacy and epistemic warrant from its relationship with the cognitive-propositional model. If the Christian doctrines that constitute the Christian Church are devoid of ontological purchase on transcendent reality, then Christianity is one vast mistake; if, on the other hand, cultural-linguistic doctrines that constitute Christian communities are simultaneously cognitive propositions that enjoy ontological purchase on transcendent reality, then Christianity is true, or at least more truthful than any available alternative. According to Gunton, the cultural-linguistic model, in focusing exclusively on the stratum of immanent culture, lacks any sustained onto-theological purchase on the reality of God, and as such constitutes 'a direct attack on the notion of revealed religion' (Gunton 1995, p. 7). However, according to Lindbeck, since the Christian creeds which constitute and regulate Christianity communities (1) recount a realistic narrative of a divinely initiated interplay between God and humanity, and, in doing so, (2) affirm propositions about ultimate reality,

> [t]here is nothing in the cultural-linguistic approach that requires the rejection (or the acceptance) of the epistemological realism and correspondence theory of truth, which, according to most of the theological tradition, is implicit in the conviction of believers.
>
> (Lindbeck 1984, pp. 68f.)

Lindbeck's position is, I would suggest, more compelling for two reasons. First, since the Triune God is simultaneously transcendent of and immanent within creation, explanations of his providential actions cannot avoid referring to events within the spatio-temporal order. Second, the modern equation of narrative with fiction and concomitant assumption that propositional truth must therefore be timeless and devoid of any spatio-temporal particularity is not sustainable. Indeed, it is a direct challenge to Christian truth claims to deny that God's historical actions are both onto-realistic *and* grounded in contingent historical events.

Eric Auerbach points out that though fictional narratives can function as sources of entertainment that seek 'merely to make us forget our own reality for a few hours', they can also constitute works of art that seek 'to overcome our own reality' by enabling readers to encounter deeper and more profound ways of making sense of their personal realities in relation to reality as a whole (Auerbach 2003, p. 15). Following Auerbach, it is possible to read the Bible as a work of realistic art: 'realistic' because it makes specific truth claims about the nature of reality, 'art' because these truth claims are embodied in a range of different literary genres. These genres include, though they are not restricted to, both mythical (e.g. the story of Adam and Eve) and historical (e.g. the Gospel accounts of the life of Jesus) narratives. The myth of Adam and Eve seeks to convey realistic theological truths about the human condition while remaining mythical ('fictional'), whereas the historical accounts of the life of Jesus in the Gospels seek to convey realistic theological truths about actual historical ('factual') events. On the one hand, Adam and Eve do not need to have been historical figures for the theological truths contained in the Genesis myth to retain an ontological purchase on reality, because the fallen nature of humanity is a universal reality. On the other hand, if Jesus was not actually crucified under Pontius Pilate and did not actually rise from the dead, then the entire edifice of Christian faith collapses, because it claims that the incarnation was a unique ontological event rather than a mere exemplification of a universal state of affairs. On this reading the myth of Adam and Eve provides insight into general *structural* truths about the human condition, while the historical life of Jesus constitutes a particular spatio-temporal event, driven by the interplay of divine and human *agency*, that cannot be de-historicised, idealised and presented as an exemplification of general structural truths without undermining the core of Christian teaching. The biblical narrative, according to Auerbach, tells the realistic story of God's creative and redemptive acts and of the varying responses of human beings to them, and in doing so employs a diverse range of literary genres, including the genres of mythical tale and historical narrative. When we read the Bible, we encounter an invitation

> to fit our own life into its world, feel ourselves to be elements in its structure of universal history.... Everything else that happens in the world can be conceived as an element in this sequence; into it everything that is known about the world ... must be fitted as an ingredient in the divine plan.
>
> (Auerbach 2003, pp. 15ff.)

This suggests that there is no necessary antithesis between the realistic narrative that constitutes and regulates Christian faith and praxis, and the propositional claims embedded, both explicitly and implicitly, within that narrative. Both combine to reveal the deep causal mechanisms, divine, natural and human, that generate and sustain particular configurations of the actual world in general and the phenomenon of Christianity in particular, and which make possible particular experiences of transcendent reality.

On this critical realist reading, the identity of Christianity is inextricably bound up with a set of retroductive explanatory hypotheses regarding the ultimate nature of reality, hypotheses that remain disputed both within and beyond the Christian Church, and which are open to critical evaluation by both insiders and outsiders. Unlike essentialism and nominalism, critical realism is not driven towards epistemic closure; on the contrary, it is intrinsically epistemically open, and as such provides a more powerful and comprehensive tool for recognising the identity of Christianity and evaluating its potential truth.

Trinitarian Christianity

The personal experience of individuals in the empirical domain who self-identify as Christian, or are identified as Christians by others, is extraordinarily diverse. Don Cupitt, Rudolph Bultmann and Karl Barth all self-identify as Christians. Cupitt's experience leads him to deny the objective reality of God and treat Christian doctrines as instrumental tools for the stimulation of self-transcendence within a naturalistic worldview (Cupitt 1980). Bultmann's experience leads him to affirm the objective reality of God and treat Christian doctrines as heuristic myths that need to be stripped away to reveal their underlying existential significance (Bultmann 1958). Barth's experience leads him to affirm the objective reality of God and treat Christian doctrines as cognitive propositions that enjoy alethic purchase on ultimate reality (Barth 1936–1952). The communal experience of groups in the actual domain who self-identify as Christian, or are identified as Christian by others, is similarly diverse. In the period of the emergence of orthodox Trinitarian Christianity the Church faced oppositions from heterodox Christian and quasi-Christian traditions such as Gnosticism, Arianism and Nestorianism. These heterodox traditions continue to exist today: Gnosticism within the New Age movement, Arianism among Jehovah's Witnesses and

Nestorianism in the Assyrian Church of the East. Once again, if we take expressions of personal experience in the empirical domain, or descriptions of communal experience in the actual domain to be exhaustive of our attempts to identify and evaluate Christianity, then we are forced to the conclusion that there is no such entity as Christianity, merely a range of disparate traditions operating under the nebulous umbrella of the 'Christian' brand, and that Christians subscribe to a range of ontologically incommensurable truth claims (Smart 1979).

Critical realism recognises that accounts of ultimate reality are generated in the transitive realms of the empirical domain of personal experience and the actual domain of communal experience. These accounts take the form of retroductive explanatory models that claim purchase on the real domain of intransitive ontological reality, and which must be subjected to rational scrutiny in the light of alternative accounts in order to evaluate their veracity. Christian truth claims about the ultimate nature of reality are not to be found in expressions of the experiences of individual Christians in the empirical domain, or in descriptions of the phenomena of Christianity in the actual domain, but rather in transitive retroductive models, generated in the domains of the actual and empirical, *that seek to provide explanatory accounts of intransitive realities in the domain of the real*. The personal experiences of Cupitt, Bultmann and Barth are each grounded in specific communal traditions that proffer distinctive accounts of ultimate reality: the Christian anti-realism of the 'Sea of Faith' movement, Liberal Protestantism and Christian orthodoxy respectively. Similarly, orthodox, Gnostic, Arian and Nestorian forms of Christianity all provide ontologically incommensurable (though hermeneutically accessible) accounts of ultimate reality. At the risk of invoking Flood's criterion of 'prototypical features', we can recognise that what unites them all is a shared concern for the theological significance of the person of Jesus of Nazareth, while what differentiates them is the ontological incommensurability of their different accounts of Jesus' theological significance. Cupitt claims that Jesus was merely a human being; Bultmann claims that Jesus exemplifies the existential possibility of living an authentic life before God; Barth claims that Jesus was the incarnate Son of God. Orthodox Christians claim that the divine and human natures of Jesus were united in hypostatic union; Gnostic Christians tend to view Jesus as the last in a chain of spiritual beings bridging the divide between the Supreme Being and the fallen material world; Arian Christians claim that Jesus was a hybrid created being subordinate to God; Nestorian Christians deny any hypostatic union between Jesus' divine and human natures.

The conviction – shared by orthodox and heterodox Christian traditions alike – that the person, teachings and actions of Jesus of Nazareth possessed some kind of theological significance has its immediate source in Jesus himself, who was abductively recognised (truly or falsely) by his disciples as embodying in some way the unique presence of the Jewish God within the created order. The tradition of orthodox Christianity responsible for the generation of the retroductive doctrines of Trinity and the incarnation can trace a historical lineage back to the first disciples and the historical Jesus. The earliest documented historical evidence of belief in the theological significance of Jesus is contained in the documents collected in the New Testament, which provide evidence of the initial preaching (Greek, κήρυγμα, kérygma) of the earliest Christian communities inaugurated by Jesus' own immediate disciples. The 'earliest biblical evocations of Jesus' significance', which 'developed in a time when eye-witnesses were still alive', took the two-fold form of brief theological narratives of Jesus' life and embryonic creedal statements regarding his person (Gunton 2002, p. 81). The narratives told the story of the fulfilment of God's salvific plan for humanity in the birth, life, ministry, death and resurrection of Jesus Christ, while the creeds identified him as the eternal Son of God worthy of worship alongside his Father. Though the New Testament contains a diverse range of accounts of Jesus' theological significance, their unity lies in the identification of the life of the human Jesus with the saving actions of God (Dunn 1977; Marshall 2004).

> Common to all the theologies and stories, in their variety, even with their tensions and contradictions of detail, is that they mysteriously converge in this unique episode in human history which they hold to be definitive for the destiny and meaning of the whole of creation.
>
> (Gunton 2002, p. 84)

Gnostic, Arian and Nestorian versions of Christianity are inconceivable apart from this antecedent belief in the unique and universal theological significance of Jesus, which was first proclaimed by Jesus' immediate disciples and later identified as 'orthodox' by the mainstream Christian Church. Regardless of their veracity, Gnosticism, Arianism and Nestorianism constitute revisionary readings of the emergent tradition of Christian orthodoxy. The Gnostic texts do not appear until the second and third centuries, and the inclusion of Jesus in their speculative ahistorical mythopoeia is, with the possible exception of a cluster of isolated verses in the *Gospel of Thomas*, and unlike the canonical Gospels, almost completely devoid of any connection to the historical Jesus. Arianism and Nestorianism emerged, in the fourth and fifth centuries respectively, as alternatives to the emergent incarnational teaching of orthodox Christianity. This is not to claim that Christian orthodoxy possesses greater alethic truth simply by virtue of its being antecedent to Gnosticism, Arianism and Docetism – to do so would be to slip into the genetic fallacy of confusing truth and origins. However, it does suggest that any account of heterodox versions of the Christianity phenomenon needs to explain how and why they diverged from Christian orthodoxy.

The label 'Christianity' covers a diversity of personal experiences, multiplicity of historical phenomena, and variety of ontologically incommensurable retroductive truth claims about ultimate reality. In the rest of this book we will restrict our discussion to the specific retroductive account of ultimate reality proffered by orthodox classical Trinitarian Christianity. The primary reason for this is that there is simply no space available to consider the variety of ontological truth claims present under the 'Christian' umbrella. However, in taking this decision we can make a virtue out of a necessity.

Critiques of Christianity frequently fail to distinguish between different versions of Christianity, and as a result tend to rely on abstract generalisations. It is increasingly clear, for example, that many Enlightenment and post-Enlightenment attacks on Christianity focus on forms of classical theism rather than Trinitarian Christianity (Buckley 1987; Babcock 1991). By centring our discussion on Trinitarian theology I hope to contribute to the differentiation of various forms of Christianity, in the belief that such differentiation is essential for the application of judgemental rationality.

Further, the decision to focus on Trinitarianism is not as restrictive as it may first appear. Orthodoxy continues to dominate contemporary Christianity, and can trace a historical lineage back to that encounter between Jesus and his disciples which first generated the notion that the historical Jesus possessed theological significance. The core tenets of Trinitarianism were iteratively honed by the seven Ecumenical Councils, starting with the first Council at Nicaea in 325 CE and running through to the second Council of Nicaea in 787 CE. Though the precise nature of the authority of these Councils is disputed, their core teachings are accepted by each of the three main traditions of Christianity, all of which affirm Trinitarian theology: Eastern Orthodoxy, the Western Roman Catholic Church, and the majority of Western Protestant denominations – including the Anglican (Episcopal), Reformed (Calvinist) and Lutheran (Behr 2004).

The 'Great Schism' between Eastern Orthodoxy and Western Catholicism in 1054 revolved around issues of ecclesiastical politics, though it did contain an important doctrinal dimension, namely the decision of the Western churches to add the '*Filioque*' clause to the Nicene-Constantinopolitan Creed. The original Creed stated that the Holy Spirit 'proceeded from the Father', while the revised Western version asserted that the Holy Spirit 'proceeded from the Father *and the Son*' (*qui ex Patre Filioque procedit*). This led to a subtle, though significant, shift of emphasis in the churches' understanding of the Holy Spirit. For our current purposes, the significance of the dispute lies in the fact that both parties assumed the basic truth of Trinitarian doctrine, and were concerned simply with an issue of nuance and emphasis – albeit a significant one – with respect to the doctrine of the Holy Spirit.

The sixteenth-century Reformation in the West, which resulted in the division between Roman Catholicism and Protestantism, was also not the result of any fundamental dispute about the basics of Trinitarian doctrine. Rather, it revolved around a disagreement concerning the nature of salvation, the authority of the institutional church, and related ecclesiastical practices that focused on contrasting interpretations of the manner in which sinful human beings are justified before God. Despite their critique of the Roman Catholic Church, all the major Protestant Reformers continued to subscribe to the doctrines of Trinity and incarnation as articulated in the Ecumenical Creeds.

The twentieth-century ecumenical movement has taken significant steps in reducing the theological distance between these three Trinitarian traditions. Torrance has sought to reduce the significance of the *Filioque* dispute by drawing attention to its departure from prior theological agreement regarding the nature and task of the Holy Spirit (T.F. Torrance 1996, pp. 186ff., 190f.; cf. Gunton 2003, p. 50). Similarly, the Roman Catholic theologian Hans Küng has argued that, when viewed with the benefit of hindsight, there is no substantial difference between Protestant and Catholic teaching concerning the Christian doctrine of justification, a view supported by the Protestant theologian Karl Barth in his Preface to Küng's book (Küng 1981). This is not to suggest that Trinitarian theology is monolithic: just as critical realism embraces a wide range of positions and perspectives within a common and relatively stable set of philosophical assumptions, so orthodox Christianity contains a diverse range of viewpoints within a relatively transparent paradigmatic framework. Such unity-in-diversity is inevitable in any tradition committed to the ongoing scrutiny and iterative refinement of its core beliefs and practices.

What, then, does it mean to be epistemically relative about Christianity? The answer depends on which particular version of Christianity we are talking about, and whether our focus is on individual spiritual experience, the socio-cultural phenomenon of particular Christian communities, or a specific set of onto-theological truth claims. For reasons outlined in this chapter – generated by the recognition that critical realism offers a more powerful tool for identifying the nature of Christianity and recognising its epistemically relative nature than either essentialism or nominalism – we will reformulate the question thus: What does it mean to be epistemically relative about the retroductive truth claims of classical orthodox Trinitarian Christianity, as subscribed to by Eastern Orthodox, Western Catholic and mainstream Protestant churches?

5 Trinitarian theology and epistemic relativism

According to the Acts of the Apostles, 'it was in Antioch that [Jesus'] disciples were first called "Christians" ' (Acts 11:26). From its original function of identifying the religious movement that was to evolve into mainstream orthodox Christianity, the label 'Christian' has expanded to become an umbrella term designating a wide range of spiritual experiences, communal traditions and incommensurable accounts of the ultimate order-of-things. Consequently, any exploration of what it means to be epistemically relative about Christianity must define its terms of reference carefully if it is not to slip into abstract generalisations. In the previous chapter we examined three different readings of Christianity: an idealist reading, that claimed unmediated experiential access to the transcendent realm; a nominalist reading, that limited itself to surface descriptions of the immanent socio-cultural phenomenon of Christianity; and an orthodox reading, that told the story of the transcendent Trinitarian God's immanent providential presence within the historical process. We elected to focus primarily on the latter orthodox reading, and its advocacy of a Trinitarian theology. We did so partly for practical reasons of space, partly because the idealist and nominalist traditions constitute, in the main, post-Enlightenment revisionary readings of Christian orthodoxy, and partly because the three major Christian traditions – Eastern Orthodoxy, Western Catholicism and Western Protestantism – all subscribe to Trinitarian doctrine. Having identified orthodox Trinitarian Christianity as the focus of our investigation, we can now move forward to explore the relationship between Trinitarianism and epistemic relativism. There is no questioning the fact that the truth claims of Trinitarian Christianity, like *all* truth claims, however puerile or refined, are open to investigation within the framework of epistemic relativism. What is at issue in this chapter is the extent to which *adherents* of Trinitarian Christianity recognise and acknowledge the epistemically relative nature of their truth claims. Failure to do so will serve to enhance the modernist notion that theological beliefs lack rational warrant and evidential support; conversely, the acceptance of epistemic relativism on the part of Trinitarian Christians will serve to enhance the critical realist claim that theological statements are, both in principle and practice, open to critical evaluation.

Epistemic relativism revisited

As we have already seen, the principle of epistemic relativism affirms that 'human intellectual enterprises are necessarily fallible, but not for that matter, necessarily mistaken' (Gunton 1983, p. 145). Since knowledge is generated by fallible human beings in particular socio-cultural contexts, it is necessarily incomplete and potentially mistaken. Because we are not omniscient beings with access to a god's-eye perspective on reality, we are right to be suspicious of claims to epistemic closure predicated on logo-centrism, onto-theology or a metaphysic of presence. However, epistemic relativism does not imply 'the doctrine of *judgemental relativism*, which maintains that all beliefs are equally valid in the sense that there are no rational grounds for preferring one to another' (Bhaskar 1986, p. 72; emphasis in original). On the contrary, the principle of judgement rationality 'gives no place to any arbitrary choice or subjective preference in assessing scientific theories', but instead affirms the possibility of adjudicating between conflicting truth claims, while recognising that such judgements will necessarily remain provisional and subject to ongoing critique and evaluation (Collier 1994, p. 57). Relativism in the transitive domain of epistemology does not imply relativism in the intransitive domain of ontology; rather, epistemic relativism presupposes the existence of a reality independent of our knowledge of it, and makes the absence of relativism in the intransitive domain a necessary condition for the generation of knowledge, however imperfect, in the transitive domain. Maintaining a proper distinction between ontology and epistemology opens up the possibility of making critical-yet-contingent judgements between competing truth claims, on the basis of the recognition of the greater and more comprehensive explanatory power of one theoretical model over another. The possibility of judgemental rationality lies in the fact that conflicting truth claims constitute transitive disagreements about a common intransitive object. Thus theists, panentheists and atheists proffer incompatible truth claims about a common object (putative or real), namely 'God'. Ontological incommensurability between conflicting explanatory models does not imply hermeneutical incommensurability:

> Just as one can switch back and forth between seeing a duck and seeing a rabbit in the famous duck/rabbit gestalt switch, so one can switch, if uneasily, between different world-views, and have no difficulty in *understanding* the one that one is currently *not* in.
>
> (Colloer 1994, p. 92; emphases in original)

The combination of epistemic relativism and judgemental rationality opens up the possibility of intellectual progress and in doing so undercuts the tyranny of the merely fashionable.

Epistemic relativism seeks to chart a course between and beyond the Scylla of modernity and Charybdis of postmodernity. Both outlooks, which are best seen as mirror images of a common tradition, are rooted in the heritage of

Cartesianism. Descartes' *Meditations* chart his intellectual and spiritual quest to make sense of the social, political, religious and intellectual upheavals that characterised much of seventeenth-century Europe (Descartes 1969).

> The terrifying quality of the journey is reflected in the allusions to madness, darkness, the dread of waking from a self-deceptive dream world, the fear of having 'all of a sudden fallen into very deep water' where 'I can neither make certain of setting my feet on the bottom, nor can I swim and so support myself on the surface'.
>
> (Bernstein 1983, p. 17)

A key source of this 'Cartesian anxiety' was the late medieval theology of William of Ockham, who dismissed the traditional scholastic synthesis of faith and reason on the grounds that it threatened to compromise the sovereignty of God by making him at least partially subservient to human reason (Ockham 1983, 1990; cf. Spade 2000, pp. 326ff. *passim*). In insisting on the absolute sovereignty of a transcendent God utterly aloof from the world of human experience, Ockham paved the way for the eclipse of the Trinitarian God, who through the incarnation was intimately engaged with creation in a gracious, loving and beneficent manner, and his replacement with a deity disengaged from creation, beyond human comprehension, and wielding absolute – and potentially arbitrary and despotic – power. Descartes' existential fear of being deceived by such a malignant deity left him with a stark alternative: '*Either* there is some support for our being, a fixed foundation for our knowledge, *or* we cannot escape the forces of darkness that envelop us with madness, with intellectual and moral chaos' (Bernstein 1983, p. 18; emphases in original). Descartes' anxiety-laden search for epistemic certainty took the route of a thoroughgoing hermeneutic of suspicion designed to strip away all fallible beliefs and reveal a secure foundation for knowledge in the immediate reflective consciousness of the thinking subject: *cogito ergo sum*.

Operating in the wake of Descartes' strictures, the philosophers of the Enlightenment,

> facing the possibility that all their given reality and perception were illusory, arbitrary and liable to negation at any moment by the Creator, decided to assert themselves and make stable their values, thereby securing a world that a perverted theology had so explicitly abandoned.
>
> (Blond 1999, p. 234)

The modern quest for epistemic certainty, predicated on the freedom to trust one's own reason and question all external authorities, took two routes: (1) the path of idealism promised secure knowledge, transcending the contingent ebb and flow of ordinary life, predicated on the immediacy of rational (Hegelian) or experiential (Romantic) awareness; (2) the path of empiricism offered access to natural laws construed on the basis of empirical observation of regular conjunctions of objects and events. Both paths promised epistemic closure through the exercise of autonomous reason, and in doing so opened the door to 'that strain of modern thought ... which sees man not as a limited and imperfect being who "muddles through", but as a superhuman being who can create the world anew through the application of his infinite will' (Gillespie 1996, p. xxiii). Consequently – and not without irony – the 'arbitrary will of the Ockhamist deity comes to be metamorphosed into the arbitrary will of the human agent' (Gunton 1995, p. 48). David Levin has traced an organic link between the modern dream of autonomous reason, the concomitant drive towards epistemic closure, and the emergence of various forms of intellectual and political totalitarianism. Cartesian subjectivity gave birth to an autonomous rationality that claimed sovereignty over the external world; however, the subjective self 'turned brutish and competitive, and a false individualism soon began to inhabit the descendants of the monadic Cartesian ego' (Levin 1988, pp. 3ff.). The will to power of the self-affirming ego convinced of its own righteousness generated totalitarian regimes that harnessed the technological resources generated by modern science: 'When reason turned totally instrumental, a function solely of power, it legitimated the construction of a totalitarian state and engineered the Holocaust' (p. 4).

Postmodernity recognised the link between epistemic closure and totalitarianism, and sought to counter the autocratic thrust of modernity by affirming a thoroughgoing judgemental relativism that identified all truth claims as moves within a modernist economy of power. In doing so it embraced as its underlying ideal 'that most invisible, because it is the most pervasive, of all modern goods, unconstrained freedom' (Taylor 1992, p. 489). Such freedom was tempered by the claim that, since truth must not and cannot be pursued, the modernist bond between autonomy and totalitarianism had effectively been severed. The exercise of autonomy avoids totalitarianism because it is driven not by any rigorous pursuit of truth but rather by a nomadic wandering around the unchartable postmodern cultural landscape. However, this merely replaces the certainty of epistemic closure with its mirror image – the certainty of thoroughgoing epistemic openness predicated on the rejection of judgemental rationality. Paradoxically, postmodernity proclaims the totalitarian truth that the pursuit of truth is morally unacceptable and intellectually unsustainable. Thus postmodernity inhabits the same Cartesian thought-world as modernity, the only difference being that where modernists fear drowning in a tide of anxiety, post-modernists throw caution to the wind in celebrating the white-water ride of epistemic ambiguity that takes centre stage in the postmodern cultural theme park. If the negative moral consequences of this postmodern move are not as immediately transparent as the modern cultivation of totalitarianism, they are nevertheless very real, not least because to reject the pursuit of truth is to affirm the status quo, thereby ruling out any suggestion that the norms of consumer capitalism, which host and nurture postmodern culture, are unjust and require transformation.

What, then, does it mean to be epistemically relative about the onto-theological truth claims of Trinitarian Christianity? Negatively, critical realism's affirmation of epistemic relativism rules out any form of modernist epistemic closure, while its rejection of judgemental relativism rules out any postmodern attempt to shield Christian discourse from the demands of rational scrutiny. Positively, to be epistemically relative about Christianity is to recognise: (1) that Christianity makes truth claims about the ultimate nature of reality; (2) that these truth claims are generated by fallible human beings in particular socio-cultural contexts, and as such are necessarily incomplete and potentially mistaken; and (3) that it is nevertheless possible to make critical-yet-contingent judgements both internally between orthodox and heterodox versions of Christianity, and externally between orthodox Christianity and alternative secular and religious accounts of ultimate reality.

Trinitarian truth claims

To be epistemically relative about Christianity is to recognise that Trinitarian theology makes alethic truth claims about the ultimate ontological ground, nature and structure of reality. Though the fact that Christianity does make such claims may appear self-evident, we must attend to the minority view that this is not actually the case. The issue is important because if Christianity does *not* proffer truth claims then there is nothing to be epistemically relative about.

The vast majority of the world's religious traditions are either implicitly or explicitly realistic in outlook (Markham 1998; Byrne 2003; Moore and Scott 2007). Despite the rise of anti-realist readings of religion in postmodern contexts, there 'is no getting around the fact that people who express their religious convictions are in so doing referring to a specific – usually divine and divinely instituted – reality and intend to assert something as true of it' (Pannenberg 1976, p. 327). This certainly seems to be the case for orthodox Trinitarian Christianity, which subscribes to what appears to be a relatively stable realistic belief system that is clearly articulated in a range of authoritative creedal statements (Kelly 1950, 1977; Patterson 1999; Marshall 2000; Moore 2003). It is sometimes suggested that Christianity is atypical of religious traditions by virtue of its highly developed doctrinal system, and that religions are on the whole best understood in terms of their social practices rather than their doctrinal beliefs, so that orthopraxis takes precedence over orthodoxy. This allows for the Liberal Protestant reading of Christianity offered by Adolf von Harnack, according to which the Church's original ethical orthopraxis was transformed into a creedal orthodoxy as a result of the Hellenisation of Christianity and its encounter with Greek philosophy (Rumscheidt 1989, pp. 108ff.). The notion that the original Christian community sought to implement the ethical teaching of Jesus and that its priests transformed it into an institutionalised dogmatic belief system based on the person of Jesus rather than his message is popular today among those who wish to hold fast to the ethical teaching of Jesus while discarding institutional Christianity and its theological and metaphysical baggage. If it is true that Christianity originally stood in line with other religious traditions in subscribing to spiritual orthopraxis rather than doctrinal orthodoxy, then Christianity in its most pristine and authentic form did not proffer realistic truth claims.

However, such a view is predicated on the modern dualistic distinction between 'act' and 'being'. Here scepticism about our ability to access knowledge of the ultimate structures of reality (being) generates reductive accounts of social behaviour (acts) devoid of any relationship to the ultimate order-of-things. It is certainly true that Christian doctrinal and creedal statements are in the main more extensive, detailed and explicit than those proffered by other religious systems. Nevertheless, the vast majority of religious traditions clearly affirm specific understandings of the ultimate nature of reality that inform the spiritual life and cultic practices of their respective communities. Jewish observance of Torah and Islamic commitment to Sharia law cannot be dislocated from the realistic theological beliefs underlying them. Judaism affirms the Shema as its basic creedal statement, while Islam proclaims the Shahadah:

> Hear, O Israel: The LORD is our God, the LORD alone. You shall love the LORD your God with all your heart, and with all your soul, and with all your might.
>
> (Deuteronomy 6:4f.)

> I bear witness that there is no God but Allah ... I bear witness that Muhammad is the Messenger of Allah.
>
> (Hamidullah 1979, pp. 44ff.)

Similarly, even Buddhists who are convinced that the present world of suffering is one vast web of illusion affirm the *reality* of Nirvana, albeit without resort to ontological language drawn from the Western philosophical tradition (Wright 2007, p. 161).

The issue, I suggest, is not so much the realism or otherwise of the beliefs of particular religious traditions, but the manner and extent to which they are articulated. Tom Wright argues that human beings inhabit worldviews that 'have to do with the presuppositional, pre-cognitive stage of culture or society' (Wright 1992, p. 122). Worldviews are bound up with our ultimate concerns and embrace 'deep-level human perceptions of reality' (p. 123). Wright identifies four characteristics of worldviews. (1) They provide the *stories and narratives* that help make sense of our place in the ultimate order-of-things. Such stories cannot be reduced to isolated observations or fragmented remarks, nor reified into abstract ontological statements. (2) These stories offer implicit answers to 'the *basic questions* that determine human existence: who are we, where are we, what is wrong, and what is the solution?' (p. 123; emphasis added). All worldviews 'cherish deep-rooted beliefs which can in principle be called up to answer these questions', since all 'have a sense of identity, of environment, of a problem with the way the world is, and of a way forward' (p. 123). (3) The stories that embody the deep beliefs of a worldview are expressed in cultural *symbols, artefacts and events*:

> In modern North America, the New York victory parade after a successful war brings together two of the most powerful symbols of the culture: the

> towering skyscrapers of business-oriented Manhattan, and the heroes of battle.... Both, in their own fashion, demonstrate, promote and celebrate The American Way.... In first-century Palestine, celebrating the Passover functioned similarly, with Jerusalem and the Temple taking the place of Manhattan, and the Passover sacrifice and meal taking the place of the victory parade. The buildings, instead of speaking of economic/ethnic goals spoke of religious/ethnic ones; instead of the celebration speaking of triumph achieved over the forces of darkness, it spoke of vindication yet to come.
>
> (Wright 1992, pp. 123ff.)

The deep beliefs of a society are brought to the surface whenever its cultural symbols are challenged in a manner that invokes fear and anger on the part of community members: witness, for example, Americans' reaction to the destruction of the World Trade Center. Cultural symbols and actions function as community boundary markers: this is seen clearly in, for example, the Conservative politician Norman Tebbit's infamous 'cricket test' which assessed the national loyalty of British citizens with roots in Asia and the Caribbean against the measure of their support for the English, Indian, Pakistani or West Indian cricket teams. (4) Worldviews encompass a *praxis*, or way-of-being-in-the-world. The future or eschatological orientation of the basic question 'What is the solution?' implies and entails action. Thus 'the real shape of someone's worldview can often be seen in the sort of actions they perform, particularly if the actions are so instinctive or habitual as to be taken for granted' (p. 124). Wright concludes that the stories, basic questions, symbols and actions that together form a worldview constitute:

> the basic stuff of human existence, the lens through which the world is seen, the blueprint for how one should live in it, and above all the sense of identity and place which enables human beings to be what they are.
>
> (Wright 1992, p. 124)

We all inhabit particular worldviews, be they religious or secular, which have embodied within them an implicit understanding of the ultimate order-of-things and of our place within it. The fact that some religious and secular communities have developed explicit creedal statements – e.g. the Christian Creeds, the Jewish Shema, the Muslim Shahadah, the democratic American Constitution, the liberal Universal Declaration of Human Rights – that identify, or come close to identifying, the realistic presuppositions underlying their particular worldviews does not warrant the anti-realist conclusion that those communities that have not done so, and whose basic beliefs remain implicit, lack a realistic understanding of their place in the ultimate order-of-things. Wright's account of the nature and role of worldviews enables us to affirm that religious traditions, including the Christian tradition, embrace realistic accounts of the ultimate order-of-things, regardless of the manner and extent of their explicit articulation. Even if it were true that the first Christians sought to implement Jesus' moral teaching rather than worship him as God, the decision to do so would still entail a realistic understanding, whether explicit or implicit, of the relationship between Jesus' teaching and the ultimate order-of-things. Thus in the early Christian era the Ebionites, a sect of Jews who recognised Jesus as the Messiah but denied his divinity and rejected mainstream Christianity, sought to implement Jesus' moral teaching within a realistic theological framework provided by Judaism.

The source of the claim that religions in general and Christianity in particular do *not* subscribe to realistic beliefs about the ultimate order-of-things is to be found in the empiricist-driven post-Enlightenment assumption that we have no epistemic access to ontological reality. Humean empiricism rejected Christian belief in miracles, and hence the entire edifice of Christian doctrine, on the ground that they could never constitute part of those regularly occurring patterns of events and configurations of objects upon which reliable knowledge of past events was deemed to depend. Logical Positivism rejected Christian doctrine on the grounds that, since its truth claims are closed to empirical verification, it is inherently and quite literally meaningless. Faced with the empiricist denial of the epistemic purchase of Christian doctrine on ontological realities, Liberal Protestant theologians increasingly sought succour in the idealist tradition. Hegel's rationalistic account of Christianity bypassed the empiricist challenge by appealing to the internal coherence of theological statements rather than their convergence with any external empirical reality. In turn Romanticism embraced a pre-rational and pre-linguistic experiential idealism that offered Christianity an epistemic foundation in the immediacy of self-legitimating transcendental spiritual experience: if we cannot see, hear or touch God, we can still be intuitively aware of his presence. This required the replacement of the received cognitive-propositional model of doctrinal statements with an experiential-expressive one: Christian doctrines were viewed, not as realistic propositional models, but as secondary expressions of primary pre-discursive transcendent spiritual experiences. This inevitably generated a general loosening of the purchase of Christian doctrines on reality. Thus Schleiermacher relegated the doctrine of the Trinity to an appendage to his systematic account of Christian faith, and Bultmann reduced the doctrine of incarnation to the status of an existentially useful non-realistic myth (Schleiermacher 1976, pp. 738ff.; Bultmann 1958). This revisionist reading of the nature of Christian doctrine opened the door to Hick's universal theology which, in affirming the possibility of direct, unmediated experiential access to the objective reality of the 'Absolute', effectively denies the doctrines of Trinity and incarnation any cognitive purchase on that reality, and thereby reduces their function to the instrumental one of aiding the stimulation of extra-linguistic spiritual experience (Hick 1989). Any affirmation of the greater alethic truth or explanatory power of one set of religious doctrines over another is ruled out *a priori*, since all are equally lacking in any epistemic purchase on reality. This denial of the cognitive-propositional nature of religious language appears to be replicated in Bhaskar's claim that meta-Reality 'is not really even a system of thought, but an intervention in the discursive process' (Bhaskar 2002c, p. xxiv).

Bhaskar presents Jesus as one of a number of inspirational gurus whose spiritual example can help stimulate awareness of the ground-state; at no stage does he consider the possibility that the Christian doctrine of the incarnation might possess alethic truth. The universal experiential-expressive theologies of Hick and Bhaskar open the door to forms of thoroughgoing anti-realistic theology that deny the ontological reality of any and all accounts of God and/or transcendence, including the Trinitarian God of classical Christianity (Cupitt 1980; Taylor 1982, 1987). If Christian doctrine functions instrumentally to express pre-linguistic spiritual experience rather than propositionally to account for the ultimate order-of-things, then Christianity does not proffer realistic truth claims, but merely acts as a vehicle for the stimulation of transcendent spiritual experience. If this is the case, we cannot be epistemically relative about Christian doctrine because Christian doctrine no longer contains any substantial realistic truth claims to be epistemically relative about.

This dualistic divide between act and being, between the instrumental and propositional function of religious language, and between religion as orthopraxis and orthodoxy, is directly linked to scepticism about the epistemic licence and cognitive warrant of Christian truth claims in the post-Enlightenment era. Such scepticism gave rise to the emergence of departments of religious studies alongside schools of theology in the modern university. If the historical, sociological and psychological study of the phenomenon of religion within a broadly naturalistic frame of reference was and remains largely uncontentious, the viability of the theological and philosophical study of those ultimate ontological realities to which religious traditions claim access, and indeed of any legitimate relationship between Christianity and the modern university, was, and in some quarters remains, a matter of fundamental dispute (Newman 1982; Fisher 1989; Griffin and Hough 1991; Marsden 1994, 1998; Diekema 2000; Cady and Brown 2002; Astley *et al.* 2004; D'Costa 2005; Hauerwas 2007). 'The dethronement of theology was directly correlated with the erosion of Christendom and the rise of the modern university rooted in Enlightenment ideals' (Cady and Brown 2002, p. 3). Prior to the Enlightenment, university education was grounded in the unity of act and being, rooted in a Christian version of the classical Greek tradition of *paideia* that posited an essential unity between the striving for truth, the flourishing of society and personal spiritual development (Jaeger 1945, 1961; Kelsey 1992; Farley 1994; Wright 2007, pp. 55ff.). This tradition began to fragment as universities drew a dualistic distinction between the pursuit of scientific 'fact' in faculties of natural and social science, and the cultivation of personal and social 'values' in faculties of arts and humanities.

This process of the disintegration of the unity of fact/value and being/act is clearly visible in the contributions of Kant and Schleiermacher to the debate surrounding the founding of the University of Berlin in 1810 (Howard 2006, pp. 130ff.). Kant's *Conflict of the Faculties* drew a distinction between the 'higher' faculties of law, medicine and theology responsible for professional training, and the 'lower' faculty of philosophy responsible for questions of truth (Kant 1992). Though he allowed the state to intervene in the higher faculties on pragmatic political grounds, the lower faculty was deemed the exclusive preserve of academics committed to the pursuit of truth. The disinterested pursuit of objective truth was thus distinguished from the political interest invested in the training of those professionals responsible for overseeing the flourishing of individuals-in-society. Kant's inclusion of theology in the higher faculties and exclusion of the discipline from the lower faculty was driven both by a concern for the professional training of the clergy in a society in which the Church remained an established part of the socio-political order, and by the belief that theology did not meet the rational standards required for entry into the lower, foundational, faculty. According to Schleiermacher, the study of theology was a 'positive' rather than a 'pure' science, rooted in historical contingency rather than universal truth (Schleiermacher 2011; see also Farley 1994, pp. 73ff.). As such, its task was 'to train professional church leadership for their indispensable social rules' (Kelsey 1992, p. 88). 'Pure' theology, concerned with the truth of Christian doctrinal claims, was excluded from the curriculum because its need to assign privileged status to the Bible as the revealed Word of God placed it beyond the scope of critical enquiry (p. 88). For both Kant and Schleiermacher the place of theology in the university must be limited to its contribution to the orthopraxis of the Church, and hence to the flourishing of society and the spiritual development of individual citizens, and excluded academic investigation of its orthodox truth claims.

Both the turn to an experiential-expressive model of Christian doctrine and the limiting of theology to an instrumental role in the modern university replicated the Enlightenment's overlapping dualistic distinctions between being/act, fact/value, truth/truthfulness and orthodoxy/orthopraxis. Apart from this dualistic framework, the notion that religious traditions and worldviews had no interest, either implicitly or explicitly, in relating their spiritual praxis and doctrinal beliefs to the ultimate order-of-things is virtually inconceivable. Insofar as the dualistic framework was generated by epistemic scepticism regarding the ability of religious traditions to provide realistic accounts of the ultimate order-of-things, it is a direct product of the epistemic fallacy. It follows that claims that Christianity does not proffer alethic truth claims about the ultimate ontological ground, nature and structure of reality are equally products of the epistemic fallacy. Once we accept the critical realist argument that theological truth claims are open to rational scrutiny, then it makes little sense to suggest that there was ever a time, prior to the modern era, when Christians accepted that their beliefs were merely instrumental tools that bore no relationship to the ultimate order-of-things. Thus Trinitarian Christianity meets the first of our criteria for being epistemically relative about Christianity: contra claims generated by post-Enlightenment empiricist, idealist and pragmatic modes of thought, both within and outside the Christian community, orthodox Christians proffer onto-theological alethic truth claims about the ultimate nature of reality that they are under an intellectual obligation to defend. Since Christianity does indeed proffer ultimate truth claims we can be epistemically relative about Christianity, since there is now something substantial to be epistemically relative about.

The fallibility of Trinitarian truth claims

To be epistemically relative about Christianity is to recognise that Christian truth claims are generated by fallible human beings in particular socio-cultural contexts, and as such are necessarily incomplete and potentially mistaken. Non-Christians, of course, have no problem in recognising that Trinitarian truth claims are the fallible products of fallible human beings. Hence the crucial question before us is whether Trinitarian Christians themselves accept the fallibility of their truth claims. Any failure to do so would effectively exempt their beliefs from the canons of epistemic relativism and judgemental rationality, and leave them with no defence against the charge of subscribing to fideistic or fundamentalist beliefs that, since they have no rational support, can only be propagated through rhetorical persuasion and/or coercive force. It is certainly true that Christianity has flirted with epistemic closure and the denial of epistemic relativism, particularly as a result of its encounter with the Enlightenment, as exemplified above all in the emergence of the distinctively modern notions of papal infallibility and biblical inerrancy. The modern quest for epistemic closure initiated by Descartes and followed through by Locke and Hume sought to distinguish between demonstrable objective fact and indemonstrable subjective opinion by utilising either the rationalist criterion of clear and distinct ideas or the empiricist criterion of observational verification. Faced with the possibility of Christian truth claims being reduced to the level of mere subjective opinion, some Christian traditions sought to demonstrate the compatibility of Christian doctrine with modernist demands for epistemic certainty. The doctrine of papal infallibility, propagated at the First Vatican Council in 1870, asserts that whenever the Pope promulgates a clear and distinct doctrine *ex cathedra* then, provided that doctrine is compatible with divine revelation, the Holy Spirit will act to preserve it from error. Similarly, claims to biblical inerrancy asserted by some Protestant groups identified the Bible as absolutely true and completely free from any form of error in *all* fields of knowledge, theological, moral, historical, sociological, scientific, etc. Neither papal infallibility nor biblical inerrancy is prototypical of classical Trinitarianism.

The doctrine of papal infallibility is contested within the Roman Catholic Church, and is rejected by other Trinitarian traditions (Küng 1971). With regard to the more general question of ecclesial authority within which the doctrine of infallibility sits, though the proclamations of the Ecumenical Councils are considered authoritative, they are not viewed as inerrant or infallible. Thus, for example, the *Thirty-Nine Articles*, which have served to ground the doctrinal teaching of the Church of England since their promulgation in 1563, recognise that the teachings of the Ecumenical Councils 'may err, and sometimes have erred, even in things pertaining unto God' since they are the products of 'an assembly of men', not all of whom may be 'governed with the Spirit and Word of God' (Bicknell 1955, p. 267). The Anglican tradition follows classical Christianity in insisting that, apart from the person of Jesus Christ, there is no direct revelation or experience of God, and that all theological knowledge is mediated through fallible human agents. The pursuit of doctrinal truth is necessarily a hermeneutical process, through which the teachings of the Ecumenical Councils are subject to theological scrutiny in the light of Scripture, tradition and reason.

Biblical inerrancy, in the fundamentalist sense outlined above, is rejected by all major Trinitarian traditions (Barr 1981). Advocates of inerrancy tend to assume a positivist hermeneutic, according to which biblical statements enjoy a direct unmediated relationship with the realities to which they refer. The notion of biblical inerrancy is virtually inconceivable, apart from the Cartesian demand for epistemic certainty and the positivist requirement of a direct unmediated connection between sense experience, language and empirical objects, and as such is a product of modernity. Prior to the modern era, orthodox Christianity had little, if any, comprehension of the option of a positivistic or literalistic reading of Scripture. Traditional exegesis of the Bible was certainly not 'critical' in the sense of being aware of, and conforming to, the norms of modern historiography. The early church drew on hermeneutical models drawn from Jewish scriptural exegesis: literalistic, midrashic, pesheric, allegorical and typological. 'Literalistic' interpretation in this context is not to be misunderstood as in any way positivistic, but rather as an exegetical method in which 'the Scriptures were understood in a straightforward fashion, resulting in the plain, simple and natural meaning of the text being applied to the lives of the people' (Longenecker 1975, p. 28). Such 'plain, simple and natural meaning' did not preclude the recognition of the presence of metaphor: literalistic exegesis had no problem recognising 'the Lion of Judah' as a metaphor for the Messiah, and 'the body of Christ' as a reference to the ontological status of the Church. Augustine, writing towards the end of the fourth century, reflected received hermeneutical assumptions when he insisted that the Bible is not a scientific text, and that 'Scripture must be interpreted in a way that is not contradicted by established scientific conclusions' (Hodgson 2005, p. 26). Theology, that is to say, does not possess authority over scientific discourse, which has its own independent object of study and method of enquiry. Indeed – and here Augustine appears unerringly prophetic – it is a

> disgraceful and dangerous thing for an infidel to hear a Christian talking nonsense on these topics, and we should take all means to prevent such an embarrassing spectacle, in which people see vast ignorance in a Christian and laugh it to scorn.
>
> (Hodgson 2005, p. 27)

Though the Bible is accepted as the revealed Word of God inspired by the Holy Spirit, it is also recognised to be the product of fallible human beings. This immediately rules out any naïve positivistic or literalistic reading of Scripture, and requires the biblical exegete to be hermeneutically astute and epistemically open (Thiselton 1980, 1992, 2007; Watson 1994, 1997). From the New Testament onward, orthodox Christianity has consistently denied the possibility of epistemic closure with regard to theological statements, on the ground that – prior, at least, to the eschatological consummation and fulfilment of creation – God is, by his very nature, always and necessarily one step beyond human comprehension: 'For now we see through a glass, darkly; but then face to face: now I know in part; but then shall I know even as also I am known' (1 Corinthians 13:12, Authorised King James Version).

True, Trinitarian Christianity claims that its distinctive understanding of ultimate reality eclipses all alternatives. However, in doing so it does not claim epistemic closure: because the Triune God is always one step beyond human comprehension, the epistemic and hermeneutical struggle to understand God more profoundly and more truthfully remains an ongoing human enterprise. All that orthodox Christianity claims is that the Trinitarian explanatory model is more powerful and comprehensive than any theological or naturalistic alternative. There is a direct parallel here with critical realism, which asserts its superiority over idealist, empiricist and pragmatic alternatives, yet makes no claim to epistemic closure.

Classical Trinitarian doctrine emerged from precisely the same abductive-retroductive-iterative process described by critical realists. The emergence of a new retroductive hypothesis is generally a result of abductive or intuitive reasoning in response to new evidence or original insight into ways of reconfiguring established evidence. Abduction is 'fundamentally innovatory and creative, generating new ideas and insights in response to "surprising facts"' (McGrath 2009a, p. 44). As such, it constitutes the 'eureka' (Greek, εὕρηκα: 'I have found') moment of insight and discovery: Archimedes' grasp of the principle of displacement as he steps into his bath, Newton's grasp of gravity as the apple falls on his head, Peter's blurted confession 'You are the Messiah, the Son of the Living God' in response to Jesus' question 'But who do you say that I am?' (Matthew 16:15f.). Contra Socrates' maieutic pedagogy, we do not learn by recollecting that which we once knew but have since forgotten, but rather by abductively projecting potential meaning onto objects and events that we encounter but do not fully understand or comprehend. This is so equally in ordinary life and academic research: when Mount Vesuvius erupted in 1944 American servicemen stationed at Pompeii first assumed they were under military attack, and then revised that assumption in the light of the evidence before them.

The moment of abductive insight paves the way for the development of retroductive models designed to offer powerful and comprehensive explanatory hypotheses in the light of the available evidence. The New Testament presents a range of distinctive-yet-complementary attempts to generate retroductive accounts of the person, teaching and actions of Jesus. All are rooted in the theological legacy of Judaism, and all affirm a direct connection between the person and work of Jesus and the culmination of God's economy of salvation. Thus, for example, the Prologue to John's Gospel alludes back to the first of the two Hebraic creation myths contained in the book of Genesis in identifying Jesus as the creative Word (Greek λόγος, logos: word, reason, wisdom) of God:

> In the beginning when God created the heavens and the earth, the earth was a formless void and darkness covered the face of the deep, while the spirit

> of God swept over the face of the waters. Then God said, 'Let there be light'; and there was light.
>
> (Genesis 1:1ff.)

> In the beginning was the Word, and the Word was with God, and the Word was God.... All things came into being through him ... And the Word became flesh and lived amongst us, and we have seen his glory, the glory of a father's only son.... No one has ever seen God. It is God the only Son, who is close to the Father's heart, who has made him known.
>
> (John 1:1ff.)

The construction of a retroductive model marks the start of an ongoing iterative process that entails 'the successive and incremental revision of how we see or understand something in the light of insights disclosed through the process of the engagement itself' (McGrath 2009a, p. 32). The iterative development of the New Testament's retroductive accounts of the theological nature and significance of the person and actions of Jesus led directly to the generation of the classical doctrines of the Trinity and the incarnation. Because classical Christian doctrines are the products of abductive-retroductive-iterative reasoning carried out by fallible human beings in particular socio-cultural contexts, they have always been recognised by the Church as human constructs that necessarily fall short of being final and definitive expressions of ultimate truth.

Traditionally, Christian theology has referred to its doctrines as 'dogmas'. In the post-Enlightenment era Christianity has frequently been accused of being 'dogmatic', in the sense that its teachings are products of closed authoritarian minds that impose premature epistemic closure upon indemonstrable opinions. Such an accusation is predicated on Locke's fact–value dualism, which assumes that theological statements possess the status, not of objective fact, but of subjective opinion. Thomas Torrance points out that in the pre-Christian classical world 'dogma' was used to refer to public decrees, judicial decisions and philosophical or scientific principles grounded in informed judgements generated by particular schools of thought. In the early church 'dogma' was used to refer to ecclesiastical ordinances (such as the sacraments of baptism and the Eucharist), and to expressions of Christian truth affirmed by the Ecumenical Councils.

> Dogmas thus came to mean doctrinal propositions formulated and given out by the Church on the ground of decisions or judgements reached by the Church through corporate or conciliar acts.... Throughout the whole Church, in East and West, the emphasis upon *dogma* came to be upon well-grounded and agreed affirmation rather than arbitrary and individual opinion, and upon positive and constructive as opposed to sceptical or merely critical thought.
>
> (Torrance 1969, pp. 338ff.; emphasis in original)

However, under the influence of the epistemological assumptions of the Enlightenment, the original meaning of Christian dogma underwent an almost complete reversal, and 'came to be understood as philosophical or theological dogmatizing without critical reference to the evidence of primordial realities' (p. 340). Torrance argues that a critically realistic perspective allows for the rehabilitation of the notion of dogma in a manner that may be legitimately applied to both the natural sciences and the science of theology: on this reading dogma refers to 'the kind of knowledge that is forced upon us when we are true to the facts we are up against, and in which we let our thinking follow the witness of those facts to their own nature and reality' (p. 341).

Despite the post-Enlightenment aberration of claims to papal infallibility and biblical inerrancy, mainstream Trinitarian Christianity has consistently recognised its doctrinal and creedal formulations as the fallible products of fallible human beings responding to the actuality of divine revelation and in some instances inspired by, though never dictated to by, the Holy Spirit.

Judgemental rationality and Christian heresy

To be epistemically relative about Christianity is to recognise that it is possible to make critical-yet-contingent judgements, internal within the Christian community, between orthodox and heterodox versions of Christianity. The dispute between 'orthodox' Trinitarian Christians and 'heterodox' Christians – e.g. Arians, Nestorians and those Gnostic sects that identified themselves as 'Christian' – led to the labelling of the latter as 'heretics', and in some cases resulted in their excommunication from the orthodox Christian community. The Church traditionally identifies truth claims generated within the Christian community that are deemed to fall short of Christian orthodoxy as 'heretical'; the term is *not* applied to truth claims that contradict Christian orthodoxy when propounded by non-Christians (McGrath 2009b, p. 33). Despite the recurrence in Christian history of profoundly regrettable and distinctly un-Christian attempts to retain the purity of orthodox faith through the active persecution of heretics, there is nothing intrinsically sinister about the notion of heresy per se. A clear distinction needs to be made between the judgement that a particular view is heretical and the response of the Church to that judgement. Just as the intellectual integrity of Marxism is not necessarily undermined by the violent totalitarian actions of those who claim to act in the name of Marx, so the intellectual integrity of Christianity is not necessarily undermined by the violent totalitarian actions of those who claim to act in the name of Christ. That said, it appears that those perpetuating violence in the name of Christ are more culpable for the dislocation of belief from action than those perpetuating violence in the name of Marx, since Christ expressly forbids his disciples to follow the path of violence, while Marx allows that in certain circumstances violence may be justified. Be that as it may, it does not weaken the claim that the veracity of a belief system is not undermined simply because some of its adherents act in ways inconsistent with it. The Christian Church is a voluntary organisation that individuals are in principle free to enter and leave at will, and as such it quite reasonably reserves the right to challenge members whose views undermine its core beliefs, initially through reasoned argument and if necessary through exclusion from the community. If the British Conservative Party were faced with an influx of self-styled Marxist-Conservatives, it would be within its rights both to challenge their political philosophy and if necessary to withdraw their party membership – however regrettable some may deem such a course of action to be.

The crucial issue for us here is whether the resolution of internal disputes within the Church between 'orthodox' and 'heterodox' positions is achieved through rational argument or through the exercise of power. Was Arius' claim that Jesus Christ was a hybrid 'superman' figure bridging the gulf between Creator and creation, rather than the incarnate hypostatic union of God and humankind, rejected for theological reasons or because Trinitarian and Arian Christians were involved in a political power struggle for control of the Christian community? In responding to this question, we need to beware the trap of imposing a false choice between binary opposites. It is palpably *not* the case that we must choose between judgemental relativism (all knowledge is the product of power) and 'pure' judgemental rationality (all knowledge is the product of brokerless intellectual debate). Critical realism's affirmation of a necessary connection between epistemic relativism and judgemental rationality means that, in all spheres of life, though the pursuit of truth can and should be a rational process, it can never be hermetically sealed from a range of socio-political and psychological power structures.

Sean Creaven argues that theological truth claims are closed to rational debate, and that Christian theology's (supposed) lack of evidentially grounded reason 'licenses inflationary flights of speculative fancy' constrained only by the 'limits of mass popular religiosity' and 'the realities of elite power within religious institutions' (Creaven 2010, pp. 295, 297). He singles out the second-century theologian Irenaeous [*sic*] as one of those 'peddlers' of Christianity responsible for 'ruthlessly suppressing beliefs and ideas ... that did not fit with a particular church-sanctioned orthodoxy' (p. 343). However, in *Against Heresies* Irenaeus offers a closely argued reading of a range of Gnostic texts which demonstrates that it was his Gnostic opponents who, in claiming access to a body of esoteric knowledge available only to a spiritual elite, produced a series of speculative spiritual systems devoid of evidential grounds and reasoned argument (Irenaeus 2004; cf. Osborn 2001). His *Demonstration of the Apostolic Preaching* set out to ground the rationality of theology on its compliance to the life of Jesus and Jewish theology by tracing a line of publicly accessible apostolic testimony back through successive generations to the four canonical Gospels and the encounter of the first Christians with the historical Jesus (MacKenzie 2002). Irenaeus was certainly not subservient to mass popular religiosity; on the contrary, he actively resisted the ascendancy of Gnostic popularism. There is no evidence whatsoever of his involvement in the ruthless suppression of ideas: as a leader of a small persecuted sect he was hardly in a position to suppress anything, and prior to the discovery of the Gnostic library at Nag Hammadi in 1945 he was the primary source of our knowledge of Gnosticism (Robinson 1996). Faced with what he judged to be a false turn in the development of the Christian tradition,

Irenaeus simply subjected the texts associated with that turn to rigorous critical scrutiny in order to defend a position he judged to be more intelligent, more reasonable, more responsible and more attentive to the evidence (McGrath 2009b, pp. 53, 122ff.). This, of course, is precisely what Bhaskar does in his wide-ranging critique of Western philosophy when developing his dialectical critical realism, and what Craven himself attempts to do in attacking the 'spiritual turn' in critical realism.

Walter Bauer sought to defend the thesis that the earliest forms of Christianity were inconsistent with later orthodox Christianity, and that they were only deemed 'heretical' by orthodox Christians once they had established political ascendancy in the Church (Bauer 1971). 'Orthodoxy', on this reading, 'is nothing more than a heresy that happened to win out – and promptly tried to suppress its rivals and silence their voices' (McGrath 2009b, p. 2). If this were the case, then we must 'recognise the existence of a group of "lost or suppressed Christianities", which were repressed and silenced by those who wished their own ideas to be acclaimed as orthodoxy' (p. 3). Bauer's case was dependent upon the modernist supposition that religious belief is not open to rational interrogation, to which we may now add retrospectively the postmodern supposition that knowledge is nothing more than a product of economies of power. Without the possibility of an appeal to reasoned argument, theological debates can only ever be resolved by blind leaps of faith that are inevitably and inextricably bound to socio-political and psychological power structures. However, Bauer's modernist assumptions, together with the postmodern supplement added here, are both dependent on epistemic suppositions that critical realism has been conclusively demonstrated to be fundamentally flawed. If theological truth claims are indeed open to rational evaluation, then adjudications between orthodox (true) and heterodox (false) truth claims cannot be summarily dismissed as unavoidable exercises in raw power. This is not to suggest that economies of power have no influence of critical judgements, nor is it to suggest that theological debates always embrace canons of judgemental rationality. Because of this, the integrity of particular theological debates vis-à-vis the relationship between force and reason need to be carefully assessed in terms of the available evidence. This is something Creaven singularly fails to do in his dismissal of Irenaeus: he does not engage with primary historical sources or with secondary academic scholarship, and his only attributed source is a semi-popular text, written by an *aficionado* of esotericism without any formal theological or historical training (Creaven 2010, p. 343; Martin 2005). Creaven's failure to provide any evidence to support his claim that Irenaeus was responsible for the ruthless suppression of ideas leaves him vulnerable to the charge of preferring rhetorical persuasion to rational argument.

The modern celebration of personal autonomy gives the heretic a certain counter-cultural attraction. As the Jewish writer Will Herberg puts it, 'people eagerly vaunt themselves as heretics, hoping that they will thereby prove interesting; for what does a heretic mean today but an original mind, a man who thinks for himself and spurns creeds and dogmas' (Herberg 1976, pp. 170ff.;

quoted in McGrath 2009b, p. 2). The heretical imperative to state a preference for Gnostic Christianity over Trinitarian Christianity, simply because the latter was once victorious over the former, appears to be driven by a power game predicated on the affirmation of the rights of the underdog over the hegemony of those in positions of power. If Gnostic Christianity does indeed offer a more powerful and comprehensive account of ultimate reality than Trinitarian Christianity, then this claim must be supported by rational argument grounded in a close reading of Gnostic and Christian texts. It is insufficient to valorise Gnosticism purely on the grounds of the hegemony of classical Christianity, since it is entirely possible that the historical victory of Christianity over Gnosticism is rooted in the greater rationality and explanatory power of the Trinitarian faith. True, an uneven distribution of power may indicate the possibility that the views of the underdog may have been misrepresented or not given a fair hearing. However, this only serves to underscore the importance of judgemental rationality, since it requires the interpreter to pay particular attention to this possibility when scrutinising the evidence. This applies as equally to Gnostic Christianity faced with the ascendancy of Trinitarian Christianity in the first few centuries of the Christian era, as it does to orthodox Christianity faced with the ascendancy of secular liberalism in the modern era. Insofar as Christianity increasingly plays the role of 'heretic' to secular liberalism's 'orthodoxy', it cannot legitimately claim possession of ultimate alethic truth simply on the grounds of its newfound 'heretical' status.

Judgemental rationality and Trinitarian theology

Trinitarian Christianity has, from its inception, been committed to rational debate. The New Testament's identification of Jesus as both divine and human remained relatively unsystematic and, at least according to the theological and philosophical norms of contemporary Jewish and Hellenistic thought, a crass absurdity: 'we proclaim Christ crucified, a stumbling-block to Jews and foolishness to Gentiles' (1 Corinthians 1:23). If Christianity was to establish intellectual credibility it was faced with the task of iteratively refining and systematising its core beliefs. In accepting the authority of Jewish Scripture, and especially of the Wisdom Literature that affirms the inherent rationality of both Creator and creation, Christianity acknowledged an imperative to subject its emergent beliefs to critical scrutiny. In following through this imperative it rejected any engagement with the raft of speculative esoteric spiritual traditions that were rife in the Roman Empire at the time, and turned instead to a critical dialogue with classical Greek philosophy. The ensuing engagement with philosophy continued through Trinitarianism's long history, and had a decisive impact upon the formation of Western Christianity, above all in Augustine's appropriation of Plato and Aquinas' requisition of Aristotle. It is significant that any substantial theological interest in the relationship between Christianity and Gnostic esotericism was effectively ended by Irenaeus' decisive exposure of the latter's intellectual poverty; from then on interest in the relationship has flourished only in contexts

– from the stagnant religious life of the late Roman Empire, through the esoteric underbelly of the Renaissance, to the contemporary New Age movement – in which the rationality of theological statements was either disregarded or placed under question.

The doctrine of the incarnation identifies the dual nature of Christ: he is simultaneously both fully God and fully human. To deny either the divinity (Ebionism) or humanity (Docetism) of Christ, or to turn him into a hybrid 'superman' figure neither fully divine nor fully human (Arianism), was to deny the core belief of orthodox Christianity that, in an act of unconditional love, God overcame the ontological gulf between Creator and creature by uniting himself with a particular human being, and through him with the entire human race and the totality of creation. The core of Christian doctrine is thus the affirmation of the essential hypostatic union (ὑπόστασις, hypostasis: sediment, foundation, substance, subsistence) of the two distinct natures of Christ. This utilisation of non-biblical philosophical language did nothing to accommodate Christianity to the theological and philosophical norms of Jewish and Hellenistic thought, both of which – albeit in significantly different ways – affirmed the utter transcendence of God. To appease these traditions by rejecting either the full divinity or full humanity of Christ would be to deny the distinctive core of Christian teaching. Orthodox Christianity's attempt to iteratively refine and systematise its doctrines by engaging in critical conversation with the language and thought forms of classical philosophy resisted any temptation to revise or adapt the substance of Christian belief by bringing it into line with prevailing philosophical and theological norms. Instead the early theologians sought to secure the intellectual integrity of their distinctive beliefs by attempting to demonstrate their greater and more comprehensive explanatory power.

Charles Cochrane argues that the iterative refinement of Christian doctrine conducted by the early Christian theologians provided the Church with a retroductive model that, in Colin Gunton's paraphrase, effectively 'out-thought the decaying classical civilisation in which it lived by providing answers to certain questions that pagan antiquity was unable to answer' (Cochrane 1944; Gunton 1998, p. 103). (1) The claim that God is essentially loving rather than essentially omnipotent, grounded in the notion of the reciprocal giving and receiving at the heart of the interpersonal relationships within the Trinitarian, provided the Church with an ontologically grounded moral vision of selfless under-labouring service (a moral vision rejected by Nietzsche as a 'slave morality') that fundamentally challenged the prevailing moral norms of the Roman Empire, rooted as they were in an ethic of honour rather than service. (2) The claim that in the incarnation God was hypostatically united with creation and humanity subverted the received onto-theological assumptions that insisted either (and predominantly) on an absolute dualistic separation between God and creation, or (and less frequently) on the pantheistic identification of God and creation. According to Trinitarian theology, the ultimate salvation and divinisation of human beings is dependent upon overcoming the ontological gulf between the Triune Creator and the created order. Consequently, the doctrine of incarnation, in affirming the hypostatic union of God and humanity in the person of Jesus Christ, rejected both a theology of absolute transcendence (Docetism: Jesus was not fully human) and a theology of absolute immanence (Ebionism: Jesus was not fully divine). The Christian doctrine of incarnation provided a means of resolving the neo-Platonic problem of how to account for the mediatory relationship between an utterly transcendent God and an utterly immanent creation: a problem Gnosticism sought to resolve by invoking a speculative series of spiritual beings, each successively less divine and more material, and Arianism countered by invoking Jesus as a 'superman' figure neither fully God nor fully man. In place of such mythopoeic speculation, Christianity offered a revised understanding of God, not as a static impersonal force, but as an active personal agent utterly committed to the flourishing of humanity. As a result, it became possible:

> to envisage the divine principle as both transcendent and immanent, 'prior' to nature, the world of time and space in which we live, and yet operative within it.... [F]rom this point of view, the panorama of human history may be conceived as a record of the divine economy, the working of the Spirit in and through mankind, from the creation of the first conscious human being to its full and final revelation in the Incarnate Word.
>
> (Cochrane 1944, pp. 367ff.; quoted in Gunton 1998, p. 103)

Cochrane's core thesis is that the remarkable success of Christianity in the Roman world was due to the fact that its retroductive claims about the ontological ground of reality generated a moral praxis (the witness of the martyrs) and intellectual vision (the witness of the theologians) that proved itself to be more powerful and comprehensive than any available alternative.

Our concern here is not directly with the truth or falsehood of Cochrane's substantial claims. Rather, I am suggesting that from the start classical Christianity set out to affirm the truth of its doctrinal claims via a critical engagement with alternative non-Christian explanatory models, and through the rigorous self-scrutiny of its core beliefs and practices. This critical process has been a constant feature of orthodox Christianity from its inception through to the present day; the occlusion of this fact is largely due to modernist and postmodernist critiques of Christianity grounded in idealist, empiricist and pragmatic modes of thought that critical realism has exposed as unsustainable – the paradigmatic model being that of Comte's oft-rehearsed narrative of the dark ages of the pseudo-sciences of theology and metaphysics giving way in the modern era to the glories of enlightened positivistic thought. This being the case, the critical realist community faces an intellectual imperative both to address the ultimate truth claims of Christianity, and to seek to adjudicate between them and the conflicting claims proffered by naturalism, the philosophy of meta-Reality and a range of other alternatives. To his credit, Sean Creaven remains the one critical realist who has sought to take this task seriously and attempted, albeit – as I have argued elsewhere – in a deeply flawed manner, to follow it through systematically (Wright 2011). One of the major problems with Creaven's work is his untested and unwarranted assumption that theological statements necessarily lack rational support and that consequently fundamentalism and literalism 'are precisely core elements of faith, hard-wired into the formal ideology of [Christianity], and embraced by millions of the faithful' (Creaven 2010, p. 243). If Creaven is correct, then Christianity eschews epistemic relativism and judgemental rationality in favour of ontological certainty and epistemic closure. Contra Creaven, I have suggested in this chapter that, far from being 'hard-wired into the formal ideology of Christianity', fundamentalism and literalism are heretical aberrations generated by the Christian encounter with a flawed modernity. The orthodox Christian tradition makes truth claims about the ultimate nature of reality that it recognises to be *necessarily incomplete and potentially mistaken* and which it subjects to critical scrutiny in pursuit of a deeper and more truthful understanding of the Triune God.

6 The 'problem' of Christian exclusivism

In the previous chapter I suggested that Trinitarian Christianity embraces epistemic relativism insofar as it makes ultimate truth claims that it recognises to be contingent and subjects to critical scrutiny in pursuit of a deeper understanding of the Triune God. This chapter seeks to test this thesis by addressing the issue of the so-called 'problem' of Christian exclusivism. Does the Christian claim to possess exclusive access to ultimate truth undermine its acceptance of epistemic relativism? We will address this question from the perspective of Bhaskar's critique of Christian exclusivism and concomitant defence of religious pluralism. In doing so, I hope to show how critical realism can shed further light on the nature of inter-faith dialogue at a meta-theoretical level.

Religious exclusivism and inter-faith dialogue

In *From East to West* Bhaskar recognises that ontological realism about God (or the absolute or transcendence) entails epistemic relativism.

> It is important to note that *ontological realism* is not only consistent with but also entails epistemological, and more generally experiential, relativism, including pluralism, diversity and fallibilism.
>
> (Bhaskar 2000, p. 22; emphasis in original)

> *Ontological realism* about God (in the [intransitive domain]) is consistent with, and indeed entails, experiential (including *epistemic*) relativism (in the [transitive domain]), including pluralism, fallibilism and diversity.
>
> (p. 40; emphases in original)

He then goes on to to affirm that epistemic relativism about God does not preclude judgemental rationality.

> This is, moreover, quite consistent with a moment of *judgemental rationalism* in the intrinsic or normative aspect of the transitive or epistemological and social, relative, dimension of science.
>
> (p. 22; emphasis in original)

> It should also be reiterated that that this epistemological relativism ... is consistent with a moment of concrete singularised *judgemental rationality* ... in, say, the assessment of the claims of specific religious practices.
>
> (p. 40; emphasis in original)

However, as we shall see, Bhaskar fails to follow through these assertions that truth claims about God are open to rational assessment. He adds an important caveat to his argument – predicated on *a priori* assumptions regarding (1) the universality of divine manifestation, and (2) the equality of all divine manifestations – that leads him to reject theological exclusivism and affirm theological pluralism.

> [W]e can allow that God, or the absolute, or the transcendent (or transcendent beings or phenomena generally) can, like nature (or ordinary material things) be accessed or experienced in *a multiplicity of different ways*.
>
> (Bhaskar 2000, p. 22; emphasis added)

The assumption of an analogy between God and nature draws the conclusion that God, like nature, can be manifest, accessed and experienced by all human beings in a variety of different ways across a plurality of different cultural contexts. The implication, which is teased out in Bhaskar's subsequent writings, is that all religious (and, indeed, secular) traditions offer equally valid routes to God. The analogy with nature is grounded in Bhaskar's account of God as a *structural ingredient* within the world, and disqualifies the Christian understanding of God as a *personal agent* other than the world, who reveals himself in particular ways in specific spatio-temporal contexts, not all of which are directly accessible to all human beings. Where Bhaskar allows that God is potentially accessible in a plurality of religious traditions, Trinitarian Christianity claims privileged access to exclusive knowledge of the definitive self-revelation of the Triune God in the life of Jesus of Nazareth. Since both arrive at their respective positions on the basis of incompatible understandings of the nature of God, the resolution of the *epistemic question* (Is God universally manifested, or manifested exclusively within a particular tradition?) is dependent upon the resolution of the prior *ontological question* (Is God a structural ingredient within the world or a personal agent apart from the world?).

Even if we bracket out the Trinitarian understanding of God as personal agent and concede a structural analogy between God and nature, this still does not necessarily warrant the pluralist conclusion that human perceptions of the manifestation of God possess cross-cultural validity. The fact that all people, in all cultural contexts, can access the natural order does not mean that alchemists, astrologers, Aristotelian philosophers, Newtonian scientists and Einsteinean physicists possess equally valid understandings of the natural world. On the contrary, judgemental rationality enables us to recognise that alchemy and astrology are deeply flawed, and affirm that the Einsteinean paradigm is more powerful and comprehensive than the Newtonian, which in turn is more powerful and comprehensive than the Aristotelian. In the same way, judgemental rationality about God should enable us, in principle at least, to adjudicate between atheism, deism, theism, pantheism and panentheism, and between the conflicting claims of Buddhism, Christianity, Hinduism, Islam, Judaism, meta-Reality, Naturalism, Sikhism, etc.

However, Bhaskar appears unwilling to concede this point. He argues that, because God is accessible in a variety of different ways in a variety of different traditions, it follows that 'God is *both absolute* (unbounded) and *relative*; that is, manifest (perhaps such as a "personal lord" such as Krishna or Christ), and accessed in a particular, and to that extent relative, form' (Bhaskar 2000, p. 40). By analogy, nature is an unbounded universal that manifests itself in particular contexts in relative ways, so that alchemists and Einsteinean physicists perceive the same reality differently. So far, so good – the problem lies in the *absence* of the next stage in the argument. We would expect Bhaskar to go on to argue that, since (say) Hinduism and Christianity proffer ontologically incommensurable truth claims about ultimate reality, their respective positions must be subject to rational evaluation. However, it becomes clear that he holds that, since the Hindu and Christian traditions offer *equally valid* paths to God, rational assessment of their incommensurable ontological claims is a redundant activity. The critical issue for the scientist faced with the choice between alchemy and Einsteinean physics is to seek to adjudicate between the two ontological incommensurable positions. However, Bhaskar shows no concern to adjudicate between the ontologically incommensurable claims of Christianity and Hinduism. The fact that Hindus attend to Krishna and Christians embrace Christ appears to have no substantial epistemic or ontological significance for Bhaskar: there are many other possible routes to God, and the particularities of Hindu and Christian beliefs have no ontological weight, since both traditions are equally capable of manifesting God. The logic demands that we draw the same conclusion regarding the debate between physicists and alchemists: since both are equally capable of manifesting nature, both must be treated as equally valid. This cuts across the grain of judgemental rationality and implies judgemental relativism. Though this interpretation is not stated explicitly at this stage of Bhaskar's argument, it becomes clear throughout his subsequent writings that he embraces a theological pluralism that affirms all religious and secular traditions as offering equally valid routes to transcendence, and rejects all claims to exclusive or privileged access.

The ensuing contrast between Bhaskarian pluralism and Christian exclusivism is fundamental to Bhaskar's identification of Christianity as 'judgementalist'.

> Judgemental rationalism in turn is not the same as *judgementalism*, which is prescriptively and abstractly universalising and derives, at least in part, from the failure to acknowledge the *concrete singularity* of the *dharma* (nature, station or position) of the individual, group or situation.
>
> (Bhaskar 2000, p. 22; emphasis in original)

Christianity stands accused of prescriptively and abstractly universalising its particular culturally bound experience of God as manifest in the life of Jesus of Nazareth, and of failing to acknowledge the concrete singularity and historical particularity of its truth claims. Because Christian adherents experience God in a particular way does not mean that all human beings must do the same; on the contrary, different traditions have their own equally valid ways of accessing God. But is Einsteinean physics not *legitimately* 'prescriptively and abstractly universalising' in claiming that its understanding of the physical world is universally valid, despite the fact that it is grounded in the concrete singularity of a particular community of scientists? Why is the path from particularity to universality open to the Einsteinean physicist but closed to the Christian theologian? On what grounds can the distinction between the judgemental rationalism of the physicist and the 'judgementalism' of the theologian be asserted? Does not the philosophy of meta-Reality proceed from the particularity of Bhaskar's thought to make abstract and universal truth claims? And on what grounds can a distinction between the judgemental rationalism of the philosophy of meta-Reality and the 'judgementalism' of Christian theology be asserted?

Bhaskar affirms pluralism on the grounds that God is potentially accessible in all religious and secular traditions, and rejects exclusivism because it follows from this that no single tradition can legitimately claim privileged access to God. Here 'exclusivism' and 'pluralism' function as binary opposites. However, interfaith dialogue has tended to adopt a slightly more complex typology that distinguishes between 'exclusivism', 'inclusivism' and 'pluralism' (Race 1994). Exclusivists hold that a single divine manifestation or revelation, owned by a single religious tradition, offers an exclusive path to enlightenment or salvation; at the other end of the spectrum, pluralists hold that many divine manifestations/revelations, accessible through a plurality of different religious and secular traditions, offer a plurality of paths to enlightenment/salvation. Inclusivists adopt a mediating position, affirming that, though there is a single definitive divine manifestation/revelation owned by a single religious tradition, other traditions may have partial access to this manifestation/revelation and thereby offer partial paths to enlightenment/salvation. A classic example of religious exclusivism may be found in the particular interpretation of Calvinism offered by the Synod of Dort in 1619: though humanity is totally depraved, God has eternally elected a chosen few to salvation, Christ died on the cross to redeem only these elect who, because they are unable to resist the power of God's grace, will necessarily persevere in the Christian faith (Pelikan 1985, pp. 236ff.). The most celebrated example of religious pluralism is found in the theology of John Hick, who, as Gavin D'Costa points out, 'argues that all religions are salvific paths to the one Divine "Real", none being better or worse, and none with a privileged or exclusive revelation, despite what some of their adherents may claim' (D'Costa 2000, p. 25; see also Hick 1989). One of the most influential examples of inclusivism is found in the Jesuit theologian Karl Rahner's notion of 'anonymous Christians': non-Christians who have never encountered Christian revelation in any meaningful way may, nevertheless, on the basis of their moral, spiritual and/or existential orientation towards life, implicitly accept the salvific grace of God offered exclusively in the incarnation (Rahner 1998, pp. 311ff.).

Despite their popularity, both the exclusivism-pluralism and exclusivism-pluralism-inclusivism typologies are increasingly perceived by theologians and scholars of comparative religion as blunt instruments, imposing simplistic solutions to complex problems. One of the central contentions of this chapter is that critical realism has the potential to generate deeper insights into the multifaceted nature of inter-faith dialogue at a meta-critical level. Specifically I will argue that, far from being polar opposites, Trinitarian Christianity and the philosophy of meta-Reality actually function in remarkably similar ways: (1) both are *ontological exclusivist*, insofar as they lay claim to specific accounts of ultimate reality that are ontologically incommensurable, both with each other and with other conflicting accounts; (2) both are *epistemic exclusivist*, insofar as the claim that the possibility, nature and extent of knowledge of God is dependent upon the prior truth of their respective ontological commitments; (3) both are soteriological inclusivist, insofar as they claim that the actuality of enlightenment/salvation is in principle universally available to all human beings, irrespective of the cultural traditions they inhabit.

Exclusivism and ontology

At the ontological level pluralists and exclusivists disagree about the nature of God. Both Bhaskar and Trinitarian Christians claim cognitive knowledge of the ultimate nature, structure and ground of reality: for Bhaskar this is the 'ground-state' or 'cosmic envelope'; for Christians this is the Trinitarian God. Bhaskar argues that God is ingrained within the categorical structures of the world, whereas Trinitarian Christians argue that God is united with creation only in the hypostatic union of God and humankind in the particular individual Jesus of Nazareth. These incommensurable ontological claims pan out at the epistemic level: for pluralists the nature of God is such that he is universally manifest and accessible; for exclusivists the nature of God is such that he reveals himself and is explicitly accessible only in specific contexts. Bhaskar claims that the ingrained God is accessible in all religious and secular traditions, whereas Christians claim that only the Church enjoys privileged access to knowledge of the incarnate God. Our focus in this section is on the ontological dimension of the dispute between pluralism and exclusivism.

As they stand, the ontological claims of Trinitarian Christianity and the philosophy of meta-Reality are mutually exclusive: God cannot simultaneously be ontologically ingrained within the totality of reality *and* ontologically united with creation uniquely and exclusively in the incarnation. It follows that Trinitarian theology, the philosophy of meta-Reality, or another (unnamed) alternative constitutes the most powerful and comprehensive explanatory account of the alethic truth of the ultimate ground of reality currently available to us. A scientific analogy will help clarify this point: the universe is either in a steady state or in process of expansion. The fact that the explanatory models of an expanding and steady-state universe are necessarily contingent does nothing to undermine their ontological incommensurability: if the universe is expanding, then it is not in a steady state, and vice versa. To claim, on the grounds of their contingent status, that both explanatory models are equally valid is to take the anti-realist path of acknowledging that neither possesses a substantial cognitive purchase on reality. The end result will be a thoroughgoing relativism in which both accounts are deemed equally valid by virtue of the (paradoxical) fact that both are equally false. If we wish to hold fast to the basic tenets of critical realism, we have no alternative other than to avoid judgemental relativism by employing judgemental rationality in an attempt to adjudicate between their respective claims.

As a theological pluralist, Bhaskar accepts the dual task of rejecting the exclusive ontological claims of Christianity while simultaneously embracing Christianity as one of many epistemic vehicles through which the ontological reality of God may be accessed.

Bhaskar rejects Christian exclusivism on the ground that belief in a unique revelation owned by a privileged community is incompatible with the fact that a God ingrained in nature must necessarily be accessible by adherents of all religious and secular traditions. However, this is an *a priori* argument, in which a pre-established understanding of God determines the means through which he may be accessed. Christians might legitimately retort by asserting their own *a priori* alternative: a universal revelation, to which all religious and secular traditions may claim access, is impossible because God, by his very nature, is fully accessible only in the incarnation. It follows that the truth of the epistemic claims of Christianity and the philosophy of meta-Reality are dependent upon the truth of their respective, and mutually exclusive, ontological claims about the nature of God. In affirming the exclusive ontological claims of the philosophy of meta-Reality, Bhaskar must reject the exclusive ontological claims of Christianity, and vice versa. This cannot be achieved by mere assertion: it requires the application of judgemental rationality, driven by an attentive, intelligent, reasonable and responsible analysis, evaluation and comparison of both retroductive explanatory models. However, there is little evidence to date of any sustained attempt by either party to engage in a critical evaluation of the relative merits of their distinctive accounts of God.

How then does Bhaskar establish Christianity as an epistemic vehicle through which the ontological reality of God can be accessed, given his understanding that Christian truth claims are devoid of ontological significance? It is instructive in this context to compare Bhaskar's response to this question with that of the theological pluralist John Hick. Both deny that Christian doctrines enjoy any substantial purchase on reality. Hick does this explicitly, arguing that the doctrines of Trinity and incarnation are essentially mythological and as such reveal nothing of ontological substance about reality. Bhaskar is more circumspect, having little to say about Christian ontology and offering no substantial engagement with the cognitive truth claims of Christianity; he appears to assume, by default, that Christian doctrines possess no ontological significance, or at least no ontological significance worthy of sustained debate. Having set aside

Christian onto-theological truth claims, both Hick and Bhaskar proceed to address the question of the epistemic value of the Christian tradition. Hick defines myth 'as story ... which is not literally true ... but which invites a particular attitude in its hearers', so that the 'truth of a myth is a kind of practical truth' consisting, not in its ability to describe reality, but in 'the appropriateness of the attitude which it evokes' (Hick 1977, pp. 166ff.). This, as Gavin D'Costa points out, assumes an instrumental view of Christian doctrine: a doctrine is deemed 'true' to the extent to which it expresses spiritual experience and helps generate further experience, and deemed 'false' insofar as it functions to articulate exclusive cognitive claims about ontological reality (D'Costa 2000, p. 25). Bhaskar follows a similar path, by interpreting Christianity in the light of Shankara's distinction between 'higher' and 'ordinary' spiritual truth (Bhaskar and Hartwig 2010, p. 151). At the higher level Christianity has shown itself capable of establishing an environment in which certain extraordinary saints and mystics can enjoy spiritual experiences of transcendence. He subsequently qualified this position in suggesting that the ground state can also be accessed in the midst of ordinary everyday life without recourse to mystical or quasi-mystical 'peak' experiences. However, while both the extraordinary mystic and ordinary Christian adherent can operate at the higher spiritual level, they must contend with the fact that, at the lower level, the Church encourages them to embrace the exclusive truth claims of Trinitarian doctrine in a manner that serves to occlude both ordinary and extra-ordinary paths to God and the ground-state. Hick and Bhaskar agree that though Christian truth claims are essentially false, the Christian tradition can function instrumentally as a means of enabling its adherents to experience transcendence, provided it is shorn of its misplaced, and potentially hegenomic, ontological claims. Hick is clear that an exclusive focus on Christian truth claims serves to undermine the spiritual potential inherent in the Christian tradition. Similarly Bhaskar, having acknowledged higher spiritual experience and ordinary dogmatic 'uniquism' as phenomenal realities, argues that the 'time for the higher truth to become the ordinary truth has arrived' (p. 151). The implication is clear: Christian Trinitarians should abandon their exclusive ontological claims and attend instead to the instrumental value and spiritual potential of their tradition.

In order to affirm the ontology of meta-Reality it is necessary to exclude the ontological truth claims of Christianity, and indeed of any religious tradition that lays claim to knowledge of the ultimate ontological contours of reality. Despite its claims to openness, inclusivity and plurality, the philosophy of meta-Reality rejects and excludes the ontological commitments of all the world's religious traditions. The instrumental inclusion of religious traditions within the framework of meta-Reality operates only at the transitive level of contingent culture; at the intransitive level of fundamental ontology, all truth claims other than those of the philosophy of meta-Reality are either excluded or subject to reductive reinterpretation. In order to accept Bhaskar's invitation of inclusion, Trinitarian Christians must first acknowledge that their core ontological beliefs have no cognitive purchase on reality and possess only instrumental value. The veneer of inclusivism thus serves to occlude a thoroughgoing exclusivism: in the name of pluralism, meta-Reality seeks to colonise and reconfigure Trinitarian Christianity.

To conclude this section: both Trinitarian Christianity and the philosophy of meta-Reality are ontologically exclusivist, insofar as they lay claim to particular accounts of ultimate reality that are ontologically incommensurate, both with each other and with other competing accounts. Whereas Christianity is transparent about its exclusivist commitments, Bhaskar appears more opaque. However, his exclusive ontological commitments become transparent once we recognise that his inclusive rereading of Christianity must necessarily exclude exclusive Christian truth claims if it is to secure its pluralist credentials.

Exclusivism and epistemology

Having considered the 'problem' of religious exclusivism in relation to ontology, we will now move on to explore the issue of the epistemic processes, procedures and methods through which individuals and groups come to assert theological truth claims. Critical realism affirms the priority of ontology over epistemology. In doing so, it challenges the once-dominant Kantian claim that to understand the world we must order phenomena according to *a priori* categorical structures embedded in the mind's cognitive make-up. Kantians argue that our minds shape reality; critical realists argue that reality should shape our minds. According to the latter view, we develop epistemic tools and procedures *a posteriori*, in response to our encounters with reality, and progressively refine them in the light of the ontological demands reality imposes on us. If this is the case, then just as the principle of epistemic relativism leads us to anticipate encountering a range of contrasting and contradictory understandings of the ontological nature and being of God, so we should also anticipate encountering a range of contrasting and contradictory epistemic means and procedures for accessing God.

In the opening section of this chapter we saw how Bhaskar effectively reversed the received understanding of epistemic relativism in suggesting that it supported the notion of a single understanding of God accessible across a range of different cultural contexts, rather than a range of competing and conflicting understandings of God requiring critical assessment. In this section I will seek to extend this argument by suggesting that, despite his pluralistic claim that God is accessible in a variety of different ways, Bhaskar actually affirms a single, particular and exclusive path to God. If this is the case, then the veneer of inclusivism once again disguises a thoroughgoing exclusivism: Bhaskar rejects the Christian understanding of a single privileged culturally bound divine revelation as exclusivist, but only by affirming his own equally exclusive account of divine manifestation. Specifically, he defends the notion that we access God through a process of self-realisation in which we recollect and become aware of our ground-state. In doing so he excludes the Christian claim that we access God definitively in his self-revelation in the person of Jesus Christ. The fact that

Bhaskar's account claims not to be restricted by any spatio-temporal or cultural boundaries cannot disguise the fact that it remains a particular and specific means of accessing God.

The tensions between the Trinitarian and Bhaskarian accounts of theological epistemology have deep historical roots. Christianity subscribes to a notion of historical revelation shared with the other two Abrahamic religions: Judaism and Islam. Broadly speaking, the Abrahamic faiths agree that, though there are intimations of God's existence in his creation, God himself is known definitively only insofar as he elects to reveal himself in specific historical events: God's historical covenant with his chosen people (Judaism), the incarnation of God in the person of Jesus Christ (Christianity), and the revelation of the Qur'an (Islam). Bhaskar's account of a universally accessible manifestation of transcendence has its roots in the natural theology of the ancient Greeks, which sought to derive knowledge of God from the universal structures of nature rather than divine agency. It also reflects a much later tradition of theological epistemology – rooted in Romanticism, generated in the main by Liberal Protestants, coming to the fore in the emerging field of the comparative study of religions during the nineteenth century, and dominating accounts of the phenomenology of religion during the twentieth century – that posited knowledge of God on the basis of self-authenticating, pre-linguistic, intuitive experiences of transcendence (Fitzgerald 2000).

Paul Ricoeur has drawn a useful distinction between a 'phenomenology of manifestation' and a 'hermeneutic of proclamation' (Ricoeur 1995, pp. 48ff.). The hermeneutic of proclamation is rooted in the Jewish belief that God chose to reveal himself primarily through the Torah, so that God is accessed by reading and meditating on Jewish Scripture. The phenomenology of the sacred, as articulated by nineteenth-century scholars, claims to identify a common pattern of divine manifestation among the world's religions in which the sacred is: (1) encountered through hierophanies linked to a plurality of natural and cultural contexts; (2) experienced as an awesome, powerful and overwhelming reality; (3) recognised as a pre-linguistic phenomenon; and (4) understood as an irruption of the noumenal realm into the phenomenal domain. Ricoeur points out how religions committed to a hermeneutic of proclamation resist the phenomenology of the sacred. Thus Israel's emergent understanding of divine revelation directly contradicted any notion that God is ingrained within, or is manifest through, the phenomenon of nature. Consequently Israel, in rejecting Canaanite nature cults, found herself in a struggle 'against the idols of Baal and Astarte, against the myths about vegetation and agriculture, and in general against any natural and cosmic sacredness' (p. 55). According to a hermeneutic of proclamation: (1) divine revelation outweighs pre-linguistic numinous experience; (2) revealed sacred texts outweigh any nature-bound manifestation of the sacred; (3) language and history take precedence over myth and ritual; and (4) ethical obedience to divinely instituted law requires the desacralisation of nature and culture. The phenomenology of manifestation and hermeneutic of proclamation offer different and incompatible routes to different and incompatible Gods: the God who manifests himself within nature tends to be an impersonal and either deistic, pantheistic or panentheistic; the God who reveals himself through proclamation tends to be personal and theistic. Though there is no direct fit between Ricoeur's account of the manifestation of the sacred and Bhaskar's philosophy of meta-Reality, they clearly share certain family traits that set them apart from the hermeneutic of proclamation espoused by Trinitarian Christianity.

The theological epistemology of manifestation and the theological epistemology of proclamation are mutually exclusive: according to the former, God can be known because he is ingrained within the *structures* of the universe; according to the latter, God can be known because he reveals himself through his actions as a *personal agent* within history. In affirming his allegiance to an epistemology of manifestation, Bhaskar makes a specific and particular epistemic claim that necessarily excludes the Christian epistemology of proclamation. This being the case, Christian theology and the philosophy of meta-Reality are both epistemically exclusivist: both claim that the possibility and actuality of knowledge of God is dependent upon the prior truth of their respective ontological commitments.

Exclusivism and morality

Having rejected Christian exclusivism in terms of ontological realism and epistemic relativism, we might have expected Bhaskar to turn his attention to judgemental rationality. However, there is no extended discussion of the viability of making rational judgements between ontologically incommensurable accounts of ultimate reality, and no attempt to adjudicate between the respective ontological claims of meta-Reality and Trinitarian Christianity. Instead, Bhaskar elects to develop a moral critique of religious exclusivism, which proceeds through three stages. (1) He distinguishes between the open universal stance of religious pluralism, and the closed parochial stance of religious exclusivism. The spiritual orientation of the philosophy of meta-Reality is a universal presupposition of ordinary life, of religion and of emancipatory projects, and as such transcends all provincial cultural boundaries. Christianity, on the other hand, embraces a parochial perspective and affirms insular dogma. By implication, universalism is concerned with the flourishing of all human beings, while parochialism attends to the needs of a particular interest group. (2) He argues that religious pluralism enhances human freedom, while religious exclusivism constrains it. Since there are multiple ways in which spiritual experience of God may be appropriated, it follows that no one way is privileged, and that individuals are free to choose their own path:

> I might be able to say that my path is the best path for me, [but] it does not necessarily follow that it is the best path for you, or him, or her. For other people have their own rhythmics, and might have very good reasons for preferring another path to the absolute.
>
> (Bhaskar and Hartwig 2010, p. 151)

Christians who are unwilling to abandon their parochial dogmas, on the other hand, limit human freedom:

> The stumbling block is of course that most people who profess a particular religion feel they have the unique, the only way to God: this is the only way that God can be described, and if your description is different you are not actually describing God.
>
> (pp. 150ff.)

(3) He contends that religious pluralism engenders tolerance, while religious exclusivism breeds intolerance. If, as pluralists claim, the absolute is manifest in different religious traditions in different ways, then 'that gives you an immediate tolerance for other religions' (p. 150). Religious exclusivists, on the other hand, tend to be intolerant of alternative positions: 'To the Christian right in America there is no way Islam can be seen as an alternative path to God' (p. 151). Indeed, 'uniquism is what lies behind "the clash of civilisations" in its religious form' (p. 151). We will consider each of these three stages in turn, under the headings 'parochialism', 'freedom' and 'tolerance'.

(1) Parochialism. Bhaskar suggests that the philosophy of meta-Reality is open and pluralist, while Christianity is closed and parochial. In doing so he brackets out his own standpoint and assumes access to a universal perspective, a privileged vantage point offering a neutral 'view from nowhere', from which he can understand parochial religious traditions better than they understand themselves. He describes how, during his initial exploration of meditation, he 'adopted a standpoint of innocence' in which he 'did not have preconceptions' and suspended his 'critical faculties' (p. 149). Similarly, he notes that, in his discussions with Margaret Archer, Andrew Collier and Doug Porpora, 'I felt that a Christian slant or interpretation was being put on things, whereas *it was my aspiration to be neutral*' (p. 151; emphasis added). This process of bracketing prior commitments in order to establish a neutral or universal vantage point appears to contradict the principle of epistemic relativism, which insists that all knowledge is necessarily contextual and grounded in parochial cultural contexts. Bhaskar's commitment to universality and neutrality marks him out, in this particular context, as representative of a specific tradition: that of the 'Enlightenment Project', which 'promised a conception of rationality independent of historical and social context, and independent of any specific understanding of man's nature or purpose' (D'Costa 2000, p. 3; see also MacIntyre 1985). Hans-Georg Gadamer suggests that claims to neutrality, objectivity and universal reason are unable to do justice to the historicity of understanding. When striving to make sense of the world, we do so from within received cultural traditions, so that understanding necessarily emerges from the revision of our pre-understanding, presuppositions and prejudices in the light of our expanding and deepening knowledge of reality. The Enlightenment's claim to access a universal perspective is itself parochial: the 'fundamental prejudice of the enlightenment is the prejudice against prejudice itself, which deprives tradition of its power' (Gadamer 1979, pp. 239ff.).

The Enlightenment project's claim to a privileged access to truth is grounded in the liberal principles of reason, freedom and tolerance. We can usefully distinguish between political and comprehensive liberalism at this point (Wright 2004, pp. 192ff.; 2007, pp. 29ff.). John Locke envisaged liberalism as an interim-ethic, or doxastic praxis, in which the values of freedom and tolerance support an 'animating vision of a society in which persons of diverse traditions live together in justice and peace, conversing with one another and slowly altering their traditions in response to their conversations' (Wolterstorff 1996, p. 246). However, these principles gradually ceased to be a means to the greater end of the pursuit of truth and justice, and instead became reified as ends in themselves. As a result, Locke's political liberalism gave way to a comprehensive liberalism in which the ideal society is not one that pursues truth and truthfulness, but one that maximises autonomy and tolerance. In such a society, rather than defend their particular exclusive beliefs and explore the exclusive beliefs of others in a common quest for truth, representatives of particular traditions must abandon their exclusive beliefs because they are deemed to undermine freedom and breed intolerance. Instead of imagining a genuinely open pluralistic society harbouring a range of different traditions and contradictory belief systems, we must imagine a closed monolithic society living in peace and harmony only because its members have abandoned their exclusive truth claims *en masse* (Rorty 1989). In such an imaginary utopia, as John Lennon once famously observed, the absence of exclusive religious and nationalistic beliefs means that there is no longer anything worth killing or dying for. However, such a utopian vision is profoundly exclusivist, since to achieve it the entire human race must abandon its received belief systems, collectively embrace the parochial truths of comprehensive liberalism, and affirm autonomy and tolerance as the ultimate goods to which humankind should aspire. Bhaskar rejects Christianity because of its parochial and exclusive truth claims. However, he does so only on the basis of his own equally parochial and exclusive commitments: parochial, because they are grounded in the culturally specific Enlightenment dream of universal reason; exclusive, because they require religious adherents to discard their core ontological commitments.

(2) Freedom. Bhaskar argues that the universal philosophy of meta-Reality enhances human freedom, while the parochial theology of Christianity undercuts it. 'Freedom' is a radically under-determined concept. Comprehensive liberalism, insofar as it embraces freedom as an end in itself, understands it primarily as freedom *from* all external and internal constraint and freedom *for* autonomous belief and action. Unrestricted autonomy frees individuals to pursue their own goals and desires, provided they avoiding harming others in the process. At first glance unrestricted autonomy is a vacuous notion, lacking any ontological foundation and perpetuating the Enlightenment-generated divide between being and value. However, on closer inspection a specific ontological assumption emerges: namely that the ontological nature of human beings and the universe they indwell is such that the ultimate human good is that of autonomous belief and action. This is highly questionable on two fronts. First, it is far from obvious that

the ontological structures of reality are such that the greatest human good is autonomy of belief and action. The Abrahamic religious traditions, for example, assert almost the polar opposite: the greatest human good is obedience and submission to the will of God. Similarly, Bhaskar suggests that giving free reign to the ego can only undermine the well-being of a person's transcendentally real self. Second, some ways of exercising autonomy are clearly better than others: the freedom to live an altruistic life is self-evidently preferable to the freedom to self-harm. On this reading, freedom *from* inappropriate relationships with self and others is simultaneously freedom *for* appropriate relationships. This begs the question: What exactly constitutes an 'appropriate' relationship? This question cannot be answered without reference to underlying ontological assumptions: for comprehensive liberalism the ontological status of human beings is such that each individual is free to answer the question for themselves in the light of their own personal preferences and desires; for meta-Reality appropriate relationships are dependent upon individuals being in harmony with their ground-state and the cosmic envelope; for Trinitarianism appropriate relationships *within* the created order are dependent upon appropriate relationships *between* the created order and the Triune God, which in turn are ultimately dependent upon the ontological primacy of love – specifically the reciprocal loving relationships between the persons of the Trinity. With these initial considerations in mind, we return to Bhaskar's dual claim that religious pluralism enhances human freedom, while religious exclusivism undercuts it.

Bhaskar claims that theological pluralism offers all human beings the freedom to access God, while religious exclusivism curtails this freedom by restricting such access to a select minority. Despite popular assumptions, the suggestion that classical Christianity restricts access to God to a 'chosen few' is significantly wide of the mark. An important distinction can be drawn here between (1) ontology, (2) implicit epistemic knowledge relationships, and (3) explicit epistemic knowledge relationships.

1 At the *ontological level*, Trinitarian Christianity affirms that, as a direct consequence of the hypostatic union of God and humanity in the incarnation, God's love, grace and righteousness is poured out on *all* of humanity, regardless of whether human beings are aware of this fact or not. The ontological reality of the incarnation has soteriological significance for all human beings: as Saint Paul puts it, 'for as *all* die in Adam, so *all* will be made alive in Christ' (1 Corinthians 15:22; emphases added).
2 At the *epistemic level* of *implicit knowledge relationships*, Christianity affirms that those who are unable to access the Christian account of reality for whatever reason are still free to enter into an implicit knowledge relationship with God. The Christian theology of nature affirms that human beings can become partially aware of God indirectly through the traces of the Creator within creation. The footprints of divinity are embossed in the inherent rationality of the natural order-of-things, and in the actuality of the moral and aesthetic strata of reality. Christ 'from the very beginning and in

every part of the world, gives a more or less obscure revelation of the Father to every creature'; consequently – and in line with the teaching of the first Christian theologians and Thomas Aquinas – 'the grace of Christ is of universal application ... no soul of good will [whether atheist, theist or agnostic] lacks the concrete means of salvation, in the fullest sense of the word' (Lubac 1988, pp. 218ff.).

> Then the king will say ... 'I was hungry and you gave me food, I was thirsty and you gave me something to drink, I was a stranger and you welcomed me, I was naked and you gave me clothing, I was sick and you took care of me, I was in prison and you visited me'.... Then the righteous will answer him, 'Lord, when was it that we saw you hungry and gave you food, or thirsty and gave you something to drink...?' ... And the king will answer them, 'Truly, I tell you, just as you did it to the least of these my brothers and sisters, you did it to me'.
>
> (Matthew 25:34ff.)

3 At the *epistemic level* of *explicit knowledge relationships*, Christianity affirms that it possesses exclusive access to an understanding of the Trinitarian nature of God: God reveals himself definitively and directly in the incarnation, witnessed to in the Bible and testified to by Christian proclamation.

This, I suggest, has direct parallels with the philosophy of meta-Reality. (i) At the *ontological level*, all human beings are already united to their ground-state and to the cosmic envelope, whether or not their awareness of this fact has been occluded by the demi-reality of the dualistic world. (ii) At the *epistemic level* of *implicit knowledge relationships*, all human beings can in principle access their ground-state, whether or not they are aware that this is what they are doing, and regardless of their knowledge of Bhaskar's philosophy of meta-Reality. (iii) At the *epistemic level* of *explicit knowledge relationships*, those who consciously embrace the philosophy of meta-Reality claim exclusive access to an alethically truthful account of the ultimate order and structure of reality.

This being the case, it is difficult to see why Christianity places any greater restrictions on human freedom than meta-Reality: in both cases the ultimate transcendent ontological order exists independently of human knowing; in both cases individuals in every cultural context can implicitly access transcendence without necessarily being explicitly aware of what they are doing; and in both cases explicit awareness of the nature of the transcendent realm, though in principle available to all, is actually restricted to a minority of human beings inhabiting particular geo-temporal bound cultural coordinates. On this reading, at the levels of ontology and implicit knowledge relationships neither meta-Reality nor Trinitarianism appears to attribute a significantly greater or lesser level of freedom to human beings. At the level of explicit knowledge relationships, it is clear that a far greater proportion of human beings, in a far greater range of geo-temporal coordinates and cultural contexts, are in a position to access the

Christian Gospel than the philosophy of meta-Reality. This is so both by virtue of the more widespread geo-temporal dissemination of Christian teaching, and by virtue of the fact that an explicit understanding of the Christian Gospel does not demand the same level of intellectual expertise currently needed to access Bhaskar's writings. That said, Bhaskar cannot possibly be held to be morally culpable for the fact that the philosophy of meta-Reality is less widely disseminated than the Christian Gospel, and – despite Bhaskar's clear intent to reach out to a wider audience – currently accessible only to an intellectual elite.

If this argument is correct, then there is no fundamental difference between the level of human freedom allowed by Trinitarian doctrine and the philosophy of meta-Reality. Even if this was not the case, and one or other position *did* significantly limited epistemic access to the ontological reality of the ultimate order-of-things, this would have no deep epistemic or moral significance. If either Trinitarianism or meta-Reality possesses alethic truth, then the ontological realities they refer to exist entirely independently of the manner in which, and the extent to which, they are accessed. If Nazi human experimentation in the concentration camps had revealed previously unknown ontological truths, they would remain ontological truths despite the horrific cessation of human freedom involved in the discovery process.

(3) Tolerance. Bhaskar contends that religious pluralism engenders tolerance, while religious exclusivism breeds intolerance. 'Tolerance', like 'freedom', is a significantly under-determined concept. British schools systematically train pupils, in the name of tolerance, to be *intolerant* of racism, sexism and homophobia. This is of course right and proper; however, it highlights the fact that tolerance is a vacuous concept without any prior understanding of that which is worthy of tolerance. This becomes clear when we consider Karl Popper's 'paradox of tolerance', according to which the tolerant rightfully 'claim, in the name of tolerance, the right not to tolerate the intolerant' (Popper 1966, p. 265). Formally, this is a mere tautology: though Popper clearly invokes it to justify intolerance of (say) intolerant racists, it could equally be invoked to justify intolerance of those who are intolerant of racists. This being the case, 'tolerance' and 'intolerance' are both morally neutral categories that assume positive or negative moral capital only in specific moral contexts. If it is true that in certain contexts (e.g. multicultural communities) tolerance can be a virtue, then it is equally true that in other contexts (e.g. Nazi Germany or apartheid South Africa) intolerance can also be a virtue. John Kekes has argued that one of the primary reasons why the liberal tradition has allowed the 'virtue' of tolerance to remain so significantly under-determined is its lack of a sufficiently robust understanding of evil (Kekes 1999, pp. 23ff.). The liberal rejection of the Christian doctrine of the fallen nature of humanity and concomitant assertion of humankind's natural goodness fails to account for the prevalence of evil in the world. The 'assumption that the attractions of liberalism are sufficient to incline people to conduct themselves according to its prescriptions' is palpably false, since 'much human conduct is evil even though its agents understand what liberals tell them' (p. 23). Clearly the boundaries of both tolerance and intolerance need to be established,

and this requires reference to the underlying ontological presuppositions underpinning any judgements that might be made, whether these be the ontological priority of personal autonomy, of the cosmic envelope, or of the Triune God. As it stands, Bhaskar's assertion of the intrinsically intolerant nature of theological exclusivism requires further explication, particularly in the light of the previous suggestion that Bhaskar's own position is itself exclusivist.

As D'Costa points out, religious pluralists 'usually argue that exclusivists have no grounds for real openness and tolerance toward other religions, nor have they anything to learn from these religions', and that the 'resultant attitude is politically translated into empire, imperialism and aggressive mission' (D'Costa 2000, p. 1). However:

> Despite their intentions to encourage openness, tolerance, and equality [religious pluralists] fail to attain these goals (on their own definition) because of the tradition-specific nature of their positions. Their particular shaping tradition is the Enlightenment ... [which] in granting a type of equality to all religions, ended up denying public truth to any and all of them.
>
> (D'Costa 2000, p. 2)

If theological pluralism is indeed a disguised form of theological exclusivism, then in rejecting the truth claims of all traditions other than that their own meta-tradition, theological universalists are themselves culpable for failing to establish grounds for genuine openness and tolerance, failing to learn from other exclusive religious traditions, and seeking to colonise such traditions within their own exclusive theological frame of reference.

Bhaskar claims that the fact that the ground-state is accessible within a plurality of different religious traditions provides grounds for the tolerance of other religions. However, he does not extend that tolerance to those who insist on remaining religious exclusivists, since their beliefs foster intolerance and contribute to the 'clash of civilisations'. Since the vast majority of religious adherents are religious exclusivists, and the vast majority of religious traditions proffer exclusivist claims, if Bhaskar is to be consistent he must be intolerant of the majority of the world's religions and their adherents, insofar as they proffer exclusive truth claims. In advocating toleration of theological pluralists, he necessarily advocates intolerance of theological exclusivists. To attempt to mitigate this conclusion by suggesting that *religious adherents* should be tolerated but not their *exclusivist beliefs* is to impose the hegemonic liberal assumption that religious beliefs are extraneous cultural accoutrements that can be discarded by religious adherents without any substantial impact upon their personal identities.

Bhaskar's culturally specific position appears to support the liberal assumption that different religious traditions advocating contradictory belief systems can only live together harmoniously if they first adopt a universal liberal morality. On this view social integration is dependent on maximising the common ground between different traditions and minimising their differences. Since there can be no shared agreement regarding the exclusive and ontologically incommensurate beliefs proffered by particular religious traditions, such common ground must transcend the particularities of specific cultures. This means that cultural particularity must be stripped away in order to expose a generic and universally valid account of human nature. Yet any account of universal human nature, including those proffered by Christians, the philosophers of the Enlightenment and advocates of the philosophy of meta-Reality, cannot – given the principle of epistemic relativity – avoid being culturally specific.

The notion that the liberal tradition possesses the only viable formula for promoting harmonious relationships between exclusivist religious traditions is unsustainable, especially when we take into account the fact that liberalism's basic orientation is antithetical to their core theological beliefs, and regards them as a major stumbling-block to cross-cultural understanding. The root meaning of 'tolerance' is the ability to accept otherness and difference, so that the practice of tolerance entails a willingness to allow the 'other' to remain genuinely 'other'. To attempt to colonise the 'other' within a single liberal frame of reference is to ignore fundamental differences in favour of an economy of 'sameness' that requires religious adherents to put aside their exclusive beliefs before entering into inter-faith dialogue. Such an economy is profoundly intolerant of the particularity of distinctive religious traditions. Indeed, if there is any shared ground between adherents of exclusivist religious traditions, it is probably to be found in their common rejection of the hegemony of a liberal economy that, in the name of pluralism, seeks to colonise them within its own exclusive dogmas.

Inter-faith dialogue between incommensurable religious belief systems does not require a third liberal party to act as intermediary between them. Stanley Hauerwas recounts how, after a lecture in which he defended Christian exclusivism, he was accosted by a professor who suggested that his 'stress on the centrality of Christian convictions provided no theory that would enable Christians to talk to Buddhists' (Hauerwas 2007, p. 58). Hauerwas took this to imply 'the necessity of a third language to mediate between two traditions', and observed that such a language 'is often said to be necessary in "pluralist" societies in order to mediate differences in the "public" square' (p. 58).

> I, however, apologized for being deficient of such theory, but asked, 'How many Buddhists do you have here in Conway? Moreover if you want to talk with them what good will a theory do you? I assume that if you want to talk with Buddhists, you would just go talk with them. You might begin by asking, for example, 'What in the world are you guys doing in Conway?'
>
> (Hauerwas 2007, pp. 58ff.)

Nor does inter-faith dialogue between exclusive religious traditions require any prior agreement between the parties. The vast majority of exclusive traditions possess their own internal theological resources for dialogue and peaceful co-existence with those who do not adhere to their particular accounts of reality: Christians can draw on their understanding of God as love, while Muslims can turn to their understanding of God as merciful. If they fail to do so, the problem lies not in their exclusivist theologies, but in their inability to translate their beliefs into appropriate actions.

Trinitarian orthodoxy affirms that the three persons of the Trinity are ontologically one, but without any loss of their distinctive individuality. The Church rejects 'modalism', the view that God is ontologically singular, and that the three persons of the Trinity are merely three contingent modes of divine activity, since this fails to do justice to the ontological singularity of the Father, Son and Holy Spirit. The Church also rejects 'tri-theism', the view that the three persons of the Trinity are three distinct and separate entities, since this fails to do justice to the ontological unity of God. The orthodox Trinitarian position is predicated on the belief that God is love, and that divine love is constituted by the reciprocal process of perfect giving and receiving between Father, Son and Holy Spirit. To draw on a human analogy, in giving himself totally in the service of his beloved, the lover seeks a relationship in which the beloved is affirmed in her particular identity and is loved as the person she actually is, rather than as a person she is not or as the person he wishes her to be. Such love is necessarily non-coercive and unconditional, and as such cannot desire the loss of the beloved's distinctive identity. To insist coercively that the beloved must change, and that the continuation of the loving relationship is conditional upon such change taking place, is precisely *not* to love the beloved. This distinctively Christian understanding of the act of love (Greek αγάπη, agape) derives from the manner in which Jesus Christ engaged with those about him – especially social outcasts, sinners and the dispossessed – in proclaiming and embodying God's gracious and unconditional love of creation and inviting listeners to reciprocate this offer of love without any attempt at coercion, rhetorical persuasion or insistence on prior conditions. In doing so, Jesus revealed the nature of God as love, and established the grounds for a retroductive theological account of the nature of God as Trinity.

The ontological nature of God as love, in which the persons of the Trinity are united in the reciprocal affirmation of their distinctive identities, pans out into a specific Christian ethic: the non-coercive and unconditional acceptance of, and agapeistic love for, the 'other-as-other'. Christians have a God-given duty to serve and under-labour for all of humanity, and this requires them to acknowledge, respect, affirm and celebrate the freedom God has given to all human beings to be the people they actually are. It follows that Christian under-labouring can never be coercive or conditional. This, I suggest, is a far richer and more substantial basis on which Christians can enter into intercommunal relationships than the liberal notion of tolerance. Tolerance, in the sense adopted by the liberal tradition, demands at best a grudging acceptance of the 'other-as-other' and at worst the negation of the identity of the 'other' and their colonisation within an economy of 'sameness', whereas Christianity is committed to the unqualified and unconditional offer of hospitality to the stranger, as if the stranger were none other than Christ himself (Bretherton 2010). This is not to deny that Christians are bound to witness to the Gospel in thought, word and deed in a manner that opens out the possibility of conversion to the Christian faith. However, the Christian doctrine of mission is primarily oriented towards the flourishing of humanity in all its rich diversity: the Church is not a self-serving 'Noah's Ark' that individuals must enter if they are to avoid drowning in a flood of divine punishment; rather, it is commanded to be the salt of the earth and light of the world, under-labouring to enrich and illuminate the whole of God's creation, and striving for the common good of all. Insofar as the Gospel invites people to enter the Christian community, any such invitation must be non-coercive. All communities that believe they have access to truth will naturally seek to pass that truth on to others: in this respect there is little difference between those in the community of critical realists who seek to persuade others of the greater explanatory power of critical realism, and those in the Christian community who seek to persuade others of the greater explanatory power of Trinitarian theology (Thiessen 2011).

Part III

Ontological realism

Christianity and truth

7 Classical theism and the Triune God

In Part III we turn from epistemic relativism to ontological realism, and consider the specific onto-theological truth claims of Trinitarian Christianity: here in Chapter 7 we will differentiate the Triune God from the God of classical theism, in Chapter 8 we will explore the Christian understanding of God as Trinity, and in Chapter 9 we will investigate the economy of salvation through which the Triune God reconciles himself to his fallen creation. We will reserve the question of the epistemic justification and rational warrant of the ontological truth claims of Trinitarian theology for Parts IV and V, when we turn to the issue of judgemental rationality.

'God' is a radically under-determined concept. Consequently, in any theological enterprise it is necessary to differentiate between conflicting concepts of God: deistic, theistic, pantheistic, panentheistic, Trinitarian, etc. In the present context it is particularly important, for reasons that will become apparent, to distinguish between the God of classical theism and the God of Trinitarian Christianity. As Pascal famously pointed out, there is a fundamental difference between the God of the philosophers and scholars, and the God of Abraham, Isaac and Jacob (Pascal 1966, p. 309). Colin Gunton has argued that the failure to properly distinguish the two is a major source of the theological crisis facing Western Christianity in the modern age (Gunton 1978, pp. 11ff.; cf. 1991, 2003). In attempting a synthesis of classical theism and Trinitarianism, Christian theologians imported a set of theological assumptions, derived ultimately from ancient Greek philosophy, that were alien to orthodox Christian faith. Classical theism's dualistic vision of an utterly simple, transcendent, timeless, omnipotent and omniscient Supreme Being set over-against the world is incompatible with the Trinitarian vision of God as Father, Son and Holy Spirit united in a bond of reciprocal self-giving love and intimately engaged with his creation. Gunton argues that the basic flaws of classical theism become apparent once it is differentiated from Trinitarian theology. In particular, classical theism is unable to account for either the ontological or epistemic relationship between God and creation: if God is ontologically timeless then he must utterly transcend our time-bound universe, and if this is the case we can have no epistemic access to him. The Trinitarian notion of God as a personal, all-loving agent, on the other hand, has no problem in asserting a mediatory relationship between God and his creation. In *Becoming and Being* Gunton suggests that once the incoherence of classical theism is accepted, the only viable *theological* options (discounting atheistic naturalism, which Gunton views to be just as incoherent as classical theism) are Trinitarianism and panentheism, as exemplified in the respective theologies of Karl Barth and Charles Hartshorne (Gunton 1978). Coincidentally, naturalism, Trinitarianism and panentheism are the three basic options currently debated by critical realists. In this chapter we will outline the evolution of classical theism, identify its basic contours, differentiate it from Trinitarianism, consider the separation of the two traditions in the modern era, and consider the crucial issue of theological mediation in greater detail.

Before proceeding, it is important to acknowledge that Gunton's desire to free Trinitarian theology from the strictures of classical theism is not shared by all Trinitarian theologians. Richard Swinburne, for example, has developed a comprehensive defence of Trinitarianism allied to classical theism within the Anglican tradition, while the Roman Catholic Church continues to uphold a synthesis of the two that was first established by Thomas Aquinas (Swinburne 1993, 1994, 1998, 2001, 2004, 2005, 2007; cf. Mößner *et al.* 2008). Both, though, would accept the priority of Trinitarianism over classical theism. Thus the account of Trinitarian theology presented here and in the following two chapters represents one move among many within the orbit of classical Christian orthodoxy, which – like critical realism – is constituted by a complex web of contrasting, and occasionally contradictory, perspectives operating within a shared paradigmatic framework.

The evolution of classical theism

Classical theism has its roots in a set of philosophical questions first posed by classical Greek thought (Gerson 1994; Guthrie 1968). One cluster of questions was oriented towards the natural order-of-things: Why is there something rather than nothing? Why is the world intrinsically rational? A further cluster took its lead from the human condition: Why is human existence morally, aesthetically and spiritually value-laden? What is the ultimate meaning and purpose of life? Responses to these questions took two main forms: first, a literary tradition, rooted in mythical and legendary accounts of the gods provided by Homer, Hesiod and others, which generated a variety of religious and cultic practices; second, a set of rational arguments concerning the existence and nature of God developed by Greek philosophers and explored in the academies. The former was concerned with a dysfunctional divine family headed by Zeus, whose power struggles and love affairs both impact upon and mirror the world of human experience. The latter focused on a Supreme Being who functions as the ultimate principle (Greek ἀρχή, archē) of all reality and source of rational order (Greek λόγος, logos) in the universe (Gerson 1994, pp. 5ff.). Parmenides distinguished between the 'way of truth' and the 'way of opinion', between 'the steadfast heart of persuasive truth' and 'the beliefs of mortals, in which there is no trust' (Parmenides 1984, p. 53). The way of truth leads to the transcendent realm of singular, timeless, necessary and unchanging 'Being', while the way of opinion leads to the demi-reality of the immanent realm of plurality, time, contingency and change. In Greek natural theology the way of truth leads to the supreme transcendent rational source of the universe, while the way of opinion leads to the dysfunctional immanent gods of myth and legend.

Plato's natural theology follows the way of truth. In the *Republic* Socrates affirms that 'God is not the cause of all things, but only of the good'; but according to Homer 'Zeus is dispenser alike of good and of evil to mortals' (Plato 1961, pp. 626ff.). Since the gods are perfect beings they would never willingly assume inferior forms, yet various myths and legends speak of the gods adopting the forms of animals and human beings, such as the account of Zeus disguising himself as a swan and raping Leda. It is clear from this that the gods of myth and legend fall short of the God of philosophical investigation, and that since Homer and the myth makers follow the way of opinion there can be no place for them in Plato's ideal Republic. According to Plato the intelligible, immaterial, unchanging and transcendent Form of the Good, which functions to unite all eternal Forms, constitutes the ultimate ground and rational principle underlying all being. It stands over-against the sensible, material, contingent and immanent cosmos, though – as the allegory of the cave seeks to show – shadowy intimations of the Forms remain available to those held captive by the demi-reality of the material world. If the Form of the Good is eternal and unchanging, then it cannot act to bring the cosmos into being. In the *Timaeus* this task is attributed to the creator God or demiurge (pp. 1161ff.). The Greek 'demiurge' (*δημιουργός*, dēmiourgos) refers to a craftsman, artisan or public worker. The divine demiurge is a lesser reality subordinate to the Forms, whose possession of intellect and intelligence (Greek *νοῦς*, nous) enables him to contemplate the Forms, and to use them as a celestial blueprint when creating the universe by imposing order upon formless matter.

According to Aristotle, it is unnecessary, and indeed misplaced, to posit the existence of ideal Forms apart from concrete material objects. Unless the ideas embedded in the Platonic forms actually contact the cosmos, we cannot explain how the cosmos came into being as a rational entity. The demiurge does not solve this problem, because Plato cannot adequately explain the origins of either the creator God or primal chaotic matter: they cannot be created by the Forms because they are unchangeable and so cannot act, nor can they exist eternally because this would give him the same status as the Forms. Aristotle rejects Platonic Forms in favour of the notion of categories. The categories of actuality (Greek *ἐντελέχεια*, entelecheia) and potentiality (Greek *δύναμις*, dynamis) constitute the basic rational ground of being. All created objects possess both actuality and potentiality: they are what they are, and have the potential to become what they will become. Actuality is antecedent to potentiality, so that all potential objects necessarily originate from prior actual objects. This leads Aristotle to invoke a causal chain of being. In the *Metaphysics* he avoids any notion of infinite regress in this causal chain by positing a 'first cause', or 'unmoved mover', which exists as pure actuality and is devoid of all potentiality (Aristotle 1984, vol. 2, pp. 1688ff.). As pure actuality, this Supreme Being does not require any antecedent cause: his essence is simply to exist. Aristotle's God is perfect, beautiful and indivisible, existing in passive contemplation of its own perfection. As such it is devoid of will, because any intentional act of will implies the potentiality to change – a possibility incompatible with the notion of an eternally unchangeable being. The unmoved mover 'acts' passively: as perfect being it is the ultimate object of desire that provokes responses from the (apparently pre-existent) heavenly bodies without any effort on its part. Thus the 'First Heaven' moves because it is attracted by the intrinsic beauty of God, just as we might be moved by a beautiful sunset. Beautiful objects do not need to act in order to generate responses – they simply need to passively 'be' what they are: 'The object of desire and the object of thought . . . move without being moved' (p. 1694).

Trinitarian theology is not grounded in philosophical reflection and argument, but in shared human responses to God's self-revelation in specific events in the history of Israel and life of Jesus of Nazareth. The decision of the first Christian theologians to seek to iteratively refine their emergent retroductive understanding of the Triune God in dialogue with philosophy meant that their revealed Trinitarian theology came face-to-face with the natural theology of the ancient Greeks. In the West this led to a synthesis of the two: initially by Augustine in dialogue with Plato, and later by Aquinas in dialogue with Aristotle. In time the distinctions between the theologies of Plato and Aristotle were more or less obscured as a generic classical theism emerged as handmaiden to Trinitarian theology. In Western Christendom the God of the philosophers and the God of Abraham entered what, I will suggest below, turned out to be a less than holy alliance.

The contours of classical theism

As already noted, the late Colin Gunton was a leading defender of the claim that Western Christianity became seriously compromised through its attempted synthesis of Trinitarian theology with classical theism. The thesis is not in itself original: the first Protestant Reformers sought to recover a pristine biblical faith uncorrupted by Aquinas' Aristotelianism. Though Gunton identified Aquinas as a prime source of classical theism, his distinctive contribution was to extend the thesis to include Augustine's Platonism (Gunton 1990). This was a highly significant move, because Augustine was and remains the dominant theological force in Western Christian theology. Increasingly Gunton and other like-minded theologians turned to Eastern Orthodoxy as an instance of classical Trinitarianism relatively uncorrupted by classical theism. Gunton identifies three major traits of classical theism: (1) a commitment to supernaturalism, (2) an affirmation of the timelessness of God, and (3) an acceptance of a hierarchical ordering of reality (Gunton 1978, pp. 1ff.). We will consider each of these in turn.

(1) Supernaturalism. Gunton points out that classical theism proceeds via a 'negative' or 'apophatic' theological method (p. 2). According to Aquinas, 'we cannot know what God is, but rather what He is not' (Aquinas 1920, vol. 1,

p. 28). This *via negativa* asserts that God is not limited in the way human beings are limited: because human beings lack power and knowledge, God must be omnipotent and omniscient; because human society falls short of goodness and justice, God must be perfectly good and just. The result is a fundamental dualism between the immanent order-of-things and the transcendent being of God, between the realms of the natural and supernatural. Gunton argues that Aquinas' attempts to demonstrate the existence of God are predicated on, and serve to reinforce, this dualism: God is unmoved, nature moves; God is causeless, nature caused; God is timeless, nature time-bound; God is necessary, nature contingent.

(2) Timelessness. According to the logic of apophatic theology, if nature is time-bound then God must be timeless: '[b]ecause the natural is necessarily temporal, and because God is the negation of the natural, his relation to time will be that of the timelessly eternal' (Gunton 1978, pp. 2ff.). As Augustine puts it, because God is perfect he cannot change, and because time depends on change God must be timeless. Since time is a sequential relationship between objects and events, the creation of the universe necessarily involves the creation of time itself. The question as to what God was doing before he made heaven and earth is redundant because God is essentially timeless:

> But in eternity nothing moves into the past: all is present. Time, on the other hand, is never all present at once. The past is always driven on by the future, the future always follows on the heels of the past.
>
> (Augustine 1961, pp. 261ff.)

Similarly, 'when Aquinas understands God to be the cause of the world ... he is always thinking of instantaneous causation'; the timeless God views reality *sub species aeternitatis*, so that the beginning and end of time, together with the temporal sequences in between, are simultaneously perceived by God at the 'moment' of creation (Gunton 1978, p. 3). 'God is then, not a deist God belonging to the beginning of time, but a timeless absolute on which all temporal reality depends ... the one changeless, infinite, perfect, abstract, absolute' (pp. 3ff.).

(3) Hierarchical order. Classical theism affirms a hierarchical ordering of reality (pp. 3ff.). Objects are located in a scale of being from the lower through to the higher. In the Platonic scheme of things, the Forms are placed at the top of the hierarchy, the natural world at the bottom, and the demiurge somewhere in between. In the Aristotelian system the unmoved mover is placed at the start of a causal chain of being. The hierarchy may be conceived on a horizontal line, either looking back to an original source of being, or forward to the final perfection of being; alternatively it may be understood on a vertical line, either looking upward to the transcendent source of being or downward to the deep ground of being. Regardless of what particular metaphor is employed, the quantitative differences in the natural order-of-things ultimately give way to a qualitative difference between the natural order and the supernatural reality that stands at the tip, base, beginning or end of a hierarchical chain of being. Typically different strata within the natural order are perceived to be linked in an economy of dependency,

with lesser objects dependent on higher ones. As a consequence, more dependent objects may be deemed to possess lower levels of 'reality' than less dependent objects, so that the Platonic Forms are more 'real' than the demi-world of the cosmos constructed by the demiurge. Classical theism tends to affirm the ultimate reality of God and the lesser reality of the created order. On one reading modern naturalism rejects this hierarchical order of being by rejecting any notion of an ultimate supernatural stratum of reality; however, on an alternative reading it simply reverses the polarity of this classical hierarchy, so that the strata described by natural scientists are designated as more 'real' than the strata described by social scientists, which in turn are considered more 'real' than the strata of aesthetics, morality and spirituality engaged with by the humanitarian disciplines. Thus where the physical world constitutes the lowest strata in the Platonic scheme, it now forms the highest strata in a naturalistic hierarchy.

Given the radical dualistic nature of the ontological distinction between nature and supernatural, time and timelessness, and ultimate reality and demi-reality natural theology – in the tradition running from Plato and Aristotle through to Augustine and Aquinas – struggled to provide a satisfactory account of the relationship between the two realms. This generated the theological problem of 'mediation': How can an utterly transcendent God possibly relate to an utterly immanent universe? Naturalism seeks to overcome this dilemma, intrinsic to classical theism, by denying the existence of any transcendent reality, while panentheism seeks to do so by partially conflating nature and transcendence. Trinitarian theology, as we shall see, offers an alternative solution to the problem of mediation between transcendence and nature, God and creation.

Trinitarian theology and classical theism

Orthodox Trinitarian theology rejects any dependency upon classical theism, and insists instead on the uniqueness of its particular *sui generis* understanding of God as Father, Son and Holy Spirit. It is not sufficient simply to seek to balance or synthesise classical theistic concepts of divine omnipotence, omniscience and omnipresence, and Trinitarian concepts of divine love, justice and faithfulness. Rather, the latter take priority over the former in Trinitarian doctrine, and do so by virtue of the fact that God reveals himself as essentially three persons, Father, Son and Holy Spirit, co-existing in the indivisible unity of reciprocal self-giving and self-receiving love. Trinitarian theology is not epistemically grounded in philosophical questions about the origins, nature and purpose of the natural world, but in God's unique historical self-revelation in the person of Jesus Christ. It follows that it is only possible to understand the nature of the Triune God through the economy of salvation flowing from the Father and administered by the Son and Holy Spirit.

> The Synoptic Gospels ... speak powerfully of the Love of God in its concrete embodiment and manifestation in the self-giving love and compassion of Jesus for the sick and the suffering, the weary and the heavy laden, the

> poor and the destitute, the sinners and the outcasts, and all who hunger and thirst for righteousness.
>
> (T.F. Torrance 1996, pp. 57ff.)

> What Jesus himself is and does in life and death is the window God has provided for us through which we may discern the Love of God, for in Jesus, the Son of God who came forth from the Father to be one of us and one with us, there is disclosed the very nature of the Love of our Father in heaven for all his children.
>
> (p. 58)

The God revealed in the person of Jesus Christ is not an abstract, impersonal and utterly transcendent unmoved mover of limited ontological import and virtually no existential significance to humankind. Though there is a vital ontological distinction to be made between the Creator and his creation, this does not mean that the Triune God is utterly transcendent of the world. Rather, he is simultaneously immanent within the world and transcendent of it. Unlike Plato's abstract and idealised Forms, divine goodness is concretely embodied in the reciprocal relationships of Father, Son and Holy Spirit within the one Holy Trinity. Unlike Aristotle's unmoved mover, the Triune God is personal and creates out of love through an intentional act of will. Unlike the deistic God of the Enlightenment philosophers, he does not cease to engage with his creation once he has brought it into being. Instead, the inner dynamic of reciprocal divine love within the Trinity is intimately engaged with an economy of salvation in which creation is free both to be itself and to reciprocate its Creator's gracious gift of love. The personal agency of the Triune God thus offers a solution to the problem of mediation that had so bedevilled classical theism.

The ontological differences between the Triune God and the God of classical theism have an epistemic dimension. The epistemology of apophatic negative theology constitutes a form of the epistemic fallacy. The *a priori* epistemic assumption that whatever God is he must be other than the natural order inevitably generates a *projective*, *particular* and *pre-established* understanding of God. Trinitarian theology, on the other hand, operates on the basis of an *a posteriori* understanding of God grounded in divine self-revelation. If God is defined in advance as supernatural, then any engagement with nature must necessarily be an act of violation. But if God is the all-loving Creator of an intrinsically good creation, then his engagement with the world can only be driven by infinite non-coercive care for its well-being and flourishing.

The division between the natural and supernatural raises fundamental problems for Trinitarian Christianity as it seeks to affirm God's gracious and loving engagement with his creation. Classical theism's commitment to supernaturalism, timelessness and a dualistic hierarchy of being leads it to view divine engagement with the world as a series of arbitrary punctiliar interventions by an all-powerful supernatural deity that necessarily violate the natural order-of-things. This makes no sense within a Trinitarian framework that speaks of a God who, in an act of unconditional love, under-labours, serves, suffers and dies for the entirety of the human race. Trinitarian Christianity understands God's engagement with creation as a coherent and meaningful process intended to affirm nature and return it to its pristine goodness. The Trinitarian account of the economy of creation, redemption and consummation assumes a God intimately engaged within the spatio-temporal coordinates of the created order, responsive to the suffering of humankind, and proactively engaged in promoting justice and nurturing goodness.

Whereas the God of classical theism is timeless, the Triune God engages personally with his time-bound creation. It is difficult to explain how a timeless God can be intimately engaged in the temporal events of his creation, and equally difficult to explain how an unmoved mover can act in a genuinely personal and loving way. As Robert Jenson puts it with reference to the doctrine of incarnation: 'if to be God is simply to be timeless, then a man with a mother and an executioner simply cannot be full-fledged and full-time God, chop the logic as we may' (Jenson 2010, p. 9). Trinitarian theology tends to understand God as everlasting and sempiternal rather than timeless: God exists eternally and as such has no beginning or end, but in doing so he experiences temporal succession, both in the inner life of the Trinity and in the economy of creation and salvation.

Though Trinitarianism affirms a basic ontological difference between Creator and creation, it does not designate creation as in any way less 'real' than its divine Creator. Creation is genuinely real not because it is self-generating, self-rationalising and self-sustaining, but because it is the environment designed by God for the establishment of reciprocal loving relationships between the persons of the Trinity and humanity. Because God is present within creation in the persons of the Holy Spirit and the incarnate Son, he is not to be found at the start or end of a temporal hierarchy, nor in the transcendent heights or immanent depths of a spatial hierarchy, but *within the created order itself*. Consequently, we 'must place ourselves theologically where the action is, because if we turn away from God's actual historical self-identification in Jesus, we simply manufacture an idol, or series of idols' (Gunton 2003, p. 26). As Paul Cumin observes, 'Gunton has developed a doctrine of God that looks less like the top portion of a vertical continuum and more like an open dynamic of personal love' (Cumin 2010, p. 79).

The parting of the waves

A fundamental problem with the medieval synthesis of Trinitarianism and classical theism lay in the difficulty of explaining how an utterly transcendent, timeless and impassable God could possibly act benevolently towards the created order by seeking to promote justice, nurture goodness and secure the salvation of humankind. The Trinitarian account of the economy of creation, redemption and consummation assumes a sovereign God intimately engaged within the spatio-temporal coordinates of the created order, responsive to the suffering of humankind, and proactively engaged in promoting the human good. However, both the

Platonic Form of the Good and Aristotelian Unmoved-Mover are necessarily impersonal static forces: a personal force requires the movement of mind and will, but an already perfect being cannot move without moving away from its perfection. Attempts to solve the problem of mediation by invoking a quasi-divine mediator required some form of myth making: Plato resorted to the demiurge, who creates the cosmos by shaping pre-existent matter according to a cosmic blueprint provided by the Forms; esoteric Gnosticism appealed to a variety of speculative hierarchies of spiritual beings linking the perfect Supreme Being to the corrupt physical world; the Christian heresy of Arianism posited Jesus Christ as a hybrid superhuman/quasi-divine figure linking the transcendent Creator and his immanent creation. From a Christian perspective, the basic problem with all such attempts is that any divine or demi-divine mediator must necessarily be at least one step removed from the reality of God himself.

The identification of God as a less-than-ultimate being was an option that the first Christian theologians categorically refused to adopt. For God to be truly God, he cannot be subservient to any greater reality (such as the Forms) or share his status to any other reality (such as pre-existent matter). Any synthesis between Trinitarianism and classical theism thus required, at the very least, the *identification* of the Triune God with the Platonic Forms and Aristotelian Unmoved Mover. This raised two fundamental problems. (1) The problem of the relationship between the divine personality, mind and will of the Trinitarian God, and the impersonal, timeless and static reality of the God of classical theism. Here Augustine's commitment to the doctrine of divine simplicity ran the risk of making the Trinitarian account of the creative, redemptive and consummatory work of the Father, Son and Holy Spirit secondary and accidental to the intrinsic being of the God of classical theism. (2) The problem of accounting for the relationship between a timeless and unchanging God and his time-bound and contingent creation. The incarnational claim that God was hypostatically united with humanity in the person of Jesus simply brought this latter problem into stark relief. How was it possible to identify a timeless, omnipotent and omniscient being with a particular time-bound human being lacking in absolute power and limited in knowledge? The three prototypical Christological heresies constitute attempts to avoid this dilemma: Docetism by denying the humanity of Christ, Ebionism by denying the divinity of Christ, and Arianism by identifying Christ as a hybrid demi-god floating midway between divinity and humanity. Given the intrinsic problems of synthesising these two very different accounts of God, Western Christianity ran the risk – or, as Gunton argues, fell into the trap – of affirming a form of classical theism overlain with an idiosyncratic and thinly spread Christian veneer.

The ambiguous nature of the relationship between classical theism and Trinitarian theology is clearly evident in the structure of Aquinas' *Summa Theologica*. He first establishes his theological benchmark by developing a theistic account of God's existence, his nature as simple, perfect, good, infinite, immutable, eternal and indivisible, and his abstract attributes of will, love, justice, mercy and providence (Aquinas 1920, vol. 1). He then goes on to overlay this classical theism with a distinctively Christian account of the Trinity, and of the Father, Son and Holy Spirit (vol. 2, pp. 1ff.). Aquinas repeats this procedure in his theological epistemology, which begins by establishing natural knowledge of God grounded in inferences from nature, and then goes on to supplement this with an account of God's self-revelation in history (vol. 1, pp. 120ff.). According to this epistemology, theological language is neither univocal nor equivocal, but analogical: as such, it is able to engage with the reality of God without either reducing God to human categories (univocity) or positing God as an unknowable entity (equicovity). Analogical language is possible because of a God-given connection between divine grace and human reason: God creates an ordered universe that reflects its Creator in a manner that is comprehensible to human beings. Thus Aquinas' turn to Aristotle partially succeeded 'in connecting theological and scientific concepts in such a way that theology and science shared together in the developing understanding of a rationally ordered universe' (Torrance 1980b, p. 22). Contrary to modern attempts to domesticate God within the bonds of human reason, this process of faith seeking understanding sought to establish a balance between 'awe in the face of divine mystery and boldness in envisioning the possibilities of grace' (Placher 1996, p. 6). Although Aquinas insists that 'we have a more perfect knowledge of God by grace than by natural reason', such positive theology consistently operates against a background of negative theology (Aquinas 1920, vol. 1, p. 147). The place of classical theism in Aquinas' thought is such that an apophatic theology in which 'divinity is readily conceived of as thoroughly inconceivable' is never far from the surface (Davies 1993, p. 42). Nevertheless, he understands positive and negative theology as dialectical correctives of one another, and thereby seeks to avoid the twin dangers of univocal and equivocal theological language (Lossky 1991, p. 26).

The danger of equivocal theological language implicit in Aquinas became explicit in the late medieval theology of William of Ockham (Ockham 1983, 1990; cf. Spade 2000, pp. 326ff.). He dismissed the traditional scholastic synthesis of faith and reason on the grounds that it ran the risk of compromising divine sovereignty by making God in some way subservient to human reason. Once Ockham established an absolute ontological dualism between God and creation, analogical language inevitably broke down into nominal equivocal language, in which human talk of God has no guaranteed purchase on divine reality. 'For Ockham the idea of divine omnipotence thus means that human beings can never be certain that any of the impressions they have correspond to an actual object' (Gillespie 1996, p. 18). In insisting on the absolute sovereignty of a transcendent theistic God utterly aloof from the world of human experience, Ockham effectively paved the way for the eclipse of the Trinitarian God intimately engaged with humankind in a gracious, loving and beneficent manner. Thus, immediately prior to the dawning of the Enlightenment, the God of classical theism took its leave from the God of Trinitarian theology, thereby extricating the deistic God of the philosophers from the Trinitarian God of the theologians.

The problem of mediation

As we have already noted, according to Gunton one of the fundamental problems with classical theism is its inability to provide a satisfactory account of the manner in which God and creation are mediated. An *absolutely* transcendent God cannot by definition engage with our immanent world, either ontologically or epistemically. The qualitative dualistic distinction between God and creation is thus a fundamental – and, Gunton would argue, ultimately irresolvable – theological problem for classical theism. In this section we will explore three contrasting ways of resolving the problem of mediation: atheism, panentheism and Trinitarianism.

(1) Atheism. The simplest way of doing away with the incoherent dualistic distinction between God and the natural order-of-things is simply to remove God from the equation. We have already noted how Ockham invoked an all-powerful and wilful God stripped of any connection with Trinitarian belief. Deism, a latter-day version of classical theism, followed a similar path, rejecting any notion of Trinitarian self-revelation in favour of a divinity conceived of as a celestial watchmaker who designs and brings into being a mechanistic universe that he then leaves to its own devices. This deistic God functioned as the ultimate principle (Greek ἀρχή, archē) of all being and source of rational order (Greek λόγος, logos) in the universe. However, since this absent deity possessed no existential significance for humanity, and given the developing understanding of the order inherent in nature generated by the emerging natural sciences, its existence was increasingly seen as an unnecessary hypothesis. At the pinnacle of the natural order stand human beings who, by virtue of their possession of rational conscience, assume responsibility for the moral ordering of the universe. From here it was but a short step to the dual claim: (i) that the universe is inherently rational, self-generating and self-sustaining and hence does not need a Creator; and (ii) that human beings can and must take responsibility for generating and sustaining their own value systems. Both the God of Ockham and the God of Deism fail to oversee, support and sustain the flourishing of human beings, who consequently find themselves thrown back on their own devices.

Michael Buckley has argued cogently that the emergence of modern atheism was largely predicated on the rejection of the classical theistic notion of God as an omnipotent and omniscient deity, rather than the rejection of Christian Trinitarianism (Buckley 1987). According to Christopher Hitchens, in the context of a public debate with Tony Blair,

> Once you assume a creator and a plan, it makes us objects, in a cruel experiment, whereby we are created sick, and commanded to be well.... And over us, to supervise this, is installed a celestial dictatorship, a kind of divine North Korea.
>
> (Hitchens 2010)

Significantly, Blair chose to attempt to rebut Hitchens from the standpoint of a classical theism, rather than from that of his own Trinitarian faith. Hitchens' theology is a version of classical theism which makes no reference to the particularities of Trinitarian doctrine. Trinitarian Christianity does not claim that human beings are 'created sick', or that they are 'commanded to be well'; on the contrary, it claims that human beings were created good, and that they are incapable of becoming well without divine aid, which is provided freely and unconditionally. Further, this divine aid comes not in the form of a celestial dictator, but in the form of an all-loving crucified God who chooses to unite himself with humanity in the hypostatic union of God and man in the incarnate Jesus Christ. The issue here is not the veracity or otherwise of Hitchens' anti-religious polemic; rather, it is with the fact that both Hitchens' argument and Blair's response replicate a long-standing and deeply rooted confusion between the God of classical theism and the God of Trinitarian Christianity.

As we have seen, the primary source of classical theism is the negative projection of human attributes onto a divine being: God is conceived to be omnipotent and timeless on the grounds that he must necessarily be 'other' than weak, time-bound human beings. According to Karl Barth, the greatest modern exponent of Trinitarianism, 'the attempt to build up a concept of God by a process of abstraction from human personality' provides 'confirmation of Ludwig Feuerbach's belief that the concept of God was the result of the projection of human characteristics (stripped of all limitations) on to a transcendent (non-existent) being' (McCormack 1997, p. 106). 'A concept of God which results from projecting human self-awareness into the realm of the transcendent cannot reach the reality of [the Trinitarian] God, let alone describe it exhaustively' (p. 106). Barth thus affirms 'the liberating truth' of Feuerbach's critique of theological projectionism and rejection of classical theism (Barth 2001, p. 522; cf. Busch 2004, pp. 57ff.). Insofar as Christian theology – in both its medieval Catholic past and modern Liberal Protestant present – has allowed itself to be compromised by an ill-judged engagement with classical theism, Feuerbach speaks to the Church with a prophetic voice in calling it to reject all theological idols of human construction. If a theology grounded on the projection of human ideals and aspirations produces a God devoid of any ontological grounding, then the problem of the mediatory relationship between transcendence and immanence is immediately resolved.

(2) Panentheism. Pantheism (*sic*) seeks to overcome the theological problems of dualism and mediation by invoking a thoroughgoing monism that simply identifies God with the totality of all that exists. The nature and extent of this monistic reality may be understood idealistically, materialistically or in terms of a mind–body dualism in which mind and matter combine to constitute the two basic interrelated substances that together form a monistic whole. In the West, pantheism flourished briefly among the Stoics and was revived in the modern era by Spinoza, from whence it impacted upon aspects of Idealism, Romanticism and most recently the New Age movement. Critics of pantheism suggest that it is a redundant move to identify the cosmos as divine, because in doing so we add nothing substantial to our knowledge of reality: at best, the identification of the cosmos as 'divine' may have a relatively superficial attitudinal impact upon the way we live our lives and engage with the natural order.

The process theology established by Alfred North Whitehead and Charles Hartshorne (Whitehead 1979; Hartshorne 1941, 1948, 1953, 1965, 1967; cf. Cobb and Griffin 1976; Sia 1990) sought to overcome the theological redundancy of pantheism by rejecting the Platonic prioritising of being over becoming and the Aristotelian prioritising of actuality over potential. On this reading reality is constituted by a dynamic process of becoming that cannot be reduced to a deterministic process, whether idealistic (e.g. Hegelian dialectic) or materialistic (e.g. biological determinism). The dynamic process of being-as-becoming is essentially holistic and open-ended: holistic, because the totality of objects and events, including both the physical and mental, interact together as part of a seamless whole; open-ended, because natural evolution generates emergent entities that are essentially contingent, and because sentient human beings possess free will and the capacity for self-determination. Because reality is monistic and all objects and events are necessarily interrelated, all human experience, however trivial it may appear, contributes to the organic and dynamic development of the whole. Process theology is a form of panentheism: God is ingrained in the dynamic process of being-as-becoming, but – contra pantheism – not completely identified with it. The panentheistic God of process theology is not omnipotent, deterministic or coercive, and interacts with and participates in emergent reality. Because of this, God changes in response to events, though his essential attributes (goodness, wisdom, love, etc.) remain constant. The dynamic progress of reality towards its future perfection is thus the result of a complex web of causal mechanisms, natural, human and divine. The affinities of process theology with Bhaskar's philosophy of meta-Reality are transparent.

Gunton recognises that panentheism overcomes the dualistic and mediatory problems inherent in classical theism, and implies that process theology offers the only viable alternative to Trinitarian theology (Holmes 2010, pp. 32ff.). However, he suggests that process theology emancipates itself from the constraints of classical theism only at considerable cost. (i) It is unable to deal with the particularity of individual objects, events and persons, which are ultimately subsumed within a monistic whole. Thus Hartshorne rejects any notion of objective immortality, and suggests instead that following our death our experiences continue to contribute to the whole by living on in the mind of God. Gunton notes that process theology is able to embrace notions of reincarnation and the transmigration of souls, and sees this as further evidence of a failure to affirm the lasting reality of the particular. Ultimately panentheism affirms the 'One' indivisible unity of all things at the expense of the 'Many' particular objects and events. (ii) It ceases to allow God to be genuinely God. Rather than the ultimate all-loving source of all being, God is reduced to one among many causal mechanisms operating within the monistic whole, albeit a supremely important one. As such God cannot qualitatively oversee the salvation of the world, merely make a quantitative contribution to such salvation: 'divine relativity in Hartshorne leaves no room ontologically for genuine divine action over against the world'

(Schwöbel 2010, p. 184). Unlike the Trinitarian God, the God of process theology cannot create the world *ex nihilo*; instead he functions in a way similar to Plato's demiurge, forming and shaping pre-existent matter, existing independently of himself and over which he does not have complete control. God thus ceases to be *the* ultimate reality, and becomes instead merely the chief object among many other objects ingrained within the ontological order-of-things. Further, process theology follows classical theism in reducing God to a projection of humanity, albeit in its own distinctive manner: if the anthropological ideal for classical Greek philosophy was an all-powerful and all-knowing creature devoid of human limitations, so the anthropological ideal for process theology is a limited human being, able to live in harmony with others and learn alongside them. Politically, God ceases to be an all-powerful sovereign ruler and becomes instead *the* ideal cosmic citizen, interacting on equal terms with other, ordinary, human citizens in a democratic ontological cosmology.

Karl Barth's discussion of the relationship between Trinitarian theology and panentheism takes as its starting point the actuality of the coexistence of God and humanity and explores its theological and ontological grounds. He argues that the Trinitarian God would be the same even without a co-existent relationship with his creatures, since God is a God who *loves* in *freedom* (Barth 1936–1952, vol. 2/1, pp. 272ff., 297ff.). God's freedom to co-exist with human persons is not dependent on a higher comprehensive force operating equally on God and humanity, as it where, 'from outside'. That would make God subservient to a greater power and limit his freedom-to-love. In Platonic terms, the freedom to co-exist would function like an eternal Form, with God adopting the role of a lesser demiurge who co-exists with human beings out of a necessity imposed by that higher Form. If God and human beings co-exist because of a comprehensive force greater than both of them, then the distinction between God and creation becomes blurred and generates panentheism, 'which tells us that everything is permeated in God or by God as an all-pervasive world soul' (Busch 2004, p. 123). This means that God is not fully divine in the Christian sense, but rather dependent on a greater force or reality in which genuine divinity ultimately resides. 'In such a case he would be bound to humanity, not because he willingly loves us but because he *has* to actualise himself because he *has* to submit to a higher constraint, namely, the supreme law of "totality"' (p. 123; emphases added). On this reading, it is only the higher law of totality, which posits a necessary relationship between God and human persons, which is genuinely divine. Barth contests the thesis that God only becomes God when he co-exists with humanity. The Trinitarian God loves and creates in freedom, and is grounded in his own being, not in any being external to himself. 'The eternal correlation between God and us, as shown in God's revelation, is grounded in God alone, and not partly in God and partly in us' (Barth 1936–1952, p. 281). God does not create out of any lack within himself, or necessity imposed from outside himself; rather, he creates freely, out of and through his own all-loving will.

The co-existence of God and human persons is irrevocable, not because of the abstract law of totality, but because God's love is eternal and unconditional. Because God will never cease to love his creatures, human persons necessarily co-exist with God. Such necessity is not, however, authoritarian, since it is grounded not in the supreme law of totality but in an eternal and irrevocable love that is non-coercive and allows human persons the freedom to be themselves apart from God. In the Kingdom of God all human beings are created equal, ruled over by an all-loving sovereign God who is simultaneously their all-loving servant.

(3) Trinitarianism. Trinitarian theology offers 'a way of escape from the paralysing effects of a false polarity: either static, unrelated divine transcendence or the subsuming of the divine into a general ontology of flux' (Webster 2010, p. 20). Within this polarity, God is either an abstract and utterly transcendent unmoved mover, or else pantheistically or panentheistically ingrained in the fabric of the universe, and thereby subservient to fate and compromised by a range of non-divine forces and causal mechanisms. In both cases we arrive at a God who does not possess the same ultimate ontological import and fundamental existential significance for humankind possessed by the Triune God of Christianity. Gunton argues that Trinitarian theology follows process theology in emancipating theological reflection from the problems of dualism and mediation. However, it differs from process theology: (i) because it holds fast to the particularity of objects, events and persons, and sees them as possessing infinite value in the eyes of God; and (ii) because it holds fast to the absolute sovereignty of God without a repetition of the problems of unconstrained omnipotence associated with classical theism, and of the limited notion of divine sovereignty associated with panentheism. The Trinitarian God is neither utterly aloof from the world nor fatally compromised by it, but rather simultaneously immanent within creation while remaining transcendent of it. The inner dynamic of reciprocal divine love between the persons of the Trinity rules out any notion of divine despotism associated with classical theism. It is divine love, in the persons of the Son and Holy Spirit, which mediates between Creator and creation. Because God is not identified with creation, he retains his absolute sovereignty over creation, though crucially this is a sovereignty of self-giving love rather than raw omnipotent power. Because divine love can never be coercive, God allows creation the freedom to be itself, and loves it with infinite care. This means that the particularity of creation, of every created object, event and person, is of eternal significance to God:

> Are not two sparrows sold for a penny? Yet not one of them will fall to the ground unperceived by your Father. And even the hairs on your head are all counted. So do not be afraid; you are of more value than many sparrows.
>
> (Matthew 10:29ff.)

God desires to draw creation into the reciprocal love within the Trinity, but for this to happen the divine offer of love must be freely reciprocated by human beings. This exchange was actualised in the hypostatic union of God and humanity in the person of Jesus Christ: in Christ God embraces humanity, and humanity embraces God, in a bond of reciprocal uncoerced love. Classical Trinitarian theology thus stands apart from both classical theism and the panentheistic God of process theology. It is precisely because God is sovereign love and not raw power that creation and creatures can be affirmed in their particularity. The polarity between the absolute 'One' and the eternal flux of the 'Many' is resolved in the Triune God of Love (Gunton 1993).

8 The Triune God

Father, Son and Holy Spirit

We turn now to the heart of the ontology of orthodox Christianity: the being of the Triune God, Father, Son and Holy Spirit. While Eastern Orthodox theologians have consistently embraced Trinitarian doctrine as the hermeneutical key to understanding the distinctive truth claims of orthodox Christianity, Western theologians have not always followed their lead consistently. Roman Catholicism's flirtation with classical theism, conservative Protestantism's suspicion of doctrinal claims not explicitly contained in the Bible, and Liberal Protestantism's modernist-derived suspicion of metaphysical questions, have at times served to obscure the centrality of the doctrine of the Trinity. Western Christianity is, however, currently in the middle of a renaissance of Trinitarian theology that is proving instrumental in the recovery and reaffirmation of the core ontological claims of Christian orthodoxy (Kärkkäinen 2009; cf. Barth 1936–1952; Mackey 1983; Gunton 1991, 2003; J.B. Torrance 1996; T.F. Torrance 1996; Cunningham 1998; O'Collins 1999; Marshall 2000; Tanner 2001; Grenz 2004). The doctrine of the Trinity is essential to orthodox Christianity: it is not a logical problem to be resolved; not a 'difficult' or 'embarrassing' doctrine to be placed to one side; not a mystery to be carefully guarded; not the product of idle speculation; not the outcome of an unnecessary and misplaced flirtation with Greek metaphysics; and not a thin Christian veneer superimposed on the God of classical theism. Rather, it is the result of a retroductive attempt to identify, describe, explore and iteratively refine an understanding of the ontological being of God, and hence the ontological foundation of the whole of creation, in the light of the history of Israel, the first disciples' encounter with Jesus of Nazareth, and the ongoing lived experience of the Christian Church. As Robert Jenson puts it, the doctrine of Trinity 'has explanatory and regulatory use in the whole of theology ... it is not a separate puzzle to be solved', but rather the 'framework within which all theology's puzzles are to be solved' (Jenson 1997, p. 31).

The one Triune God

> And the Catholic Faith is this: That we worship one God in Trinity and Trinity in Unity, neither confounding the Persons, nor dividing the Substance. For there is one Person of the Father, another of the Son, and another

> of the Holy Spirit. But the Godhead of the Father, of the Son, and of the Holy Spirit, is all one, the Glory equal, the majesty co-eternal. Such as the Father is, such is Son, and such is the Holy Spirit.
>
> (Anglicans Online 2011)

The Athanasian Creed, established in the West since the sixth century, offers perhaps the most substantial formal statement of Trinitarian belief. The doctrine of the Trinity affirms that the Triune God exists as three persons in one being, and one being in three persons. The Father, Son and Holy Spirit exist as three distinct persons, entities or individual instances (Greek, ὑπόστᾰσις, hypostases; Latin, *persona*). The three persons are united in one substance, subsistence or essence (Greek, οὐσία, ousia; Latin, *substantia*), so that the Son and the Holy Spirit are consubstantial, or of one substance (Greek ὁμοούσιος, homoousios; Latin, *una substantia*), with the Father.

Unlike classical theism, belief in the Trinitarian God is not the result of abstract philosophical reflection, though such reflection certainly played a noteworthy, albeit limited, role in the iterative development of the doctrine. Rather, Trinitarianism arose out of the concrete historical events of the life, death and resurrection of Jesus of Nazareth, as interpreted through the dual lens of Jewish history and Scripture, and the ongoing experience of the Christian Church. The belief that this particular human being was God incarnate, coupled with Christian experience of the ongoing presence of the Holy Spirit in the life and worship of the Church, inevitably led to attempts to account for the relationships between Jesus Christ and the Holy Spirit, and between each of them and the Father-God worshipped by Jew and Christian alike. In Hebrew Scripture aspects of God's divine being and action in the world are personified as God's word, wisdom and spirit: he 'sends out his word' to Israel, his 'wisdom cries out in the street', he promises 'to put my spirit within you' (Psalm 147:18; Proverbs 1:20; Ezekiel 36:27). Though the New Testament is replete with references, both implicit and explicit, to God as Father, as Son and as Holy Spirit, there are only two explicitly Trinitarian formulae: 'In the name of the Father and of the Son and of the Holy Spirit'; 'The grace of the Lord Jesus Christ, the love of God, and the communion of the Holy Spirit be with all of you' (Matthew 28:19; 2 Corinthians 13:13). The doctrine of the Trinity cannot be simply read off the surface of Scripture; rather, it is the result of an extended process of theological reflection within a hermeneutical circle in which Christian theologians struggled to interpret the Bible in the light of the Church's communal life and worship, and the Church's communal life and worship in the light of the Bible. Their concern was to make sense of the ontological reality of the God who generates the historical encounters reported in the Bible and sustains the ongoing life and worship of the Church. In the language of critical realism, the theological task was to generate and iteratively refine a retroductive explanatory model of God's past and present dealings with the created order in general, and with Israel and the Christian community in particular. If God was indeed active in the history of Israel, if Jesus was indeed the unique embodied presence of God within his creation yet prayed continually to his Heavenly Father and acknowledged the presence and power of the Holy Spirit in his life, if the Spirit of the transcendent God is indeed immanent within the life of Israel and the Christian community, what then must God be like?

The Father, Son and Holy Spirit are three persons, three distinct individual entities. Greek philosophy drew a dualistic distinction between the outward appearance and actions of persons (Greek πρόσωπον, prosopon) and their underlying ontological ground (ὑπόστᾰσις, hypostases) (Zizioulas 1985, pp. 27ff.). In Platonic thought this underlying ground, the source of a person's being, was the abstract impersonal Forms. Because persons lack their own individual ontological support, their existence is ultimately ephemeral: they inhabit a demi-reality that falls short of the enduring reality of the Forms, is under constant threat of deterioration into non-being, and consequently binds them to the deterministic forces of the cosmos. Greek tragedy is predicated on a drama of heroic but ultimately futile self-assertion: however hard the tragic hero seeks to assert his identity and freedom, since he lacks any secure ontological foundation particular to his unique identity he can never escape the ravages of fate. The role adopted by the tragic actor, the mask (προωπεϊον) worn on stage, simultaneously identified his self-asserted personhood and obscured his lack of any stable and enduring identity. The Roman world embraced a similar understanding of persons: *persona* referred to an individual's social relationships and the role he played in the civic order, rather than his essential identity.

The Latin theologian Tertullian's identification of God as three persons (*persona*) in one substance (*substantia*) ran the risk of reducing the persons of the Trinity to the role of part-time actors, transient modes of divine action whose adopted 'masks' (the Son and Holy Spirit) obscure rather than reveal the essential underlying reality of the Father-God. Working from their conviction that their experience of the Son and Holy Spirit was experience of the very *essence* of God rather than of two ephemeral masks, the Greek Cappadocian Fathers elected to deconstructed and recast the establish Graeco-Roman understanding of personal identity. By translating the Latin *persona* with the Greek *hypostases* (ὑπόστᾰσις) they asserted that the persons of the Trinity existed as distinct entities and possessed their own distinct ontological foundations. By insisting on the essential identity of the persons (πρόσωπον) with their distinctive foundations (ὑπόστᾰσις), they effectively undercut the dualistic distinction between act and being, between the outward appearance of a person and his underlying ontological ground. The persons of the Trinity are thus not mere actors acting out roles, but three distinct personal entities whose appearance and actions are indivisible from their essential being. Because they possess their own ontological ground the persons are self-subsistent and self-identifying and thus essentially free from subservience to the Platonic Forms (contra classical theism) and the cycles of fate in the demi-reality of the cosmos (contra panentheism). Zizioulas suggests that the modern ideal of personal identity-in-freedom 'both as a concept and as a living reality is purely the product of patristic [early Christian] thought' (p. 27; cf. A.J. Torrance 1996).

This affirmation of the distinct ontological identities of the persons of the Trinity raises the chimera of Tri-theism and hence the inevitable question of the unity of the Triune God. As we have already noted, the three persons are perceived as united in one subsistence, substance or essence (Greek, *οὐσία*, ousia; Latin, *substantia*), so that the Son and the Spirit are consubstantial, or of one substance (Greek ὁμοούσιος, homoousios; Latin, *una substantia*), with the Father. In Greek philosophy *ousia* refers to the essence, ontological ground or underlying substratum of an existing object: its 'substance ... is that which is not predicated of a subject, but of which all else is predicated' (Aristotle 1984, p. 1625). As a metaphysical concept, *ousia* is static and impersonal, and may be applied to 'animals and plants ... natural bodies such as fire and water and earth. ... Forms and numbers ... the heavens and [all] sensible objects' (p. 1624). Just as the Cappadocian Fathers sought to recast philosophical language by identifying the essential unity of the personal acts and ontological ground of the persons of the Trinity, so they followed Athanasius in recasting the metaphysical notion of *ousia*. Athanasius insisted that because God is ultimately beyond human understanding, his unity can only be understood through his self-revelation, rather than through categories imported from Greek metaphysics. Hence in generating a revised understanding of *ousia* he was 'governed by the revelation of God's redemptive activity as recorded in the Scriptures of the Old and New Testaments' (T.F. Torrance 1996, p. 116). As a result *ousia* was understood 'not as static but as *living being* and not as dumb but as *speaking being*, and hence as *personal being*' (p. 116, emphases in original). The persons of the Trinity are thus not bound together in unity by some underlying metaphysical substance, but through their interpersonal relationships. The 'substance' that unites the Trinity is 'Love', understood not as an abstract static Form or underlying metaphysical force, but as a concrete dynamic interrelationship of the three divine persons. God is Love by virtue of the loving bond of reciprocal love that unites the persons of the Trinity. Each person of the Trinity gives himself totally to the other two persons in an everlasting act of unconditional loving service, and each in turn receives and accepts the gift of unconditional loving service offered by the other two persons. There is thus a dialectic of interpersonal love at the heart of the Trinity (and hence at the foundation of all created being) that sets the Christian God apart, both from the static impersonal God of classical theism, and from the dynamic God of panentheism whose consciousness exists only as an as-yet-unrealised potential. (1) On the one hand, the reciprocal self-giving of the persons unites them in a single substantial loving reality. The concept of perichoresis (Greek περιχώρησις) refers to the mutual indwelling, co-inherence and interpenetration of the persons of the Trinity, and 'serves to hold powerfully together in the doctrine of the Trinity the identity of the divine Being and the intrinsic unity of the three divine Persons' (T.F. Torrance 1996, p. 102). It is precisely this relationship of mutual indwelling that God desires to establish with his creatures: 'As you, Father, are in me and I am in you, may they also be in us' (John 17:21). (2) On the other hand, the same reciprocal self-giving affirms and upholds the distinctive identities of the three persons. Just as the bond of love that unites the Trinity affirms and upholds the distinctiveness of each divine person, so God creates out of love and allows creation the space and freedom to be itself apart from its Creator, and so Jesus accepts people unconditionally, just as they are.

God the Father

The Christian understanding of God the Father is not derived from philosophical speculation, but from theological reflection on the person and work of Jesus Christ in the light of Jewish tradition. Knowledge of God the Father is revealed by God the incarnate Son. Jesus reveals God as his Father and as the Father of the human race, not as an omnipotent deity who kick-starts the rational universe into motion. This means that categories such as 'omnipotence' drawn from classical theism can be applied to God the Father only if they are subjected to substantial revision and qualification in order to bring them into conformity with this revelation. Though the notion of divine fatherhood is analogous with notions of human fatherhood, our understanding of human fatherhood cannot provide the basis for the Christian understanding of divine fatherhood. God cannot be projected as an idealised human father writ large; rather, the Fatherhood of God revealed by the Son constitutes the standard against which human fatherhood itself should be measured. Saint Paul makes this explicit: 'For this reason I bow my knees before the Father [Greek πατέρα, patera], from whom all fatherhood [πατριά] in heaven and on earth takes its name' (Ephesians 3:14f.). (It should go without saying that the metaphorical nature of theological language means that talk of divine Fatherhood is not to be taken literally: God the Father is not gendered, and does not procreate as human beings procreate.)

The classical Christian understanding of the Fatherhood of God has its roots in the Old Testament: 'Have we not all one father? Has not one God created us?' (Malachi 2:10). Israel is understood collectively as a child of God, often in the context of God's redemptive plan for his chosen people: 'When Israel was a child, I loved him, and out of Egypt I called my son' (Hosea 11:1). In the New Testament this understanding of divine Fatherhood is intensified and rooted in the unique relationship between Jesus and his heavenly Father: 'All things have been handed over to me by my Father; and no one knows the Son except the Father, and no one knows the Father except the Son' (Matthew 11:27). As the divine Father, God is essentially love, and as such desires the salvation of all humankind: 'God's love was revealed among us in this way: God sent his only Son into the world so that we might live through him' (1 John 4:9). God is eternally the Father of the Son, and invites human beings to share in that relationship: 'He destined us for adoption as his children through Jesus Christ, according to the good will of his pleasure' (Ephesians 1:5). In the New Testament 'Father ... is the personal Name of God in which the form and content of his self-revelation as Father through Jesus Christ his Son are inseparable' (T.F. Torrance 1996, p. 56).

The retroductive accounts of Christian faith presented in the classical Creeds promulgated by the Ecumenical Councils sought to hold fast to this biblical vision. The Nicene Creed (325 CE) proclaims:

> We believe in one God, the Father [πατέρα] almighty, maker of all things visible and invisible; And in one Lord Jesus Christ, the Son of God … light from light, true God from true God, [eternally] begotten not made, of one substance with the Father, through Whom all things came into being, things in heaven and things on earth.
>
> (Kelly 1950, p. 215)

God is the almighty Father – the 'maker of all things visible and invisible' and Creator *ex nihilo* of an inherently rational, intrinsically beautiful and fundamentally good universe – *only insofar as he is first and eternally the loving Father of the Son.* As such he cannot be reduced to the level of the first cause or rational principle of the universe without substantial theological qualification. The eternal loving Fatherhood of God constitutes 'the inner ontological framework of the Gospel message' (T.F. Torrance 1996, p. 138). This rules out any compatibility between Christianity and both classical theism and Deism: the theistic God creates human beings for an arbitrary – and, for Hitchens and others, apparently demonic – purpose utterly beyond human comprehension; the deistic god functions as a demi-personal first cause of creation that acts out of raw power to establish the rational order of the universe. The Triune God, on the other hand, creates as the loving Father of the Son, acting in and through the reciprocal love that unites and binds the persons of the Trinity. He creates in order to bring all of humankind into an uncoerced filial relationship with God the Father, in the name of God the Son and in the power of God the Holy Spirit.

(1) God the Father is love (1 John 4:8). The Father's love manifests itself in Jesus as selfless service of the 'other' (Greek αγάπη, agape) rather than erotic love (Greek ερως, eros) or brotherly/sisterly relationships (Greek φίλος, philos). Out of agape Jesus washes his disciples' feet and dies on the Cross. Agape is a relational concept: just as one serves the 'other', so one must allow the 'other' to serve you. Just as reciprocal giving-and-receiving-in-love binds the three persons of the Trinity into an indivisible unity, so in order for human beings to participate in this divine love it is necessary for God to love humanity, and for humanity both to accept God's love and to begin to learn to love God in return – if Peter is to serve Jesus, he must first allow Jesus to wash his feet (John 13:3ff.). In the incarnation the love of the Father is poured out on humankind, and because God's love is unconditional this outpouring remains an eternal ontological constant. However, human beings are unable to fully accept and reciprocate God's love through their own merits; only the man Jesus, in the hypostatic union of God and humankind in the incarnation, is able to achieve this. Consequently human beings learn to enter into a reciprocal loving relationship with God by participating in Christ's achievement through the power of the Holy Spirit. Such participation is not conditional on the merit of any individual human act, but is

rather grounded in the achievement of Jesus of Nazareth. Because in Christ God and humankind enter into a perfect relationship of reciprocal love, God gracefully judges all human beings to be worthy of entering into such a relationship, and embraces them in such a relationship *despite* the fact that their actions do not merit this judgement. This forensic understanding of grace preserves human freedom and undermines the quasi-magical notion that we are somehow *made* to be good against our will.

(2) God the Father is omnipotent. As we have already noted, the use of this concept drawn from classical theism required substantial qualification before it could be applied to God the Father. Without such qualification, omnipotence refers to raw unconstrained power, precisely the kind of power Hitchens attributes to his 'celestial dictator'. Christian theology rejects all

> abstract notions of divine omnipotence, for omnipotence is not to be understood in terms of what we think God can do, defining it as potence raised to the nth power, i.e. as omni-potence, but in terms of what God actually *is* and actually *has done*.
>
> (T.F. Torrance 1996, p. 204; emphases in original)

God's power is restricted by his essential nature: if this were not the case his essential nature would *be* raw power. Though he has the power to do whatever he wants, because he is essentially love he can only act in and through love. This means that God can only exercise his omnipotence by acting in and through love in seeking to draw human beings into the reciprocal loving relationships eternally present within the Trinity. If God were to cease to do so, then raw power would replace love as the ontological foundation of all that has being. The early Christian Creeds identify God as 'Almighty' (Greek παντοκράτω, pantokrator). Though normally translated literally as 'ruler of all', a less literal but more accurate translation is 'sustainer of all'. The emphasis here is 'not so much on the power to act as on the capacity to *embrace and contain*, that is, *to establish a relationship* of communion and love' (Zizioulas 2006, p. 116; emphases in original). The Latin word used to translate παντοκράτω in Western Christendom was *omnipotens*, which carries the primary connotation 'all-powerful' rather than 'sustainer of all', and could equally serve to translate the Greek παντοδύναμος (pantodunamos, all-powerful, omnipotent). Once classical theism's notion of divine omnipotence became the default theology onto which Trinitarian theology was overlain, the risk of turning an all-loving God into an all-powerful deity became a constant and real danger.

(3) *God the Father loves in freedom.* Because God is God, there are no restrictions on his actions other than those that necessarily flow from his intrinsic nature. This means that God is eternally God the Father. Though Trinitarian theology speaks of the Son as 'begotten' by the Father and the Spirit as 'proceeding' from the Father, this generation is understood as eternal, so that there was not a time when the Father existed apart from his Trinitarian relations with the Son and Holy Spirit. The ontological significance of the notion

of divine generation lies in the fact that it allows for God to freely and eternally choose to be Triune, rather than for his essential nature to be imposed by some extra-divine power or substance. Though God is eternally God the Father by virtue of the eternal Trinitarian relationships of reciprocal love between Father, Son and Holy Spirit, he is *not* eternally the Creator of all things 'visible and invisible' that possess being outside of the Trinity, since there was a time when the created order did not exist. The act of creation *ex nihilo* was not the product of any ontological necessity: God did not need to create, and did not create out of any lack or absence in the Godhead. Rather, God created freely, purely out of love for his creation. This means that creation is intrinsically meaningful – it exists for a purpose – and that it derives its meaning directly from the loving intentions of its Creator. God continually holds and sustains creation in being in an ongoing act of love, and desires the whole of creation to enter into a reciprocal loving relationship with its Creator. Because there is no *necessary* causal connection between Creator and creation, God the Father is free to enable creation to exist as a reality other than its Creator. Out of love God allows his creatures the freedom to be themselves. This derives directly from the fact that love can never be coerced, since any such coercion would mean that raw power replaces love is the ontological ground of all being. Reciprocal loving relationships require the agreement and commitment of both parties. Though God loves his creatures unconditionally and works proactively to secure their salvation, they remain free to choose for themselves whether to reciprocate the gift of divine love or not. If human beings are not the passive puppets of a divine puppet master, neither are they abandoned by their Creator and left to their own devices.

(4) God the Father is not coercive. The parable of the 'Lost (Prodigal) Son', who squanders his inheritance before returning repentantly to the loving arms of his forgiving father, to the chagrin of his elder brother who dutifully remained at home, is rich in theological meaning (Luke 15:11ff.). In a very real sense the true 'prodigal' is not the younger son but the father, on two counts: first, because he allows his younger son to inherit before his own death, thereby ignoring received custom and by implication disinheriting his elder son; second, because in welcoming his returning son he effectively heaps shame upon himself, his family and his village. God, the parable suggests, is similarly prodigal in allowing humankind extraordinary freedom while continuing to love them unconditionally. Jesus' own association with outcasts and sinners identifies him with the prodigal father, while the indignation of the elder brother reflects the self-appointed righteousness of the Pharisees and religious leaders in their opposition to Jesus' ministry. Tom Wright suggests that in the parable 'Jesus was retelling the story of Israel's return from exile, and doing so in a sharp and provocative manner' (Wright 1996, p. 242). Just as the prodigal son journeys into 'exile' of his own free will rather than being sent away by his father as a punishment for earlier misbehaviour, so the exile of Israel is self-imposed rather than a result of divine coercion in response to Israel's sins.

> Exile, as some of the greatest prophets had seen, was itself part of the strange covenant purposes of Israel's father-god. Israel could be allowed to sin, to follow pagan idolatry, even to end up feeding the pigs for a pagan master, but Israel could not fall out of the covenantal purposes of her god.... When, therefore, Israel comes to her senses, and returns with all her heart, there is an astonishing, prodigal, lavish welcome waiting for her.
>
> (Wright 1996, p. 129)

In the incarnation God the Son unites himself with humanity, thereby enabling human beings to respond to God's gift of love freely, rather than through an economy of power and coercion. This makes us responsible for our own destiny: we are free to reject God's gift of love, even though the gift is unconditional and will never be rescinded. In Trinitarian theology 'hell' – despite the persistence of wooden literalistic readings of a rich cluster of metaphors – refers simply to separation from God, and any decision to live 'in hell' apart from God, the ultimate ontological foundation of all creation, is entirely self-imposed. Christ's 'descent into hell' following his death means that God paradoxically enters the realm from which he is by definition excluded: the ontology of love is ultimately victorious, even over death and hell. Orthodox iconography depicts the resurrected Christ standing astride the broken gates of hell, eternally holding out his hand to the lost souls below him: all they need to do is reciprocate Christ's actions by reaching out and grasping his hand. However, Christ cannot coerce them to do so without replacing an economy of divine love with an economy of raw power and thereby becoming a celestial dictator. This would appear to be a case of the irresistible force meeting the unmoveable object, were it not for the fact that *non-coercive* love is ontologically prior to human desire in the Christian scheme, so that ultimately the unmoveable object of human resistance will give way – not because it must, but because it may.

(5) God the Father suffers. Here we arrive at the point at which Trinitarian Christianity is at its furthest remove from both classical theism and Islamic theology. Out of love God identifies himself with humanity, becomes one with humanity, becomes vulnerable to humanity, and is put to death by humanity. The inner life of the Trinity reveals that love is a non-coercive reciprocal relationship. As such, the lover is necessarily vulnerable to the possibility that the beloved will not reciprocate the love that is offered. The doctrine of perichoresis refers to the mutual indwelling and interpenetration of the persons of the Trinity, and allows anything predicated of one person to be simultaneously predicated of all three. Hence Christian theology can legitimately speak of the 'crucified God' (Moltmann 1974; 1992). On the Cross God freely allows himself to become vulnerable to the demonic forces that seek to undermine divine love; in the resurrection God defeats these forces because divine love, the ontological ground of all that has being, is ultimately stronger that evil and death. Mel Gibson's 2004 film *The Passion of the Christ* depicts a tear falling from heaven at the moment of Christ's death: presumably God the Father weeping for his dead Son. The doctrine of perichoresis does not allow this theological interpretation, just as it

does not allow any suggestion that a despotic Father watched passively as his Son was nailed to the Cross and allowed his Son's sacrifice to appease his anger at the sin of humankind. Through the death of the Son the one Triune God, Father, Son and Holy Spirit, freely elects to bear the full weight of the forces of evil unleashed within his creation as a result, in part at least, of the inhumanity of humankind. The suffering Triune God experiences the eclipse of the very bonds of divine love that constitute the essence of his divinity: 'My God, my God, why have you forsaken me' (Mark 15:34). God the Father suffers, yet 'God's foolishness is wiser than human wisdom, and God's weakness is stronger than human strength' (1 Corinthians 1:25). 'This is my commandment, that you love one another as I have loved you. No one has greater love than this, to lay down one's life for one's friends. You are my friends if you do what I command you' (John 15:12ff.).

God the Son

Christian ontology affirms the eternal being of the Trinitarian God and the contingent reality of the created order. Just as God the Father is eternally the Father of the Son, so God the Son is eternally the Son of the Father. The breakdown of the reciprocal loving relationship between God and humankind, and alienation of creation from its Creator led to the instigation of a divine plan of redemption and reconciliation. This economy of salvation culminated in the incarnation of the eternal Son of God in the historical person of Jesus of Nazareth: 'God was in Christ reconciling the world to Himself' (2 Corinthians 5:19). The Nicene Creed, promulgated by the first Ecumenical Council of Nicea (325 CE), proclaims faith in the Lord Jesus Christ, the eternal Son of God: 'Light from Light, true God from true God ... of one substance with the Father.... Who because of us men and because of our salvation came down and became incarnate, becoming man' (Kelly 1950, pp. 215ff.). There is no difference between the essential eternal Son and the economic incarnate Son: 'what he is towards us, with us and for us in his incarnate mission from the Father he is antecedently and eternally in himself, the eternal Son of the eternal Father' (T.F. Torrance 1996, p. 142). In the incarnation God establishes a new ontological relationship between himself and his creation in the person of Jesus Christ.

> Far from the Gospel being empty of reality the fact that Jesus Christ is the incarnation within our alienated being and perishing existence of the eternal Son of the eternal Father means that the message of reconciliation, salvation and redemption does indeed have divine content and eternal validity.
>
> (T.F. Torrance 1996, p. 142)

Initial attempts to describe the theological significance of Jesus Christ employed categories drawn directly from Judaism: teacher, prophet, High Priest, Messiah, Son of God, Son of Man, etc. The confession 'Jesus is Lord' (Greek κύριος, kyrios) was particularly significant, both because the title was claimed by the

Roman Emperor, and – more significantly – because it is used in the Greek translation of Jewish Scripture to render the four Hebrew letters used to indicate the unpronounceable name of God (the tetragammaton 'YHWH'). The advent of Jesus Christ constituted the fulfilment of the divine plan initiated in God's covenantal relationship with the Jews: 'Long ago God spoke to our ancestors in many and various ways by the prophets, but in these last days he has spoken to us by a Son' (Hebrews 1:1f.). However, the nature of Jesus Christ transcended anything anticipated by the Jewish tradition: 'He is the reflection of God's glory and the exact imprint of God's very being' (Hebrews 1:3); 'In the beginning was the Word [Greek λόγος, logos], and the Word was with God, and the Word was God.... All things came into being through Him ... the Word became flesh and lived among us' (John 1:1ff.). The Nicene affirmation of both the divinity ('true God from true God') and humanity ('was incarnate and was made man') of Christ was iteratively unpacked in greater detail at the Ecumenical Council of Chalcedon (451 CE): 'perfect in Godhead and also perfect in manhood; truly God and truly man ... consubstantial [of one essence/being] with the Father according to the Godhead and consubstantial with us according to the Manhood' (Early Christian Texts 2011). This is the very heart of the Christian Gospel: Creator and creation are ontologically reconciled in the person of Jesus Christ.

(1) Christ is fully human. The full and perfect humanity of Jesus is fundamental to Christian ontology: 'He worked with human hands, he thought with a human mind. He acted with a human will, and with a human heart he loved' (Pope Paul VI 1996, §22). In assuming human nature, the Son of God was subject to hunger, thirst, fatigue, pain and death. If the source of this belief was simply the historical experience of the first disciples, the motivation for affirming it was soteriological. The humanity assumed by God in the incarnation became axiomatic for Christian accounts of salvation, since if Jesus was not fully human then God could not have been fully reconciled with the human race. As Gregory of Nazianzus put it, the 'unassumed is the unhealed, but what is united with God is also being saved' (Gregory of Nazianzus 2002, p. 158). God's unlimited compassion for humanity is ontologically grounded and epistemically revealed in the incarnation: 'Because he himself was tested by what he suffered, he is able to help those who are being tested' (Hebrews 2:18; cf. 4:15). This identification of God with human suffering possesses an ontological depth that cannot be reduced to mere empathy: it is not that Jesus Christ allows us to *imagine* that God empathises with our suffering, but that God in Christ *actually* suffered as a fully human being. Contra modern forms of existentialism, nominalism and liberal individualism, according to Christian theology we are the people we are by virtue of our shared interpersonal relationships: human society is a substantial ontological entity that cannot be reduced in nominalist fashion to the sum of its individual constituent parts. Consequently, in the incarnation, 'the Son of God, has in a certain way united himself with each individual', thereby restoring to the *whole of humanity* 'that likeness to God which had been disfigured ever since the first sin' (Pope Paul VI 1996, §22). If Jesus is not fully human the ontology of salvation disintegrates. Because of this, Trinitarian orthodoxy opposes all forms of the

Docetic heresy that denies or qualifies the full humanity of Christ, whether this be the Gnostic claim that because the material world was inherently evil, Jesus' body must have been ethereal rather than physical and his suffering and death apparent rather than real, the Apollinarian claim that Jesus possessed a human body but not a human mind, or the Monophysite claim that Christ's human nature was taken up and absorbed by his divinity. For orthodox Christians any theological statement that denies or compromises the full humanity of Christ constitutes a rejection of a genuinely incarnational ontology.

(2) Christ is fully God. If Christ was fully human, he was also fully divine. It was the eternal Son of God, the second person of the Holy Trinity, who became incarnate in Jesus of Nazareth. In 107 CE Pliny the Younger, Governor of Bithynia, wrote to the Emperor Trajan seeking advice on how to deal with the sect of Christians who persisted in meeting before daybreak to 'recite by turns a form of words to Christ as a god' (Stevenson 1960, p. 14). The Christian belief that a specific human being, Jesus of Nazareth, was essentially and uniquely God ran against the grain of both Jewish and Graeco-Roman theology. Recognition of the spiritual and moral significance of Jesus does not require an affirmation of his divinity: the Ebionites, an early Jewish-Christian sect, revered Jesus as the Messiah but declined to acknowledge his divinity; Islam acknowledges Jesus as a divinely inspired prophet in the line of Abraham, Moses and Mohammed; many Liberal Protestants recognise Jesus as a great spiritual teacher; many atheists are happy to acknowledge the lasting value of at least some of Jesus' moral teaching. Where Greek onto-theology viewed the essential identity of an omnipotent and omniscient God and a fallible human being as a logical contradiction, Judaism regarded the notion that God would subject his holiness to contamination by the created order as anathema. The two basic Christological heresies sought to overcome this theological conundrum by denying the incarnation: Ebionism by proclaiming a true man devoid of divinity, Docetism by proclaiming a divine being devoid of humanity. The Arian heresy sought to mediate between these two extremes: whereas the Father is an eternal being without origin, the Son is a created being with a point of origin. As such, the Son can engage with creation without logical contradiction or theological affront. Thus the ontological bridge between Creator and creation is established by a demigod or superman figure essentially inferior to God the Father. The condemnation of Arianism in the Nicene Creed was designed to affirm the full divinity of Jesus: those who say that 'the Son of God … is created', that there was a time 'when He was not', that he is 'subject to alteration or change', or that he is of a different 'hypostasis or substance' than the Father, 'these the Catholic Church anathematizes' (Kelly 1950, p. 216).

(3) Christ is fully God and fully human. Jesus' first disciples experienced themselves as being in the simultaneous presence of a man and of God. Acknowledging the alethic truth of this experience meant resisting the epistemic fallacy, in which reality is forced to conform to our prior epistemic assumptions, by refusing to deny it on *a priori* rational or theological grounds: in the face of an apparent logical contradiction it was the logic that had to give way; in the face of an apparent affront to divine holiness it was the theological understanding

of the nature of God that had to be revised. How can an omnipotent deity become a weak and vulnerable human being? As we have already noted, the Triune God is essentially love rather than raw power, so in choosing to become human for the sake of humanity God acts in accordance with his essential nature, and in doing so is not subject to any *a priori* logical or spiritual constraint other than that of the divine theo-logic of unadulterated love. The Ecumenical Council of Chalcedon proclaimed Jesus as truly God and truly man, and explained this in terms of the classical incarnational doctrine of the two natures (Greek φύσεσιν, phusesin) of Christ, 'the peculiar property of each nature being preserved and being united in one Person [πρόσωπον, prosopon] and one Subsistence [Greek ὑπόστασις, hypostasis], not separated or divided into two persons' (Early Christian Texts 2011). The two natures remain distinct: Jesus prays to his Father yet is himself worshipped, is without sin yet subject to temptation, brings eternal life yet suffers execution. Beneath the outward appearance of each nature lies a substantial hypostatic union: the two natures subsist in a single foundational reality, the unique personhood of Jesus Christ. Just as the Trinity is made up of three persons united together in reciprocal love, so the incarnate Christ's humanity and divinity are united together in precisely the same way. The Creator remains Creator and the creature remains created, yet the two are ontologically united, thus opening the door to the perfection of humanity and the fulfilment of Jesus' command 'Be holy, therefore, as your heavenly Father is holy' (Matthew 5:48). Maintaining this apparent paradox in the face of human logic and received theological wisdom secured the distinctiveness of the Christian vision of reality: the ontological distinction between Creator and creation remains in place, yet the ontological divide is effectively annulled through God's gracious loving act. Divine love is thus both revealed to be, and actualised as, the fundamental causal mechanism undergirding all that has being, both eternally divine and contingently created. Consequently, the view of Nestorius that the two natures co-exist in perpetual separation was rejected, as is the view of Eutyches that the two natures merge into a hybrid reality within which both divinity and humanity are dissolved. Neither perspective allows for the actuality of the Creator–created relationship alongside the actuality of the dissolution of the ontological divide. Recognition of the hypostatic union of the two natures of Christ generated the doctrine of *communicatio idiomatum* (communication of properties), which affirmed the mutual exchange of the attributes of the divine and human natures of Christ: God wept, this man healed the sick; God was tempted, this man was without sin; God died on the Cross, this man rose from the dead – and vice versa: this man wept, God healed the sick; this man was tempted, God was without sin; this man died on the Cross, God rose from the dead.

God the Holy Spirit

In classical theism God is generally considered to be both holy and spiritual. In identifying and naming the Holy Spirit as one of the three persons (ὑπόστᾰσις, hypostases) of the single divine Godhead (*οὐσία*, ousia), Christian theology took a further crucial step in establishing the identity of its particular and distinctive *sui generis* Trinitarian theism. Influenced by Platonic dualism, Hellenistic theology tended to stress the absolute transcendence and ultimate unknowability of God. In Plato's *Timaeus* the gulf between the transcendent eternal Forms and immanent contingent creation was bridged by a demiurge that orders the world out of pre-existent matter and lacks the attributes of omnipotence, omniscience and holiness. In Gnostic speculation the demiurge falls so far short of the supreme, spiritual and holy God that in some Gnostic systems he was considered demonic, and in other anti-Semitic Gnostic schemes he was identified with the reviled God of Judaism. The doctrine of incarnation rejected any ultimate dualistic separation between God and creation while refusing to identify the two in any way, apart from in the hypostatic union of God and humankind in the particular person Jesus Christ. Similarly, the doctrine of the Holy Spirit affirms the immanence of God within creation, while refusing to identify the Holy Spirit in any way – whether wholly (pantheism) or partially (panentheism) – with nature, the human spirit, or the evolutionary process. The Holy Spirit is utterly transcendent, holy and wholly 'other' than the created order, yet simultaneously fully immanent within it as 'the immediate presence of God in unreserved self-giving to his people' (T.F. Torrance 1996, p. 64). The heresy of Sabellianism sought to limit the Holy Spirit by affirming 'modalism', the doctrine that the work of the Holy Spirit was merely a series of temporary intrusions of God the Father into the created order. In sharp contrast, Trinitarian orthodoxy identifies the Holy Spirit as *eternally* and *essentially* the third person of the one Godhead who, from the creation through to the final consummation of the universe, is personally immanent in the world and constantly working to fulfil God's divine plan to reconcile Creator and creation. This assertion of the distinctive task of the Holy Spirit vis-à-vis creation is such that, rather than affirm the Spirit's immanence within creation, it is more appropriate, without for a moment undermining the fundamental ontological distinction between Creator and creation, to follow Saint Paul in identifying the Holy Spirit as that divine reality within whom 'we live and move and have our being' (Acts 17:28).

The Nicene-Constantinopolitan creed, promulgated by the second Ecumenical Council of Constantinople (381 CE), affirms belief in the Father, in the Son,

> And in the Holy Spirit Ghost,
> the Lord and life-giver,
> Who proceeds from the Father,
> Who with the Father and the Son is together worshiped and together glorified,
> Who spoke through the prophets.
>
> (Kelly 1950, p. 298)

There are many forces in the created order, not all of which are holy, and some of which are positively demonic in their effects. The Spirit is 'Holy' because he is sacred and set apart from all those forces and realities, be they material or spiritual, that fall short of the glory of God. The title 'Lord' is also applied to the Father and Son, and identifies the Holy Spirit as divine. The Spirit's procession from the Father identifies him as part of the Trinity, and the references to worship and glorification affirm that the Spirit participates in the inner life of the Triune God. The Holy Spirit's distinctive nature is that of 'life-giver'. The creative task of the Holy Spirit vis-à-vis the created order begins in the initial act of creation and ends with the final eschatological consummation of God's redemptive plan. The Holy Spirit is intimately involved in the economy of salvation from its very beginning, when he 'spoke through the prophets'. The role of the Holy Spirit is to actualise the reconciliation of the Triune God with his creation:

> the freedom of the Spirit is revealed in the way in which God achieves the end for which his world was created ... the Spirit is the one who gives freedom through community ... the God-given freedom to be like God in learning to love even the unlovely ... the Spirit is the one who perfects all the creation.
>
> (Gunton 2002, p. 156)

In the two Hebraic accounts of creation God creates through his divine Word and through his Holy Spirit. In the first, the Spirit of God sweeps over a formless void, and each step in the creative process opens with the refrain 'And God said' (Genesis 1:1ff.). In the second, when Adam is created the Spirit of God breathes life into him (Genesis 2:7). Creation is the result of the activity of the Triune God: in the second-century theologian Irenaeus' famous metaphor, God the Father creates through his two hands, the Son and the Holy Spirit (Irenaeus 2004, pp. 487ff.). God's creative will and design was 'to bring into being a creaturely realm ... within which he may share with others the Communion of Love which constitutes his inner Life as Father, Son and Holy Spirit' (T.F. Torrance 1996, p. 218). Christian theology denies that creation is in any sense divine: because it is absolutely dependent on God for its existence, creation must be continually sustained in being. It also rejects any notion that the contingent order is a demi-reality: rather, creation is both fully real and intrinsically good, because it constitutes the environment in which God and his creatures can enter into reciprocal loving relationships. Because love is not coercive but must be freely given and received, creation is not a closed deterministic system but an open structure that allows humanity the freedom to reciprocate God's offer of the gift of love. The drama of the divine economy of salvation is predicated on the failure of humanity to accept this gift. Thus as the 'life-giver' the Holy Spirit works to create, sustain and bring to eschatological fulfilment God's divine plan.

In the Old Testament God's spirit (Hebrew רוּחַ, ruach: wind, breath, spirit) breathes life into creatures: in Ezekiel's vision of the valley of dry bones God says, 'I will cause *ruach* to enter you, and you shall live' (Ezekiel 37:5). The divine plan for Israel is actualised by the dynamic and unpredictable spirit of God: when the Israelites escape from slavery in Egypt Moses stretches out his hand over the sea and a strong *ruach* drives back the waters and allows the people to cross to safety (Exodus 14:21). The prophets speak the word of God because the spirit of God enters into them: 'I am filled with power, with the *ruach* of the LORD' (Micah 3:8). The life-giving and providential acts of God's spirit are eschatologically oriented to the future accomplishment of God's plan:

> I will pour out my *ruach* on all flesh;
> Your sons and your daughters shall prophecy,
> Your old men shall dream dreams,
> And your young men shall see visions.
>
> (Joel 2:28)

The expected Messiah will be endowed with God's spirit:

> The *ruach* of the LORD shall rest on him,
> The *ruach* of wisdom and understanding,
> The *ruach* of counsel and might,
> The *ruach* of knowledge and the fear of the LORD'.
>
> (Isaiah 11:2)

In the New Testament the creative and redemptive role of God's spirit is confirmed, intensified and given Christological focus. The Holy Spirit descends on Jesus at his Baptism and immediately drives him into the wilderness where he confronts Satan, the personification of the forces of evil that separate humankind from God. However, this is not a further instance of God's spirit inspiring a human prophet; rather, the Holy Spirit is an agent of Jesus' very conception and creates the particular humanity that the Son of God receives from Mary as he becomes incarnate in the created order. Throughout his ministry Jesus drives out evil spirits from the sick and suffering in the power of the Holy Spirit. At the same time he nurtures those who are poor in spirit and offers them spiritual rebirth and the gift of the Holy Spirit (Luke 11:13). On the Cross he commends his Spirit into the hands of his Father, and experiences the absence of God as the full force of the demonic powers weighs down upon him. He rises from the dead in the power of the Holy Spirit. In John's Gospel he teaches his disciples about 'the Advocate [Greek παράκλητος, Paraclete: companion, comforter, counsellor], the Holy Spirit, whom the Father will send in my name' following his departure from this world (John 14:26). The Paraclete will guide the disciples into truth and glorify God in the world. In the Acts of the Apostles the prophecy of Joel referred to above is realised as, on the day of Pentecost, the disciples are filled with the Holy Spirit (Acts 2:1ff.). The task of the Holy Spirit is to re-create God's fallen creation and bring God's divine plan to fruition. The Spirit constitutes the Church as the Body of Christ, establishing fellowship, generating faith, revealing God's will through the reading of Scripture, and facilitating worship through preaching and the sacraments.

The Holy Spirit actualises the reciprocal loving relationship between God and humankind made possible by the incarnation. Since God accepts persons into the

Church unconditionally, the Church is a community of sinners in the process of transformation. The renewal of the hearts and minds of Christians, and through them of humankind as a whole, is a slow and often painful process, so that the gifts of the Holy Spirit, faith, hope and love, await full realisation in the eschatological consummation of all things. This eschatological emphasis gives a cosmic dimension to the work of the Holy Spirit: 'For the creation ... was subjected to futility ... [but now] groaning in labour pains ... [it is being] set free from its bondage to decay and will obtain the freedom of the glory of the children of God' (Romans 8:19ff.). As Gunton puts it,

> God the Spirit is ... the perfecting cause of creation. God the Father perfects his creation in and through time, through his Son and Spirit enabling particular acts, events and things, from bits of pottery to noble self-giving actions, to be themselves, and thus to be particular anticipations of the final perfecting of all things.
>
> (Gunton 2002, p. 119)

9 The economy of salvation

Christian theologians draw a distinction between the 'immanent' and 'economic' Trinity, between God-in-himself and God as he creates the world, reconciles it to himself, and brings it to perfection. Epistemically, humanity has access only to the economic Trinity, to God as he elects to reveal himself in the drama of salvation. Any positive affirmations regarding the immanent Trinity can only take the form of retroductive inferences from the divine economy, grounded in the understanding that since God reveals himself as faithful, trustworthy and incapable of deception, his inner being and outer acts must be identical. It follows that, to all intents and purposes, the immanent Trinity *is* the economic Trinity (Rahner 1970). The only qualifications that need to be drawn here are (1) that since God creates freely and not out of any compulsion he would have remained God without the act of creation, and (2) that if humanity had elected not to disobey God the economy of God's dealings with humanity may possibly have taken a different course. Such speculation cannot, however, take away from the actuality of the economy of salvation within which the created order now participates: God's creation of all that exists apart from himself, the disobedience and fall of humanity, the divine decision to prepare for the incarnation by establishing a covenant relationship with Israel, the conception, life, death and resurrection of Jesus Christ, the establishment of the Church in the power of the Holy Spirit, and the future consummation and perfection of *all* creation as it comes to share in the reciprocal bonds of love that unite the three persons of the Holy Trinity. Since there is no space here to explore every aspect of the divine drama of creation-reconciliation-sanctification, we will focus on three selected areas: the Christian doctrine of creation, the Christian account of the nature of humanity, and the Christian explanation of the reconciliation of Creator and creation.

The theology of creation

The Christian doctrine of creation has its roots in Scripture. The Old Testament is replete with references to creation (e.g. Psalm 8, 104; Proverbs 8:22ff.). The opening eleven chapters of Genesis recount a 'Primal History' that opens with two expansive accounts of creation: the creation of the world, and the creation of Adam and Eve (Genesis 1:1ff., 2:4bff.). This Primal History is rooted in Hebraic oral tradition and folklore, and has links with other ancient Near Eastern primordial myths. It is important to avoid two common misunderstandings of the opening chapters of Genesis.

First, the Primal History is misunderstood if it is read in fundamentalist terms as a primitive contribution to science or historiography. Rather than attending to the strata of nature and human history, it is concerned with the theological stratum of reality and explores questions of the nature of God and God's relationship to nature and the evolving story of humankind. As Colin Gunton points out, and critical realists affirm, '[r]eductionist claims that this or that science, or science as a whole, will ever be able to explain *everything* are simply arrogant and untenable', as are claims that the human condition can be fully explained through reference to historical events alone (Gunton 2002, p. 34; emphasis added). Natural science is not *the* meta-discipline, but rather 'the discipline, or set of disciplines, that investigates the [natural/physical] structure of this universe, as it is given to be understood': just as the disciplines of metaphysics and theology cannot legitimately encroach into the domain of natural science, so natural science 'is not qualified to move outside the universe ... in order to determine its limits and conditions' (p. 18, n. 19).

Second, the Primal History is misunderstood if it is seen to be merely dependent upon, and passively reflective of, other ancient Near Eastern primordial myths. Though the book of Genesis shares 'a common outline of primeval history with its neighbours ... the similarities between biblical and non-biblical thinking, however, are overshadowed by the differences' (Wenham 1987, p. xlviii). It is clear that the Hebraic Primal History is in part designed to challenge and recast similar primeval histories. Thus, for example, the Mesopotamian *Enuma Elish* recounts the story of the building of a temple-tower (probably a reference to the distinctive Babylonian Ziggurats) that, to the approval of the gods, reaches up to paradise and establishes a route to heaven (Pritchard 1958, pp. 37ff.). In the Hebraic retelling the Tower of Babel fails to reach heaven, functions as a source of folly and confusion, and meets with divine displeasure (Genesis 11:1ff.). The Genesis version of a familiar ancient Near Eastern myth thus functions negatively as 'a satire on the claims of Babylon to be the center of civilisation and its temple tower the gate of heaven', and positively as a warning against all attempts to undermine and displace the sovereignty of God (Wenham 1987, pp. xlviiiff.).

The transition from oral to literary tradition took place over an extended period of time: modern scholarship tends to locate the initial production of the literary form of the myth of the creation of Adam and Eve in the ninth century BCE, and the myth of the creation of the world in the sixth century BCE, with the final redaction of the received texts taking place towards the end of the fifth century BCE. The deliberate stress on the absolute sovereignty of God over creation sets the two Hebraic myths apart from their ancient Near Eastern counterparts. The Ugaritic creation myth concerning the fertility God Baal is grounded in the cycle of the seasons: in a recurring sequence Baal's death leads to infertility and chaos on earth, and his return to life under the auspices of his sister and consort Anat brings about the return of fertility and restoration of order (Pritchard 1958, pp. 92ff.). In the *Enuma Elish* Marduk creates the world from the dismembered corpse of the female principle Tiamat (pp. 31ff.). The Hebrew myth is of an altogether sparser affair, entirely devoid of any notion of a cosmic struggle between a pantheon of gods: 'In the beginning … God created the heavens and the earth … [and] God saw everything he had made, and indeed, it was very good' (Genesis 1:1, 31). The implication that the one sovereign God created the world *ex nihilo* – out of nothing – later becomes explicit: 'look at the earth and sky and everything in them, and consider how God made them out of what did not exist' (2 Maccabees 7:28).

The New Testament takes up the Old Testament understanding of creation and gives it an explicitly Christocentric focus:

> There is one God, the Father, from whom are all things and for whom we exist, and one Lord, Jesus Christ, through whom are all things and through whom we exist.
>
> (1 Corinthians 8:6)

> [In Christ] all things in heaven and on earth were created, things visible and invisible, whether thrones or dominions or rulers or powers – all things have been created through him and for him. He himself is before all things, and in him all things hold together.
>
> (Colossians 1:16f.)

> In the beginning was the Word, and the Word was with God, and the Word was God. He was in the beginning with God. All things came into being through him, and without him not one thing came into being.… And the Word became flesh and lived among us.
>
> (John 1:1–3, 14)

Building on New Testament teaching, the second-century theologian Irenaeus developed a Trinitarian understanding of creation that was to become normative for classical Christianity. God the Father creates through his 'two hands' God the Son and God the Holy Spirit: 'For with Him were always present the Word and the Wisdom, the Son and the Spirit, by whom and in whom, freely and spontaneously, He made all things' (Irenaeus 2004, pp. 487ff.; cf. Gunton 2002, p. 10).

The classic statement of the Christian doctrine of creation, affirmed alike by Eastern Orthodoxy, Roman Catholicism and mainstream Protestantism, was produced by the first Ecumenical Council held at Nicea in 325 CE. The Nicene Creed proclaims faith 'in one God, the Father almighty, maker of all things visible and invisible' (Kelly 1950, p. 215). God creates *ex nihilo*, bringing all that exists into being out of nothing rather than imposing order on pre-existent primal matter in obedience to some pre-established Platonic Form, Aristotelian principle, Newtonian natural law or Kantian moral scheme. Prior to the act of creation there was *only* the pure being of God: not God surrounded by a void of nothingness, or existing alongside any other positive reality. The actuality of our time-space continuum, of the universe we inhabit, and of any prior, future or parallel universes, have their source and origin in the creative act of the one Triune God. As such, created being is utterly contingent: generated by God, not self-generating; imbued with rational order by God, not self-rationalising; constantly held in being by God, not self-sustaining. The doctrine of creation *ex nihilo* serves to affirm God's absolute sovereignty and supreme freedom over all creation. According to Trinitarian faith, to affirm any material substance or idealistic principle as co-eternal with God is to deny that God is God. The primary ontological concern of the Hellenistic milieu in which this doctrine was formulated was to identify and explain the first principle, the *archē*, of the rational order inherent in the universe. The more basic ontological question, of the being-of-being-itself – Why is there something rather than nothing? – received relatively little attention. The common assumption that 'nothing can come from nothing' implied the eternity of all being, ideal, material and divine, and suggested that God was merely a limited part of a greater whole. Thus in Homer's pantheon of gods, even Zeus, the highest of all divinities, was subject to the primal forces of fate, chance and necessity. In Plato's *Timaeus* the demiurge creates by fashioning pre-existence matter according to the pre-established blueprint provided by the eternal Forms. The crucial issue for classical Greek onto-theology was not the source of being itself, but the relationship, within the great chain-of-being, between eternally necessary being and time-bound contingent becoming: in neo-Platonic terms, the relationship between the unchanging, transcendent and utterly simple 'One', and the contingent, immanent and multifaceted 'Many'. The Christian doctrine of creation *ex nihilo*, in proclaiming the absolute sovereignty and freedom of God, thus constituted a direct challenge to the prevailing philosophical theology of the ancient Greeks.

Christians affirm creation *ex nihilo* as the purposeful, intentional and free act of a personal agent: the Trinitarian God. The Nicene Creed affirms that all things came into being through the Lord Jesus Christ. Contra Greek philosophy, it is not the product of any impersonal force: neither Plato's abstract Forms, nor Aristotle's self-contained Unmoved Mover, nor Plotinus' absolutely simple 'One'. God's personal creative activity is neither necessary nor capricious. Since the Trinity is a perfect communion of persons existing in loving relationship, there can be no lack, need or unfulfilled desire within the Godhead. Consequently, God creates freely rather than out of necessity, not for his own sake but for the sake of creation itself. Creation flows directly from God's love, but is not a necessary product of that love. Because God creates out of love, the created order is imbued with ultimate meaning and purpose: in theological terms, the economy of creation is inextricably linked to the economy of salvation. Thus, according to the Nicene Creed, it was for us and for our salvation that Christ was incarnate of the Holy Spirit and the Virgin Mary and became truly human (p. 215). The claim that, for the sake of humanity the Creator of the universe lived among us and suffered death on a cross – 'Hands that flung stars into space/To cruel nails surrendered' (Kendrick 2011) – was, as Saint Paul observed, pure foolishness when viewed through the eyes of Greek philosophy (1 Corinthians 1:23). That the key to the mystery of the world lies not in any *abstract structure* or impersonal force, whether transcendent of, or embedded within, the universe, but rather in the *personal agency* of its Creator constitutes the heart of the Christian Gospel, a Gospel grounded in a divine foolishness that transcends and eclipses all human wisdom (1 Corinthians 1:25).

Creation is real, not illusory or deceptive. The Christian theology of creation directly challenged the Platonic distinction between the immaterial, intelligible, transcendent and eternal realm of the Forms, and the material, sensible, immanent and contingent realm indwelt by human beings. For the neo-Platonist only the former is genuinely real, since the latter constitutes a demi-reality that in various ways and to varying degrees falls short of ultimate reality. For the Christian God did not create a deceptive chimera, a cave of illusion from which humanity must struggle to escape, but rather a tangible environment in which the human race could flourish: 'God saw everything he had made, and indeed, it was very good' (Genesis 1:31).

This affirmation of the intrinsic goodness of creation constituted a direct challenge to the Gnostic claim that the material realm is inherently evil, formed by a malevolent demiurge located at the bottom of a hierarchy of divine beings, whose function was to protect the Ultimate Being from contamination by the world of matter. To be sure, Christianity recognises the presence of evil in creation; however, it understands evil as extraneous to the created order, a parasitical and cancerous perversion of an intrinsically good reality. Such goodness is not to be confused with divinity: created reality is not divine, since God gives it the freedom to be other than himself. The freedom of creation flows directly from the ontology of divine love: a truly loving relationship, as perfectly and eternally actualised in the Trinity, requires the reciprocal free gift of each person in selfless service of the other. Without such freedom, relationships become coercive and their ontological grounding disrupted. Since God lovingly created the world for the world's sake, the proper end of creation is to enter freely into a reciprocal loving relationship with its Creator. God's gracious gift of love thus requires that creation be *free* to reciprocate in kind. To achieve this, creation must be free to be what it truly is: genuinely *created* being, rather than a divine, or partially divine, emanation from the being of God. Such freedom is not to be understood as the exercise of absolute autonomy, since this would lead directly to the dislocation of the self from any meaningful relationship other than that of solipsistic and egocentric self-contemplation. Rather, as relational creatures our freedom lies in the freedom to relate, whether appropriately or inappropriately, to ourselves, to others-in-community, to the natural order-of-things and to our Creator.

In the act of creation God sanctioned otherness and raised it to ontological status (Zizioulas 2006, pp. 13ff.). If this were not so, if God had not given creation the freedom to be other than divine, then creation would have been forced into a necessary, and hence totalitarian, relationship with its maker. However, the divine gift of freedom does not mean that God has simply abandoned the world to its own devices; if this were the case, then the act of creation would have sanctioned an equally totalitarian 'absolute "abysmal" otherness' (p. 19). Just as God's immanence does not require the divinisation of creation and its dissolution into the Godhead, so God's transcendence does not require the absenting of divinity from creation and the disruption and dislocation of the relationship between Creator and creation. Instead, 'in the biblical perspective divine transcendence and immanence are convergent, both movements being united in the conception of a God who paradoxically reveals his majestic greatness through his liberating and beneficent involvement in the world' (Anatolios 1998, pp. 6ff.). For the authors of the Nicene Creed this apparent paradox undercuts both the prevailing neo-Platonic assertion of the absolute transcendence of the Supreme One, and the Stoic affirmation of the pantheistic identity of the divine with the sensible material world. Similarly, for the contemporary Church the doctrine of creation challenges both deistic notions of an absent celestial designer and neo-pagan notions of a divinised nature. The Trinitarian God is immanently engaged with his creation, constantly holding a contingent reality incapable of self-sustenance in being and, through the incarnation of Christ and creative activity of the Holy Spirit, working pro-actively yet non-coercively to bring about the reconciliation of creation and Creator. Indeed, for Paul the intimacy of God's engagement with creation is such that it is *in* God that 'we live and move and have our being', since all things have their being *in* Christ, and all things hold together *in* him (Acts 17:28; Colossians 1:16f.). The notion of divine immanence is thus reinforced by the metaphorical suggestion that creation is present within God, rather than God present within creation. At the same time, however, the theo-logic of the Christian understanding of love means that God's intimacy with creation can only be predicated on an ontological otherness in which God so utterly transcends his creation that there is a qualitative distinction between the two. The Christian ontology of love thus combines the irresistible desire of the lover for intimacy with the beloved with the equally irresistible desire to affirm and uphold the otherness of the beloved's unique and particular identity.

The glory and poverty of humankind

Secular anthropology addresses questions of the nature and characteristics of human beings in relation to themselves and other animate creatures. We are complex, multifaceted and stratified beings, open to investigation from many different-yet-complementary perspectives. Our psychosomatic make-up resists reduction to biological behaviourism and demands recognition of the social, cultural, moral, aesthetic and spiritual dimensions of our being. The search for a common essence of humanity culminated in the Enlightenment's celebration of autonomous rationality: *cogito ergo sum*; *sapere aude*. On this reading it is our ability to reason that separates us from all other sentient beings. The romantic valorisation of emotional sensibility, the existential insistence on the priority of the singular individual as active agent, and the psychoanalytic identification of the subconscious mind collectively challenged this rationalistic consensus and paved the way for the postmodern deconstruction of essentialist accounts of human nature: 'you may have killed God beneath the weight of all that you have said; but don't imagine that, with all that you are saying, you will make a man that will live longer than he' (Foucault 1989, p. 211). In the postmodern order it is our unconstrained freedom to become the people we wish to be, rather than to strive towards some pre-established, essentialised and restrictive norm, that identifies us as human beings. Despite this tension between the modern essential 'One' and the postmodern nominal 'Many', both traditions hold in common an affirmation of the capacity for self-transcendence, whether that be the transcendence afforded by the rational ordering of self and society supported by various technological advances, or that promised by the unrestricted creative freedom of individuals to construct their own identities in accordance with their own desires.

Christian theological anthropology recognises the significance of humankind's sense, sensibility, creative acts, unconscious drives and thirst for transcendence. However, in doing so it places them in a very different ontological framework: the key issue is not humankind's distinctiveness vis-à-vis the natural order and other sentient beings, but rather humankind's distinctiveness vis-à-vis the Triune God. Here the rational and emotive drive towards self-transcendence constitutes an ontologically grounded search for God: 'The thought of you stirs him so deeply that he cannot be content unless he praises you, because you made us for yourself and our hearts find no peace until they rest in you' (Augustine 1961, p. 21). Since God is ontologically prior to creation and epistemically prior to human understanding, our striving for self-transcendence is dependent on divine being (ontology) and divine revelation (epistemology) for its ultimate fulfilment.

'God created humankind [Hebrew: Adam] in his own image' (Genesis 1:27). God is a relational being: Father, Son and Holy Spirit are distinct persons whose 'otherness' is irreducible, whilst at the same time 'the unbreakable *koinonia* [Greek: fellowship] that exists between the three persons ... means that otherness is not a threat to unity but a *sine qua non* condition of it' (Zizioulas 2006, p. 5). The distinctiveness of each person of the Trinity lies precisely in their relational unity of perfect self-giving love. As God the Father exists as Father to the Son and Holy Spirit, so I exist as husband to my wife, father to my children, teacher to my students, son to my parents, citizen of my nation and child to my God – the fundamental difference being that whereas the relationality within the inner life of the Trinity is perfect, my relationships are frequently wounded, fractured and in need of healing and restoration. Ontologically, I am a relational being because I am created in the image of God; epistemically, I am able to understand myself as a relational creature because of God's self-revelation in the person of Jesus Christ. Since God, in the internal relationships of his Triune being, is essentially love, love constitutes the ontological ground of all being, both created and uncreated. We are created *imago Dei*, in the image of God, and as such exist ontologically as relational creatures whose ultimate end is the eternal bliss of loving and sanctifying communion with our Creator. As such, we are neither isolated atomistic individuals nor anonymous parts of a greater whole. Both the individualism of the 'Many' and the collectivism of the 'One' imply ontological violence: either we stand over-against our fellow creatures in a dynamic of self-assertion, or we are stripped of our distinctive identities and immolated into a greater whole. In sharp contrast, the Christian ontology of love affirms our distinctiveness-in-communion. If our ultimate end is communion with our Creator, then that relationship is necessarily asymmetrical. As finite creatures, created *ex nihilo* and thus utterly dependent for our existence on divine grace, we possess no spark of divinity within us: 'you are dust, and to dust you shall return', since apart from God 'all is vanity and chasing after wind' (Genesis 3:19; Ecclesiastes 1:14). However, as finite creatures held in being by an all-loving God we are destined for immortality:

> What are human beings that you are mindful of them,
> mortals that you care for them?
> Yet you have made them [only] a little lower than God,
> and crowned them with glory and honour.
>
> (Psalm 8:4f.)

It follows that our relationship with our Creator should be one of obedient gratitude, manifested in righteous working, resting, feasting and praising: 'Go, eat your bread with joy and drink your wine with a merry heart ... let your garments always be white' (Ecclesiastes 9:7f.). As relational creatures we are destined to come to know ourselves as we are already known by God: 'Now I know only in part; then I will know fully, even as I have been fully known'; 'Without knowledge of self there is no knowledge of God ... without knowledge of God there is no knowledge of self' (1 Corinthians 13:12; Calvin 1960, vol. 1, pp. 35, 37).

'Your kingdom come, your will be done, on earth as it is in heaven': Jesus' petition to his Father recognises that creation has fallen from the state God intended for it (Matthew 6:10). Christianity teaches that all human beings experience, to a greater or lesser extent, estrangement from themselves, from others-in-community, from the natural order-of-things and from the Triune Creator. Such experience is not delusional, but grounded in actuality. The bond of love that unites the persons of the Trinity constitutes the ontological bedrock of reality. God desires to extend the bond of love within the Trinity outward to embrace the whole of creation and draw it into the inner life of the Godhead. However, his unbounded love for creation is not fully reciprocated. To inherit eternal life, to be in proper relationship with self, others, world and God, it is necessary to follow the core moral precept of the Judaeo-Christian tradition: 'You shall love the Lord your God with all your heart, and with all your soul, and with all your strength, and with all your mind; and your neighbour as yourself' (Luke 10:27). Sin, the failure to fully adhere to this precept, generates estrangement and the disruption of proper relationships within creation itself (love of neighbour) and between creation and the Creator (love of God). The word 'sin' 'expresses what is not implied in the term "estrangement", namely, the personal act of turning away from that to which one belongs' (Tillich 1978, p. 46). This act of sinful 'turning away' cannot be reduced to individual moral failings judged according to petty bourgeois standards. Rather, Christian teaching follows Paul in placing sin in a broader cosmological framework: the fragmentation and absence of love of God and love of neighbour means that the pristine harmony of creation itself disintegrates and falls into a futile bondage of dislocation and decay (Romans 8:18ff.). Sin is thus primarily an ontological category: hubris, the sin of pride, leads a person to seek to 'elevate himself beyond the limits of his finite being'; concupiscence, the striving for impersonal sexual gratification, constitutes 'an unlimited desire to draw the whole of reality into oneself'; idolatry, the worship of created objects rather than the Creator, entails the rejection of the actual ontological order-of-things (Tillich 1978, pp. 49ff.). 'Christian theology diagnoses the ill as the disruption or distortion of the relation of personal beings with the personal creator God ... that... incorporates the whole created world in its structures' (Gunton 2002, p. 60). Since finite creation cannot possibly be self-sustaining, the ontological displacement of creation from its Creator generates a downward spiral into decay, death and non-being. However, because God's love for creation is eternal and unbounded, and despite the failure of creation to properly and fully reciprocate God's love, he not only continues to hold creation in being but also works proactively to overcome the estrangement that divides creation from itself and from its Creator.

The Primal History in Genesis, Chapters 1–11 tells a mythological story of this evolving estrangement: the failure to love God (Adam and Eve's self-imposed exile from the Garden of Eden); the failure to love neighbour both individually (Cain's murder of his brother Able) and collectively (the scattering of the dislocated and fragmented human race across the surface of the earth); and the failure of human beings to overcome estrangement through their own efforts (the futile attempt to construct a 'Tower of Babel' in order to gain access to heaven and establish a common unifying language for humanity). The myth of Adam and Eve lays the ultimate responsibility for the fallen state of creation on humankind's sinful disobedience of God. Contrary to received understanding, the Garden of Eden inhabited by Adam and Eve is not identified in the biblical text as 'heaven' or 'paradise', but rather as the 'very good' environment in which God chooses to enter into reciprocal relationship with his creatures (Genesis 1:31). As Gunton points out, 'perfection' is not a static concept: a newborn baby is perfect *both* as a newborn child *and* as a person with the potential to become a mature adult (p. 19). Adam and Eve are created to enter into a reciprocal loving relationship with their Creator, but fail to do so: beguiled by the serpent – whose association with the devil, identified elsewhere as a fallen angel, suggests that the original cosmic disruption of creation took place in another created sphere and pre-dates the fall of the human race – they allowed themselves, both individually and collectively, to participate in an economy of power that placed them in direct opposition to the love of God. The disobedience of the human race, personified in the mythical story of Adam and Eve, is possible – though perhaps neither necessary nor inevitable – because God lovingly gives his creation space to exist apart from himself, and offers humankind the freedom to reciprocate his gracious offer of reciprocal communion without compulsion. A loving relationship must always be entered into freely: self-giving love poured out for the benefit of the 'other' can never be coercive, since to coerce is to oppress and do violence to the integrity and distinctiveness of the beloved. Creation, in failing to reciprocate God's love, falls short of the harmonious relationship desired by God and thereby collapses into decay. If the *possibility* of this fall from grace can be traced back to the ontology of love from which the created order receives its being, its *actuality* is neither willed nor generated by God, since unconditional love can never desire to be dislocated from its object.

The Christian claim that 'all have sinned and fall short of the glory of God' is an offence to modern liberal sensibilities (Romans 3:23). According to Calvin,

> There is no doubt that Adam, when he fell from his state [of grace], was by his defection alienated from God. Therefore, even though we grant that God's image was not totally annihilated and destroyed by him, *yet it was so corrupt that whatever remains is frightful deformity.*
>
> (Calvin 1960, vol. 1, p. 189; emphasis added)

In affirming the earlier teaching of Augustine regarding the doctrine of the Fall, Calvin's language of 'frightful deformity' paved the way for the later Calvinist doctrine of the 'total depravity' of humankind. Taken at face value, the suggestion that the human race is utterly depraved is significantly lacking in empirical support. Most of us, under normal circumstances, live relatively good lives, and few of us could truthfully be described as *totally* depraved: Nelson Mandela and Adolf Hitler are not moral equals. It is no surprise, therefore, to find Rousseau wholeheartedly rejecting the Christian doctrine of the fall of humankind: 'The first impulses of nature are always right; there is no original sin in the human heart' (Rousseau 1986, p. 56). At the same time, he is not blind to the reality of evil: 'Everything is good as it comes from the hands of the Maker of things; everything degenerates in the hands of man' (p. 5, following the translation of Bowen 1981, p. 187). Rousseau reconciles the two statements by drawing a distinction between human agency and social structure: though human agents are naturally good, they are inevitably corrupted by social structures. Like the Calvinist doctrine of total depravity, Rousseau's position is not supported by empirical evidence: many social structures are clearly forces for good, and many human agents clearly corrupt social structures for their own selfish, and at times demonic, ends. Despite this, Rousseau's doctrine of the natural goodness of humanity has been imported into mainstream liberal thought: the unconstrained exercise of personal freedom uncontaminated by fallen social structures is seen as the basic prerequisite for human flourishing, and the exercise of tolerance is deemed sufficient to hold back the forces of evil. John Kekes develops a cogent case for the poverty of liberalism's position here, arguing that it 'pays almost no attention to the prevalence of evil' in the world (Kekes 1999, p. 23). Liberalism attributes evil to non-autonomous actions, and claims that the promotion of autonomy can make evil less prevalent. However, it is unable to explain: (1) how increasing the autonomy of evildoers to do evil can reduce the prevalence of evil in the world; (2) why increasing autonomy would *not* result in non-autonomous evil action becoming autonomous evil action; and (3) 'why reasonable people would adopt the suicidal policy ... of removing curbs on non-autonomous evil actions in order to increase autonomy' (p. 44f.). It is palpably not the case that 'the attractions of liberalism are sufficient to incline people to conduct themselves according to its prescriptions' (p. 23).

Unlike Christianity's affirmation of the fallen (but not necessarily totally depraved) nature of all human beings, liberalism's assertion of the natural goodness of humanity and lack of a robust understanding of wrongdoing and evil leaves little room for the acknowledgement of personal responsibility, and makes remembrance and acknowledgement of guilt, and therefore reconciliation and restorative justice, extremely difficult to achieve (Wyschogrod 1998; Volf 2006). The Truth and Reconciliation Commission in post-apartheid South Africa required perpetrators of violence to acknowledge and testify to their wrongdoing prior to requesting amnesty from civil or criminal prosecution; in practice such testimony frequently proved a necessary prerequisite for reconciliation between victims and perpetrators (Gruchy 2002). Nevertheless, the limitations of the liberal understanding of evil cannot gloss over the failings of the Calvinist doctrine of 'total depravity'; they can, however, enable us to understand more deeply the original Christian vision of the nature of sin and evil prior to its later distortion in Western Christendom. In the rest of this section we will explore four dimensions of Christian teaching of the universality of sin that have been effectively obscured by the later stress on 'total depravity'.

(1) The Christian doctrine of the universality of sin, from which the distorting notion of 'total depravity' emerged, was grounded, in part at least, on attacks on religious hypocrisy. Paul's assertion of the universality of sin forms the conclusion to an extended rhetorical argument aimed at hypocrisy among Christians in Rome (Romans 1:18ff.). He first courts the approval of his audience by listing the moral failings of first pagans and then Jews, only to turn his argument back onto his Christian hearers, who cannot claim moral superiority: 'For there is no distinction [between pagans, Jews and Christians], since all have sinned and fall short of the glory of God' (3:22f.). Paul here replicates Jesus' own attack on the hypocrisy of the religiously self-righteous:

> Woe to you, scribes and Pharisees, hypocrites! For you are like whitewashed tombs, which on the outside look beautiful, but inside are full of the bones of the dead and of all kinds of filth. So you also on the outside look righteous to others, but inside you are full of hypocrisy and lawlessness.
>
> (Matthew 23:27f.)

If the motivation of Augustine's infamous doctrine of 'original sin' is identical to that of Jesus and Paul, namely to attack religious hypocrisy and spiritual self-righteousness, its immediate roots lie in two specific ecclesiastical issues. In the first place, Augustine was responding to the ongoing Donatist schism in the North African church that followed the 'Great Persecution' of Christians under the Emperor Diocletian (313–303 CE). The Donatists were religious rigorists who refused to receive back into church membership those Christian apostates who had denied their faith and given up their Bibles to be burnt during the persecution. In asserting the doctrine of original sin, Augustine was seeking to reject the self-righteous hypocrisy of the Donatists by defending the claim that all Christians, apostate and non-apostate alike, fall short of the standards desired of them by God. In the second place, Augustine was responding to the claim of the British monk Pelagius that Christians are able to make themselves righteous before God through their own moral efforts. Augustine held that since all human beings are tainted by original sin and fall short of perfection, those Christians who claim to have become perfect through their own efforts are necessarily charlatans. Augustine responded to the challenges of Donatisn and Pelagianism by invoking a doctrine of 'original (or universal) sin' designed to cut to the very heart of the Christian Gospel: God loves all human beings unconditionally, just as they are, regardless of their faults, failings and sins – you do not have to be a good person to be a Christian. Augustine's doctrine, like Paul's, is grounded in Jesus' teaching: 'You hypocrite, first take the log out of your own eye, and then you will see clearly to take the speck out of your neighbour's eye' (Matthew 7:5).

It is important here to recognise that, because 'sin' is essentially a theological category, the affirmation of the universality of sin is always predicated on a comparison between humanity and God rather than between human beings themselves. When measured against one another, human beings clearly possess different levels of moral virtue, but when measured against the glory of God, all human beings, from the very best to the very worst, fall short of perfection.

(2) If sin is simply that which undermines proper relationships between God and creation, it cannot – contra Protestant pietism – be limited to individual moral culpability. Because we inhabit a fallen world, anything that cuts us off from God and from the fulfilment of God's desire for our flourishing and well-being, including all that might stand beyond the orbit of our personal moral responsibility – from cancer and AIDS, through abuse and persecution, to unjust social and economic structures – are inherently sinful. Our participation in a fallen world does not mean that we are morally culpable for all the world's evils and ills, only those over which we have some level of control. We are victims of sin as well as perpetrators of sin. This is precisely the issue Augustine's doctrine of original sin set out to address: we are all born equally into a fallen world from which we are unable to free ourselves. In retrospect, it is clear that Augustine's failure lay in his attempt to explain our unavoidable participation in a world dislocated from God by using Platonic categories. The Platonic dualism between the higher spiritual realm and lower material realm led Augustine to locate agential and structural sin in the material world, specifically in human sexuality. As a result he saw our universal participation in a fallen world to be a result of the

transmission of sin through the physical act of sexual intercourse. When the transmission of sin in this manner became linked to the pietistic reduction of sin to moral culpability, Western Christendom arrived at the absurd conclusion that newborn infants inherit the moral guilt of their parents, grandparents and ancestors, rather than inevitably participate, through no initial fault of their own, in an ontologically fallen, fragmented and dislocated world.

(3) The Christian claim that the 'wages of sin is death' needs to be put in the context of divine grace, in which 'the free gift of God is eternal life in Christ Jesus our Lord' (Romans 6:23). The 'teaching of the election of a limited number and the rejection in hell of the majority' is not that of orthodox Christianity, despite the fact that many Christian churches embraced it from the medieval era through to the present day (Gunton 2002, p. 66). 'Far from something to be feared and hated, the doctrine of election says that God has chosen the human race, *all of it*, in Christ, to come into loving relationship with its Creator' (p. 66; emphasis added). The notion that we are born into original sin and are morally culpable for the failings of our ancestors, and that as a result we will be consigned to hell unless we happen to be among the chosen few arbitrarily selected for salvation by a despotic and demonic God is not a distorted parody of the Christian Gospel, but a fundamental perversion of it.

(4) God's unconditional love for every human being means 'there will be more joy in heaven over one sinner who repents than over ninety-nine righteous people who need no repentance' (Luke 15:7). Despite the joy of reconciliation, the consequences of the ontological dislocation of creation cannot be glossed over by liberal sentimentality. Restorative justice on a cosmic scale means that evil must be exposed and eradicated rather than plastered over and left to fester. Consequently, God's love is simultaneously a divine 'yes' and a divine 'no': a divine 'yes' to the infinite worth of every human being, and a divine 'no' to the ontological disruption of creation and all that denies humanity access to the reciprocal loving relationship with their Creator. In the incarnate Christ the light of God's love shines in the world:

> For all who do evil hate the light and do not come to the light, so that their deeds may not be exposed. But those who do what is true come to the light, so that it may clearly be seen that their deeds have been done in God.
>
> (John 3:20f.)

God's love can be experienced as divine wrath and judgement: 'And this is the judgement, that the light came into the world, and people loved darkness rather than light because their deeds were evil' (John 3:19). Divine wrath is not that of a capricious tyrant king seeking revenge and satisfaction for the wrongs dealt him by his subjects; rather it is a response to 'the rebellious act of a prodigal, provoking the burning anger of a Father's heart which is, nonetheless, never other than a form of his holy love' (Hart 1997, p. 202). God's wrath is not coercive, seeking to re-establish a relationship with creation through fear, force or violence; rather it is educative, seeking not the destruction of creation but its renewal and transformation. A lover may respond to the infidelity of her beloved in different ways: through rage, in order to bring the relationship to an end once and for all; through anger, in order to heal, restore and recover the relationship; through silence, in order to seek to rescue the relationship by perpetuating the fiction that no infidelity actually took place. God's love is expressed as righteous anger because he seeks to restore his relationship with humanity, not to end it or ignore it. An all-loving God who shows no righteous anger at the evils perpetuated on and by his creatures is either a God who does not care, a God in denial, or a God willing to perpetuate a pious fiction. Genuine restoration requires both parties to be truthful to one another, by giving and receiving openly and honestly. This demands acknowledgement of personal agential guilt and responsibility as well as recognition of extenuating structural circumstances. In order to achieve reconciliation, the lover will need to express to her beloved the pain and anger caused by his infidelity, and the beloved will need both to recognise such expression as an act of love inviting repentance and to respond by amending his ways, confessing his infidelity and begging forgiveness.

The drama of reconciliation

The drama of the salvation of humankind and reconciliation of creation to its Creator is possible only because God acts providentially in the world. The Latin *providentia* refers to the foresight to look ahead and act prudently by making provision for future needs. In Christian theology God 'actively directs and involves himself in the day to day life of his creatures' (Gunton 2002, pp. 20ff.). 'God's providence is his action both within and alongside the structures of the world he has created so as to both uphold and shape the direction of things' (p. 34). Specifically, God makes provision for the future perfection of creation through divine and human action. At a general level God sustains fallen creation and holds it in being: 'your Father in heaven … makes his sun rise on the evil and on the good, and sends rain on the righteous and on the unrighteous' (Matthew 5:45). Such general providence provides a supportive backdrop to particular, historically specific, providential acts that together constitute the divine economy of salvation that runs from the covenantal election of Israel, through the redemptive and transformative events of the death and resurrection of the incarnate Son, to the future eschatological sanctification of all creation.

Trinitarian accounts of providence rejected Gnostic and neo-Platonic dualism. The divine plan does not entail the emancipation of immaterial souls from captivity in the material world, but rather the transformation of the created order and of persons in their material relationships. According to Irenaeus' theology of recapitulation, the disobedience of the 'first man' Adam leads to the disruption of the created order, while the obedience of the 'last man' Jesus leads to the reconciliation of the world to God and sets in motion a pilgrimage towards the perfection of creation. The first Christian theologians also rejected Stoic pantheism and its monistic vision of a deterministic reality that humanity can accommodate itself to only through stoic, and occasionally heroic-tragic, resignation. Gunton views the mechanistic world of Deism and the genetically determined world of some forms of neo-Darwinism as modern instances of a closed deterministic monism. 'The weakness of all pantheistic and mechanistic conceptions of providence is that they turn the richness of the created order into a homogeneity: there is only one form of causal regularity, and all must conform to it' (Gunton 2002, p. 34). Such causal regularity makes it impossible to envisage God's providential action in the world as anything other than a series of miraculous punctiliar interventions that violate the natural deterministic order-of-things. The mirror image of a deterministic economy governed entirely by the natural order is a deterministic economy governed exclusively by God. According to Aquinas, 'all things are subject to Divine government' (Aquinas 1920, vol. 5, p. 3). God works directly and miraculously from beyond the created order and indirectly and non-miraculously within nature. As the supreme final cause, the God of classical theism is ultimately responsibility for the chain of efficient, formal and material causation in the world. Because he is omniscient, omnipotent and timeless, this God's providential action is ultimately deterministic, so that all that happens in the world is divinely pre-ordained, pre-destined and pre-determined.

Trinitarian theology offers an alternative to the envisaging of providence as emancipation from the material world, naturalistic determinism tempered by miraculous intervention, or divinely driven predestination: because God creates a world that is 'very good', divine providence cannot involve any dualistic rejection of the material world; because God is sovereign over creation, providence cannot be limited to a series of miraculous punctiliar interventions that violate an otherwise deterministic natural order; because God is love, providence cannot involve any divinely imposed celestial determinism that undermines human freedom. Gunton points out that modern physics has rejected Newton's mechanistic universe in favour of an open, organic, dynamic, emergent and multifaceted natural order – a fact that that the proponents of genetic determinism, with their reductive privileging of biological science, have yet to recognise.

> Scientific and philosophical reductionism do not begin to plumb the richness and variety of our world, which is increasingly being revealed to be a compound of vastly varied patterns of energies, corresponding both to the variety of the creator's mediations through Son and Spirit and to the richly various world of Genesis 1.
>
> (Gunton 2002, pp. 34ff.)

This resonates strongly with critical realism's affirmation of a stratified universe in which emergent higher strata are dependent upon, but irreducible to, lower strata. This means *both* that any given object or event is open to explanation at different levels (physical, chemical, biological, psychological, sociological, moral, aesthetic, spiritual, theological, etc.), *and* that explanations at different levels are complementary rather than mutually exclusive. If the universe is indeed dynamic and emergent, as both contemporary physics and Trinitarian theology claim, then there is no fundamental problem in envisaging divine providence at work *alongside* other natural, socio-structural and personal-agential causal mechanisms. Further, if the ontological ground of reality is the dynamic of reciprocal love within the Trinity, rather than the natural order, then the miraculous need no longer be seen as a punctiliar violation of the immutable laws of nature, but rather as an integral part of God's holistic providential plan through which nature is freed from its fallen state and transformed into the reality God intended and desires. God thus works from within creation (in the person of the incarnate Son) and from without creation (in the person of the Holy Spirit) to perfect creation by reconciling it to himself. This providential action does not override human actions: we remain free to either obstruct of participate positively, creatively and pro-actively in this transformational process. God seeks to enable and empower – but never to coerce – humanity to recognise, accept and reciprocate his gracious offer of love. Without this freedom – including the freedom for refusal, resistance and rebellion – to interact with God, the establishment on the human side of a reciprocal loving relationship between God and humanity is impossible. This interplay between God and his creation preserves both the freedom of all parties and the ultimate sovereignty of that reciprocal love that constitutes the ultimate ground of all being. If we play chess with a grandmaster, it is inevitable that the grandmaster will win the game, but we nevertheless remain free throughout to move our pieces as we wish.

> As playwrights devise characters that self-develop according to the internal logic and structure of the play, so does God bring the world into existence and allow its inhabitants the necessary freedom for them to be what God wants for them.
>
> (Wright 2010, p. 150; cf. Gunton 2002, p. 64)

As we have already noted, God's providence operates generally by sustaining fallen creation and holding it in being, and particularly in a drama of reconciliation and sanctification that proceeds from the election of Israel, through the incarnation and resurrection, via the life of the Spirit-inspired Church as fallible witness to God's providential actions, to the final eschatological sanctification of all creation. The incarnation and resurrection of the Son of God are the decisive focal points of this drama: the incarnation constitutes the ontological reconciliation of God and humankind, and the resurrection both actualises and prefigures the future eschatological perfection of creation. Because of limitations of space, we must restrict ourselves here to an exploration of just two moments in this drama: the theological significance of Jesus' death and resurrection, and the final consummation of God's providential plan.

Before briefly considering the cluster of theological accounts of the significance of Jesus' death generated by Christian theologians, we must pause to address a methodological issue. I have suggested that Christian doctrines are best seen as retroductive models that are open to ongoing iterative refinement. Colin Gunton has subjected the traditional accounts of the theological significance of Jesus' death to just such a process (Gunton 1988). His core argument is that, though the use of metaphor is a necessary feature in all retroductive modelling, including that undertaken by the natural, social and theological sciences, there is a constant danger of the over-rationalisation of metaphor, so that what starts out as a heuristic tool becomes a procrustean constraint that occludes rather than illuminates our understanding of reality.

The natural sciences are replete with metaphor: in physics an electron can be envisaged as a 'particle' or a 'wave', and biologists refer to 'factory' cells and 'chaperone' proteins. To push any of these metaphors further, for example, by asking whether factory cells operate on capitalist of socialist lines, is to misunderstand their hermeneutical function in scientific discourse. Metaphorical description functions as a heuristic tool that is necessarily closed to rationalistic and literalistic interpretation. The rationalistic unpacking of a heuristic metaphor runs the risk of committing the epistemic fallacy by allowing the epistemic tool of metaphor to determine our understanding of the ontic reality to which the metaphor is attached. This is not to suggest that metaphor lacks cognitive purchase on reality. The standard account of metaphor as mere rhetorical decoration lacking in any positive epistemic significance has its roots in Aristotle, and flourished during the Enlightenment. Locke goes a stage further in suggesting that such rhetorical decoration is positively detrimental to understanding: the 'artificial and figurative' application of metaphor functions 'to insinuate wrong *Ideas*, move the passions, and thereby mislead the Judgement' (Locke 1975, p. 508; emphasis in original).

Against Aristotle and Locke, Paul Ricoeur insists that metaphorical language possesses powers of disclosure unavailable to rationalistic discourse. In particular, religious metaphor is able to 'eclipse the objective, manipulative world' of brute fact and 'make way for the revelation of a new dimension of reality and truth' (Ricoeur 1976, p. 68). As Gunton puts it, in theological discourse a 'metaphor or family of metaphors takes its shape from the divine and human story it seeks to narrate, and so enables aspects of the meaning of an unfathomable mystery to be expressed in language' (Gunton 1988, p. 113; cf. Terry 2010, pp. 134ff.). Just as the strength of a chain is limited to its weakest link while the strength of a rope lies in the interweaving of many strands, so the power of a retroductive explanation of reality is not dependent upon any single metaphor, but upon the interweaving of a cluster of complementary metaphors. Thus Gunton looks to a web of related metaphors to shed light on the salvific significance of the crucifixion, and seeks to avoid their reduction to a single brittle rationalistic account. Such rationalism, he argues, tends to turn acceptable metaphors into unacceptable myths. Thus, if taken literally, the metaphor of Christ's victory over Satan and the forces of evil ceases to refer to the overcoming of repressive political, social, economic and religious power structures embedded within human culture, and becomes instead a mythological account of a supernatural event that takes place 'in a sphere outside the course of concrete divine–human relations' (Gunton 1988, p. 63). Myth, in the sense employed by Gunton, is generated whenever a metaphor is used rationalistically to dictate realty, rather than heuristically to illuminate reality. The theological resistance to literalistic myth making is ingrained within the biblical text itself. Thus references to Christ's death as a 'sacrifice' are presented in terms of the explicit use of the word as a metaphor for moral action rather than a literal designation of cultic practice: 'The sacrifice acceptable to God is a broken spirit; a broken and contrite heart' (Psalm 51:17). The power of the metaphor derives not from cultic practice per se, but from the fact that in cultic sacrifice 'those who bring an animal to the alter or place a coin in a collecting box give something of themselves, of their substance, as we say, and in that way alter, be it ever so slightly, the shape of our world' (Gunton 2002, p. 70). It is this self-giving in order to help reshape the world that is at the heart of the New Testament's use of the metaphor of sacrifice. Just as God sacrifices by giving himself unconditionally to human beings, so Christians in turn are called to sacrifice by giving themselves unconditionally to God and to their fellow human beings, thereby replicating the reciprocal bond of love that unites the Trinity. Just as Christ's self-sacrifice led him to the Cross, so Christians are called to 'carry their cross' and offer their lives 'as a living sacrifice, holy and acceptable to God' (Luke 14:27, Romans 12:1).

The Christian doctrine of atonement seeks to explain how, within the economy of the incarnation, the crucifixion and death of Jesus functions to reconcile God and creation. According to Christian orthodoxy the various theories of atonement developed by Christian theologians are necessarily grounded in the primary doctrine of the incarnation, which affirms the hypostatic union of God and humankind in the person of Jesus Christ. Four retroductive models are prominent in the multifaceted doctrine of atonement: (1) ransom, (2) satisfaction, (3) penal substitution, and (4) moral influence.

(1) The ransom model. Gustaf Aulén in his classic study of the atonement, *Christus Victor*, argues that the paradigmatic model of atonement that dominated Christian theology in its formative years affirmed the crucifixion as the actualisation of Christ's victory over all those forces that undermine the sovereignty of divine love (Aulén 2010). The resurrection reveals that divine love continues to reign even after the forces of evil have done their worst, so that Jesus' death served to liberate humanity from the bondage of sin, death and the devil. This established the ground for the 'ransom' theory, which draws upon the ancient practice of paying ransom to military opponents to ensure the safe return of prisoners of war. In Isaiah the Israelites returning from exile in Babylon are described as 'redeemed': 'the ransomed of the LORD shall return/and come to Zion with singing' (Isaiah 51:11). According to Mark's Gospel Jesus 'gave his life as a ransom for many' (Mark 10:45). In both cases the word 'ransom' is used metaphorically: like ransomed prisoners of war, the people of Israel are delivered from their captors and restored to their homeland, and the entire human race is delivered from sin and repressive social structures and restored to a proper relationship with God, nature, community and self. The metaphor *qua* metaphor serves to illuminate the actuality of such restoration, and is misappropriated if understood literally. Thus to suggest that God paid a ransom to Satan in order to free humanity from the captivity of sin is to over-literalise and over-rationalise the metaphor. As the medieval theologian Anselm pointed out, the

suggestion that 'the devil had a rightful ownership of man', and hence power over God, is absurd because 'neither the devil nor man belong to any but God, and neither can exist without the exertion of Divine power' (Anselm 1962, p. 187). The price of taking the metaphor of ransom too literally is the generation of a myth of an exchange between God and Satan that is not warranted or licensed by New Testament teaching. Similarly, as we have already seen, a literalistic reading of the metaphor of 'victory' over the forces of repression generates a myth of a cosmic struggle against Satan.

(2) The satisfaction model. Anselm's development of the satisfaction model in the eleventh century, driven by a concern to challenge the rationalisation of the ransom model, was the first major iterative development of the doctrine of atonement in the history of the Church. It is important to recognise that 'satisfaction' in Anselm's theology refers to legal restitution rather than personal gratification. In medieval jurisprudence the perpetrator of a crime had a legal duty to restore justice by making recompense to their victim. The notion that the perpetrators of evil had a moral duty to make restitution to God, that any failure to do so would result in an unjust settlement, and that since human beings are unable to make proper restitution Jesus Christ paid the debt on their behalf by dying on the Cross offered a powerful metaphor that, like the established metaphors of victory and ransom, enhanced the Christian understanding of atonement.

(3) The penal substitution model. Literalistic readings of the metaphors of satisfaction and restitution generated a cosmic myth of legal exchange predicated upon the absurd assumption that God was subject to the canons of medieval jurisprudence. Such rationalistic readings of Anselm were to provide a basis for the model of penal substitution, which envisages the crucifixion as a legal transaction 'in which Jesus is conceived to be punished by God in place of the sinner' (Gunton 1988, p. 90). Gunton follows many orthodox theologians in viewing this model as a theological abhorrence, rooted, once again, in a naïve literalism that collapses on its own terms. An irate God who requires his Son to die on the cross in place of sinful humanity in order to satisfy his anger is incompatible with the Trinitarian understanding of an essentially loving God. Further, since, according to orthodox doctrine, Jesus *is* God, then according to this model God punishes himself on the Cross. This, however, provides a clue to the deeper meaning of Anselm's theory of restitution. 'Therefore, since he himself (Jesus Christ) is God, the Son of God, he offered himself for his own honour to himself, as he did to the Father and the Holy Spirit' (p. 92). Since the sovereign God cannot be accountable to any external canon of justice, such as the medieval notion of restitution, restitution for evils perpetuated against God is made through a transaction *within the Trinity itself*, so that on the Cross God-in-Christ takes upon himself the burden of the sins of humanity and in doing so reorders creation in a way that takes the issues of justice and evil seriously while remaining grounded entirely in God's gracious love. '*Satisfaction* is therefore Anselm's way of speaking of that which took place as a result of the good God's being unwilling to allow his creatures to destroy themselves' (pp. 91ff.). Anselm is not comparing God to a feudal lord whose desire for restitution for wrongs done to

him leads him to demanding that his son be executed because the actual perpetrators cannot afford the cost of reimbursement. Rather, and at the risk of extending the metaphor beyond its proper boundaries, it is God himself who makes restitution, paying the price of sin by taking on his own shoulders the burden of the consequences of sin, namely the collapse of the created order into chaos, non-being and death. 'The framework of the theology of satisfaction is thus only secondarily human fallenness; the primary focus is the goodness of God' (p. 91). On human terms restitution requires the perpetrator to compensate his or her victim in a manner proportionate to the crime itself; on God's terms, however, it is the victim – acting out of love and blind to human notions of justice – who freely chooses to make restitution on behalf of the perpetrator.

> You have heard that it is said, 'An eye for an eye and a tooth for a tooth'. But I say to you, Do not resist an evildoer. But if anyone strikes you on the right cheek, turn the other also.... Love your enemies and pray for those who persecute you.
>
> (Matthew 5:38ff.)

On the Cross God-in-Christ turns the other cheek and, in Anselm's extended metaphor, makes restitution for the debt owed to him.

> [U]nderlying this theology is the conviction that evil can be overcome only by good, for attempting to overcome it by a further evil simply reduplicates the ill.... A real change in the situation can come about only if the evil is *overcome*, that is to say taken seriously in such a way that its reality is displaced by something stronger.
>
> (Gunton 2002, p. 76)

(4) Moral influence. According to this model, the moral and spiritual example of Jesus in undergoing crucifixion inspires Christians to greater love of God and neighbour. Though popular among liberal theologians suspicious of the metaphysical 'trappings' of Christianity, it places the incarnate Christ on the same footing as Christian saints, whose lives can be equally inspirational, and makes reconciliation with God a product of human effort and achievement. Though the moral influence of the example of Christ is not entirely rejected by Christian orthodoxy, it plays only a minor role in the doctrine of atonement.

The liberal Protestant suggestion, that Jesus' death was nothing more than an example of sacrificial service that humanity would do well to attempt to replicate, is frequently accompanied by the claim that his resurrection is no more than a symbol of the persistence of the human spirit in the face of adversity. On this reading, though Jesus died once and for all on the Cross, the myth of resurrection inspired his disciples to continue to live out his moral vision. Orthodox Christianity, on the other hand, insists on the material actuality of the resurrection: 'If Christ has not been raised [from the dead], your faith is futile' (1 Corinthians 15:17). David Jenkins, former Bishop of Durham, once famously courted public controversy by suggesting that the resurrection was not merely a divine conjuring trick with bones. If what he meant by this is that God chose to confirm the identity of Jesus and reveal his own power through a dramatic intervention in the ordinary course of natural events, then Jenkins' suggestion is entirely orthodox. The Gospels record a number of occasions when Jesus raises individuals from the dead; if these serve to confirm Jesus' divine identity and reveal the power of God, then they would possess just as much theological significance as the resurrection of Jesus himself. But this runs contrary to the orthodox understanding of the resurrection. Those whom Jesus raised from the dead continued in their earthly lives and eventually died, whereas the resurrected Christ lives on eternally. As we have seen, Trinitarian Christianity affirms the absolute sovereignty of God over the natural order, so that natural causal mechanisms are his creation, are sustained by him, and remain subject to him. Decay, death and the collapse into non-being are not intrinsic to creation, but rather a cancerous travesty of it. In reciprocating divine love as a human being, Christ establishes the ontological unity of God and humankind, and once that bond is established the collapse into non-being is halted and the natural order is returned to its original pristine state. This is the theological significance of the resurrection: in the absence of a reciprocal relationship between God and humanity 'creation was subject to futility', with the establishment of a reciprocal relationship creation is liberated from 'its bondage to decay', and the human race obtains 'the freedom of the glory of the children of God' (Romans 8:20f.). The resurrection signals the transformation of material creation rather than the liberation of disembodied souls from their decaying physical bodies. Hebraic thought had no notion of a Platonic mind–body dualism, so that the resurrected life is always an embodied life. 'The resurrection is not a taking of the human soul out of the world, but the perfecting of the whole human being along with it' (Gunton 2002, p. 153). On the Cross Christ brings to fulfilment his establishment of a reciprocal loving relationship between God and humankind, so that human beings are justified before God by faith in the achievement of Christ despite their sinful natures. In the present, 'those who allow their lives to be shaped by the worship and fellowship of the church ... are becoming [often slowly and frequently painfully] like God in the way purposed in the creation, shaped by and conformed to the manner of Christ's life and death' (pp. 154ff.).

Christians look to the incarnation as the event through which creation was ontologically reconciled with God, and to the sequence of crucifixion–resurrection as the first fruits and anticipation of its future universal actualisation. Christian eschatology has both a realised and future dimension. The hope of the perfection of creation is currently being realised through the work of the Holy Spirit, who shapes the worship and fellowship of the Church, and through the Church – acting as the 'salt of the earth' and 'light of the world' – the whole of humanity (Matthew 5:13f.). However, the final consummation of the mission of the Holy Spirit remains a future hope, grounded in the promise that through the Holy Spirit 'God will complete at the end the holiness that is in process, so that the justified will finally become the perfectly just' (Gunton 2002, p. 155).

From second-century Montanism through to contemporary millennial movements, Christianity has been bedevilled by speculation about the future. Such speculation is fuelled by the presence of apocalyptic literature, in particular the books of Daniel and Revelation, in Christian Scripture (Russell 1964). Literalistic readings view apocalyptic texts as predictions of a final cosmic cataclysm that will lead to the end of the world, the last judgement, and the allocation of souls to either heaven or hell. Such literalism is at odds with the texts themselves. The reference in the apocalyptic chapter of Mark's Gospel to 'the Son of Man coming in clouds', for example, is not a reference to a literal 'second coming' of Christ from heaven at the end of time (Mark 13:26). Rather, following the symbolic convention established in Daniel, from which the phrase is taken, it would have been understood by first-century readers as doing no more than 'predicting great events in and through which God would vindicate his true people after their suffering' (Wright 2000, p. 32; Daniel 7:13). Since such events were understood to take place *within* the created order, there 'is no justification for seeing "apocalyptic" as necessarily speaking of the "end of the world" in a literally cosmic sense' (Wright 1992, p. 298). Rather, apocalyptic literature was an established literary genre in the ancient Jewish world whose primary function was to express political theology. It used striking imagery and metaphor within established stylistic conventions to refer to the working out of divine providence in relation to social and political power structures perceived to be opposed to the will of God. In the Book of Revelation the 'beast rising out of the sea' who 'was allowed to make war on the saints and to conquer them' is a reference, not to some supernatural or demonic being appearing at the end of time, but to the persecution of the Church by the Roman Empire at the time when the book was written (Revelation 13:1, 7). The author of Revelation employs the familiar apocalyptic device of 'gematria', through which letters are transposed into numbers, to identify the 'beast' with a person bearing the mark '666' (Revelation 13:18). Scholars are divided as to whether this is a reference to the Emperor Nero himself or to another agent of persecution; if it is a reference to Satan, as popularly assumed, it is so only to the extent that the persecutors might legitimately be described metaphorically as 'satanic'.

Theological reflection of the future hope for the perfection of creation was of an entirely different order to literalistic and speculative readings of apocalyptic texts. A number of New Testament passages refer to the future participation of humanity in the being of God: the gifts conferred upon the world by Jesus Christ are given 'so that through them you might escape from the corruption that is in the world because of lust, and may become *participants in the divine nature*' (2 Peter 1:4; emphasis added). The doctrine of 'theosis' refers to the divinisation or deification of humanity through participation in God's nature. According to Athanasius, the Son of God 'assumed humanity that we might *become God* … manifested Himself by means of a body in order that we might perceive the Mind of the unseen Father … [and] endured shame from men that we might inherit immortality' (Athanasius 1996, p. 93; emphasis added). In the Eastern Orthodox tradition the doctrine was developed by Gregory Palamas, and in

Western Catholicism it was picked up by Aquinas. According to the latter, 'the gift of grace surpasses every capability of created nature, since it is nothing short of a *partaking of the Divine Nature*, which exceeds every other nature' (Aquinas 1920, vol. 8, p. 369; emphasis added). Just as the persons of the Trinity are united in perfect reciprocal love, and just as the human and divine natures of the incarnate Son were similarly united in hypostatic union, so human beings will enter into just such a relationship with their Creator. This does not imply the dissolution of human identity within the Godhead: just as the bond of love that unites the Trinity simultaneously upholds the unique identities of the three persons, so the perfect bond of love between the Creator and his creatures will uphold the individual identities of both God and individual human beings. With the perfection of creation, God's perfect love of humanity will finally be reciprocated by humankind's perfect love of God. The ontological achievement of the hypostatic union between the human and divine natures of Christ prefigures the participation of humanity in the divine nature, and the perfection of their status as creatures created in the image of God.

Part IV

Judgemental rationality and Christian epistemology

10 The epistemology of divine revelation

Having outlined the contours of Christian ontology, our task in the next two sections is to address the question of their rational justification and warrant. Christian theologians draw a distinction between natural and revealed theology. Natural theology seeks to derive knowledge of God from the natural order-of-things via inductive and deductive reasoning. Thomas Aquinas' five 'demonstrations' of the existence of God take the form of inductive inferences drawn from observation of nature (Aquinas 1920, vol. 1, pp. 19ff.; Kenny 2008). The first three are variations on what became known as the 'cosmological argument': objects in the natural world are in a constant state of motion, and since motion cannot extend into infinity there must be an 'unmoved mover'; nature consists of a network of efficient causes, and since there can be no infinite regress there must be a 'first cause'; since the contingent world has the potentiality not to exist, its continuing existence must be due to a 'necessary being'. The fourth 'moral' argument proceeded from the gradation of goodness in the universe to the necessity of an 'infinitely good being'. The fifth 'teleological' or 'design' argument proceeded from the order and purpose inherent in the universe to an intelligent creator. Anselm's ontological argument seeks to deduce God's existence from the suggestion that an absolutely perfect being would be less than perfect if it did not actually exist (Anselm 1962, pp. 7ff.; Plantinga 1968). A renewed interest in these scholastic arguments at the start of the Enlightenment quickly gave way to widespread scepticism regarding their viability. Aquinas' *a posteriori* arguments faced criticism from Humean empiricists, who rejected the Aristotelian understanding of causality on which they were dependent, as well as from Darwinian evolutionists. Anselm's *a priori* argument faced the dual challenge of the anti-idealist rejection of deductive reasoning from first principles and the Kantian denial that 'existence' is a predicate. Despite the subsequent appearance of a series of robust defences of natural theology, the modern era saw a relative decline in its influence, particularly among Protestant theologians (Mackie 1982; Swinburne 1993, 2004). In the case of the latter, this was not primarily due to a perception that natural theology was struggling to sustain its intellectual credibility, but rather to the fact that the God whose existence natural theology sought to demonstrate was the God of classical theism rather than the God of the Trinitarian faith.

Whereas natural theology appealed directly to *structures* inherent in the natural order, revealed theology appealed to the concrete and particular acts of divine *agency* within the contingencies of the historical process. Specifically, revealed theology found knowledge of God in his covenantal relationship with the people of Israel, in the life, death and resurrection of Jesus Christ, and in the ongoing experience of the Christian community. The claim that God elects to reveal himself in this manner is grounded not in any *a priori* assumption, but in *a posteriori* reflection on the actuality of these historical events. Christianity could not have emerged apart from the historical figure Jesus of Nazareth who, according to Christian doctrine, reveals unique knowledge of the being of God unavailable through any other epistemic route. The Christian epistemology of revelation constitutes a disclosure model, according to which 'A *discloses in the situation* B *the content* C *for the recipient* D *with the result* E' (Gunton 1995, pp. 18ff.; Schwöbel 1992, p. 87; emphasis in original). Thus, Jesus Christ (A), discloses in the historical context of first-century Palestine (B), salvific knowledge of the Triune God (C), intended for the whole of humanity (D), with the result that the Church emerged as witness of this revelation (E). Jesus Christ simultaneously reveals (epistemically) and actualises (ontologically) the hypostatic union of God and humankind in the incarnation. The truth disclosed in this revelation has ontological significance regardless of the epistemic effectiveness of the revelatory process: if the actuality of the hypostatic union means that 'as all die in Adam, so all will be made alive in Christ', then this constitutes an ontological reality entirely independent of human knowing, so that it would remain true even if all knowledge of Jesus Christ were wiped from the collective memory of humanity (1 Corinthians 15:22). Historically, the doctrine of revelation played an underlabouring role in Christian theology, functioning as a second-order retrospective means of accounting for the Church's possession of first-order ontological truth. The epistemic basis of revealed theology is thus not deductive or inductive reasoning, but rather inference from human experience. Such experience is not to be mistaken for direct 'peak' or 'mystical' experience of the numinous or transcendent occasioned, for example, by aesthetic encounters with nature, particular meditative practices, moments of personal crisis or elation, or chemical stimulation of the brain (Otto 1931). Though a few Christian mystics may indeed enjoy direct unmediated experiences of God, Christianity places relatively little weight on their epistemic significance. Instead, it understands God's self-revelation to be mediated through contingent historical events, so that assertions of alethic theological truth constitute claims that particular historical events are most powerfully explained within the retroductive framework of Trinitarian theology.

Revelation as divine agency

The distinction between revealed and natural theology is so central to Christian epistemology that it warrants teasing out further. Johannes Climacus, the pseudonymous author of Søren Kierkegaard's *Philosophical Fragments*, is an unbeliever on the brink of becoming a Christian, and hence is concerned with the possibility of divine revelation rather than its actuality. He begins with a question: 'How far does the Truth admit of being learned?' (Kierkegaard 1967, p. 11). 'Truth' is capitalised, because it refers to the quest for ultimate truth and eternal happiness. The assumption of the need to *learn* the Truth raises Meno's Paradox: 'For what a man knows he cannot seek, since he knows it; and what he does not know he cannot seek, since he does not even know for what to seek' (p. 11; Plato 1961, p. 363). Socrates' solution is found in his doctrine of recollection, which holds that knowledge is already latent in the mind, and understands 'all learning and inquiry … as a kind of remembering' (Kierkegaard 1967, p. 11). This assumes a Platonic ontology in which the soul possesses within itself vestiges of the eternal Forms, knowledge of which has been forgotten as a result of the soul's temporal physical embodiment. The role of the Socratic teacher is not to transmit positive knowledge, but to act as a 'midwife who brings others' thoughts to birth', or, in the case of reluctant learners, 'a "gadfly", stinging the sophists into an encounter with truth' (Pattison 1999, p. 68). Climacus claims that the relation of a maieutic teacher to her students 'is the highest that one human being can sustain to another' (Kierkegaard 1967, p. 12). Despite this, 'the fact that I have been instructed by Socrates or by Prodicus or by a servant-girl' is of no lasting significance, since the teacher is merely an instrument enabling me to recollect the Truth that alone possesses eternal value (p. 14). Similarly, the moment in time when I recollect eternal Truth 'is *eo ipso* accidental, an occasion, a vanishing moment': 'The temporal point of departure is nothing; for as soon as I discover that I have known the Truth from eternity without being aware of it, the same instant this moment of occasion is hidden in the Eternal' (pp. 13, 15).

Climacus goes on to distinguish between 'Religiousness A' and 'Religiousness B'. The word 'religiousness' here is best understood in terms of the contemporary use of the word 'spirituality', and has no necessary link to institutional religion. Spirituality, in Kierkegaard's sense, has to do with the individual's relationship to ultimate reality and eternal happiness. The aesthetic and moral stages of life – concerned respectively with sensual gratification and the acceptance of civic responsibility – can be engaged in without asking questions about our eternal destiny. The transition to the religious or spiritual stage of life takes place when we come to understand that neither aesthetic pleasure nor moral convention enables us to make sense of our lives *sub species aeternitatis*. Religiousness A is found among pagans and those nominal 'cultural' Christians whose faith is 'not decisively Christian', whereas Religiousness B is the exclusive preserve of authentic Christian faith (Kierkegaard 1968, p. 495). The former embraces a spirituality of immanence in which God is deemed immediately accessible to human experience, since 'the tie which binds the individual to the divine is still, in spite of all tension, essentially intact', while the latter affirms a spirituality of transcendence predicated on an infinite qualitative distinction between God and humankind (Lowrie 1974, p. 173f.).

Religiousness A is rooted in the human craving for transcendence and the desire to give temporal existence eternal significance. Consequently, it presupposes that eternal Truth is accessible to human beings, and that the dividing line between immanence and transcendence is relative rather than absolute. Eternal Truth is thus placed within the grasp of human understanding and seen as latent within the human soul. As a result, human beings are able to self-activate their dormant relationship with eternity:

> truth is an immanent and eternal possession that *can* be made actual, if a man will but shed his temporal conditions and return to an original state of being and knowledge, outside of time and the existent order.
>
> (Collins 1983, p. 152; emphasis in original)

At its lowest level, Religiousness A seeks to retrogressively domesticate transcendence within immanent human thought and action. The chief villains here are philosophical idealists and 'cultural' Christians: Hegel's 'System' seeks to assimilate transcendence within the bounds of human reason and, in similar vein, the theology of Bishop Jakob Mynster, primate of the Danish Lutheran Church, asserts a synchronicity between 'human effort and divine effort', thereby reducing Christianity to 'a force which corresponds to human striving and creates harmony within the self' (Sponheim 1968, p. 73). At its highest level, Religiousness A is represented by Socrates who, unlike Hegel and Mynster, resists the temptation to follow 'the lure of recollection back to an unparadoxical eternity, which turns out to be the true being and essence of man', and consequently experiences transcendence in terms of guilt, resignation and suffering (Collins 1983, p. 152; Kierkegaard 1968, pp. 347ff.). Climacus attributes Socrates with a dual role: positively, he invites the learner to seek the Truth through introspection and recollection; negatively, and in direct contradiction of Hegel and Mynster, he shows the learner just how far removed from the Truth she is.

Climacus presents Religiousness B as a 'hypothetical' alternative to a spirituality of immanence and its accompanying pedagogy of recollection. If transcendence and immanence are ontologically distinct, then knowledge of the Truth and eternal happiness must entail an epistemology of revelation and pedagogy of proclamation. Further, if a vestige of transcendence is not latent in the soul, then the learner is in a state of fundamental error rather than relative ignorance, and as such 'must be destitute of the Truth up to the very moment of his learning it' (Kierkegaard 1967, p. 16). This being the case, the best that a maieutic teacher can do is to follow the path of Socrates, whose genius was to enable us to realise that we actually are in a state of error. It follows that 'if the learner is to acquire the Truth, the Teacher must bring it to him', and that the paradoxical 'Moment' when time and eternity *do* intersect 'must have a decisive significance' for the individual's eternal happiness (pp. 16ff.). Climacus describes this philosophical hypothesis in theological terms: the divine teacher, driven by his love of the learner, overcomes the error of sin and brings salvation to the seeker after truth. By (thinly veiled) implication, the incarnate Christ, in whose person time and eternity, immanence and transcendence, intersect, is paradoxically both ephemeral (bound to the contingencies of history) and eternal (grounded in the loving will of God), and simultaneously human teacher and divine saviour.

'Suppose', Climacus speculates, 'there was a king who loved a humble maiden' (p. 32). This might appear to be a cause for celebration, since 'love is exultant when it unites equals, but it is triumphant when it makes that which was unequal equal in love' (p. 33). However, what if 'there awoke in the heart of the king an anxious thought'? (p. 33). Could the maiden be truly and eternally happy, as long as she was able to recollect that the king is her sovereign lord and she merely his humble servant? Their love would always disguise the secret grief that the king had conferred a favour upon her 'for which she can never be sufficiently grateful her whole life long' (p. 33). Because the king and maid can never be equals, their reciprocal love can never be perfectly harmonious, and a note of discord will be perpetually present in their relationship. 'Our problem is now before us ... [our] task will be to find ... some point of union, where love's understanding may be realized in truth' (p. 35). The union might be achieved by the king appearing before the humble servant in all his royal power and sovereignty, but this would risk *coercing* her into reciprocating his love and would not achieve the desired end of *her* glorification. Alternatively, the union might be brought about by elevating her to royal status, but this would make her *other* than the person she actually is and *other* than the person the king actually loves. Both of these proposed solutions follow the path of Religiousness A in seeking to bypass the ontological distinction between king and servant, transcendence and immanence, God and humanity. Any viable solution must take seriously the ontological divide between these polarities, and consequently the 'union must therefore be brought about in some other way' (p. 37).

'There once lived a people who had a profound understanding of the divine': a people who perceived the infinite qualitative difference between God and humankind; a people who understood that no human can gaze into the face of God and live; a people who discerned God's dilemma – 'not to reveal oneself is the death of love, to reveal oneself is the death of the beloved!' (p. 37; Exodus 33:20). With the experience of the people of Israel in mind, Climacus turns to his proposed solution: 'Since we found that the union could not be brought about by an elevation [of the servant] it must be attempted by a descent [of the king]' (p. 39). If the king is to lower himself to his beloved, if God is to descend to the human level, then he must do so fully and genuinely. 'In order that the union may be brought about, the God must ... appear in the likeness of the humblest', and the humblest human being is the servant of others (p. 39). Further, his appearance must be real and not deceptive in any way, so that 'this servant-form is no mere outer garment, like the king's beggar-cloak, which therefore flutters loosely about him and betrays the king' (39). The 'unfathomable nature of love' is such 'that it desires equality with the beloved, not in jest merely, but in earnest and truth'; any other form of revelation 'would be a deception from the standpoint of the divine love' (pp. 39, 41). Climacus recognises that even the greatest philosopher or poet could not have dreamed this hypothetical scenario: it could not 'enter his mind that the God would reveal himself in this way' (p. 42). Climacus himself, standing as he does on the verge of embracing the Christian faith, acknowledges that his thought experiment – in rejecting natural theology 'from below' in favour of revealed theology 'from above', and requiring an epistemology of revelation rather than recollection – 'is the most wretched piece of plagiarism ever perpetuated' (p. 43).

Revelation as an epistemic problem

The epistemic turn in modern philosophy, with its valorisation of universal reason, generated widespread suspicion regarding the veracity of the Christian account of revelation. Natural theology, for all its flaws, at least retained the virtue of seeking to follow the path of reason; revealed theology, on the other hand, appeared to abandon reason in favour of a blind leap of faith. We must content ourselves here with identifying four issues surrounding the legitimacy of an epistemology of revelation, namely the problems of (1) historical contingency, (2) divine agency, (3) mediated knowledge, and (4) testimonial authority.

(1) Historical contingency. The claim that the person of Jesus Christ uniquely reveals the Triune God runs counter to Lessing's previously mentioned dictum that *'accidental truths of history can never become the proof of necessary truths of reason'* (Lessing 1957, p. 53; emphasis in original). Both idealism and empiricism resisted any substantial connection between contingent events and eternal truths. In the Hegelian idealist tradition, a historical event can neither generate nor perfectly embody an eternal truth. At best, it may exemplify or reflect an eternal truth, so that the life of Jesus Christ may be seen as *representative* of some eternal ideal. In the Humean empiricist tradition, a historical event takes on a meaning beyond itself only insofar as it is one of a regular conjunction of similar events. At best such conjunctions might act as indicators of universal laws or exemplifiers of sociocultural conventions. Since the life of Jesus Christ was a unique historical occurrence, there could be no possibility of it *alone* constituting one of a regular conjunction of similar events, and consequently no possibility of inferring from it *alone* any universally valid law or legitimate social convention. It follows that any suggestion that Jesus Christ possessed eternal significance by ontologically embodying and epistemically revealing God's nature inevitably transgressed the established boundaries of both idealistic and empiricist rationality.

(2) Divine agency. In both the idealist and empiricist traditions reason functions to identify common structures, in the form of eternal principles and natural laws, that leave only limited room for personal agency. Human agency is severely limited: it can manipulate nature but not transgress natural law, and can influence social structures but only change them with great difficulty and normally only by way of collective agreement. If the natural order is seen as a sovereign and immutable structure driven by impersonal rational principles and laws, then divine agency is limited to the initial generation of the rational order inherent in nature, and can engage in the historical process only through miraculous violations of the natural order – violations that are apparently arbitrary and potentially malignant. This being the case, the suggestion that divine agency is sovereign over the natural order-of-things, and that God reveals himself through his historical acts, appears intrinsically irrational.

(3) Mediated knowledge. There was a tendency among Enlightenment philosophers to differentiate between knowledge and opinion, and to identify the former with the possession of clear and distinct ideas grounded on either idealistic intuition (Descartes) or empirical experience (Locke). This led in turn to the positing of knowledge on either direct unmediated access to clear and distinct ideas or a direct unmediated relationship between the knower and the object of empirical knowledge. Where knowledge is grounded in direct, immediate, unmediated apprehension of an idea of experience of an object, opinion is derived from indirect, deferred, mediated apprehension and experience dependent upon third-part testimony. Thus Hegel appeals to an intuitive apprehension of God 'beheld sensuously and *immediately*', and Schleiermacher appeals to the '*immediate* self-consciousness' of being absolutely dependent on God (Hegel 1949, p. 758; Schleiermacher 1976, p. 5; emphases added; cf. Gunton 1995, p. 3f.). Such appeals to direct unmediated experience of God run counter to the claims of Trinitarian theology that God is known uniquely in the person of Jesus Christ. Given that only the first disciples enjoyed unmediated access to Jesus Christ, and that later generations of Christians are necessarily wholly reliant upon mediated testimony, Christian claims to possess secure knowledge of God appear to lack epistemic warrant when judged against the criterion of the epistemic primacy of direct unmediated experience.

(4) Testimonial authority. To accept the authority of revelation mediated through third-party testimony is to reject Kant's call to have the courage to trust one's own autonomous reason, and runs counter to Descartes' insistence on the epistemic imperative to emancipate ourselves from the hegemony of tradition. Appeals to revelation thus not only fall short of the canons of universal reason, but in doing so reveal themselves to be inherently authoritarian. Post-Enlightenment attempts to buttress mediated testimony by appealing to biblical inerrancy or ecclesiastical infallibility merely serve to reinforce perceptions of the authoritarian nature of revelation. Since universal reason affirms that 'truth as something lying within control of the autonomous human agent', any appeal to revelation, insofar as it invites us to attend to a divine reality beyond our complete control and full understanding, 'takes away our autonomy and leaves us in thrall to the authority of others' (Gunton 1995, p. 21). If 'reason is autonomous and self-sufficient', it follows that 'we do not need revelation' (p. 21). This is the central thrust of Fichte's *Critique of All Revelation*, in which he argues that since appeals to reason necessarily take precedence over appeals to revelation, any revealed knowledge of God must in principle be knowledge that could otherwise be established by the exercise of autonomous reason (Fichte 2010). Fichte does not spell out the obvious conclusion: since autonomous reason, once disconnected from the authority of third-party historical testimony, cannot possible generate knowledge of the life of Jesus Christ, Christian claims to revelation must be dependent upon the irrational authority of a hegemonic tradition.

Viewed from the perspective of this dominant strand of Enlightenment thought, there is a necessary polarity between reason and revelation, so that it is both irrational and authoritarian to claim that contingent historical events can depose the eternal truths of reason, that personal agency can override the structures inherent in nature, and that mediated testimony can countermand unmediated experience.

Revelation as an epistemic necessity

The distinction between reason and revelation is a false one because all knowledge, whether it is arrived at through rational or irrational means, is dependent upon revelation. The Latin 'revelare' refers to the unveiling of that which was previously hidden. Given that knowledge is constituted by the relationship between the knower and the object of knowledge, then the ontological reality of the object must somehow be made known, manifested or revealed to the knower. If natural science is predicated on belief in the existence of an external reality independent of the human mind, then the task of the natural scientist is to unveil nature, or enable nature to unveil itself, in order to reveal its truths (Torrance 2001, p. 2). This is the case in all spheres of human knowing. In the field of aesthetics, for example, we cannot reduce knowledge to the subjective response of the knower in a manner that ignores the ontological reality of the object of aesthetic awareness. Instead, we must learn 'to see the object represented in terms of the truth that the representation reveals ... we learn to see the sea as a Turner seascape teaches us to see it just as Rembrandt's work teaches us the depths of character a human face can reveal' (Warnke 1987, p. 59). A 'painting illuminates aspects of its subjects that were not previously apparent but that, once revealed, can be seen to divulge something essential about those subjects' (p. 59). Similarly, in our knowledge relationships with other persons we 'come to know both others and ourselves as we enable ourselves to be known, as we reveal ourselves and are granted revelation in return' (Gunton 1999, p. 24).

Modernity views reasoning as the autonomous exercise of the mind: *Sapere aude*! Have the courage to trust your own reason! However, if the exercise of rational autonomy fails to recognise and respect the ontological reality of the object of knowledge by attending to its self-revelation, but instead seeks to make the object conform to the presuppositions and prejudices of the mind, then the knowledge relationship will be fundamentally distorted. Critical realism's commitment to the primacy of ontology over epistemology requires the knower to strive to recognise the object of knowledge for *what it is* and *what it reveals itself to be*, and thereby to bring the mind into conformity with it. The proper exercise of reason is thus predicated on the subservience of the mind to the object of knowledge, rather than on the exercise of a free-floating autonomous rationality dislocated from the object. The basic form of the epistemic fallacy is thus to ignore or bypass revelation. Even Kant, despite his commitment to autonomous rationality and insistence that reality must be brought into conformity with the cognitive structures of the mind, accepted that reason 'must approach nature in order to be taught by it' (p. 25). However, he insisted that it must do so not as 'a pupil who listens to everything that the teacher chooses to say', but as an 'appointed judge, who compels the witnesses to answer questions which he himself has formulated' (p. 25). The crucial issue here is the extent to which the inquisitor's questions serve *either* to impose a set of presuppositions onto reality *or* to enable reality to reveal itself. Hans-Georg Gadamer has shown that realistic knowledge emerges when investigators allow questions formulated on the basis of their prior assumptions to be reformulated in the light of the nature of the object under investigation (Gadamer 1979). As the mind is brought into conformity with the object, so misapprehensions are stripped away and the true nature of the object is revealed.

If all knowledge is dependent upon both (i) an object's self-revelation and (ii) our recognition of such revelation, it follows that, insofar as the latter is dependent upon the exercise of forms of judgemental rationality that are necessarily contingent and open to contestation, all knowledge assumes the structure of faith seeking understanding. It follows that the basic epistemic distinction is not between reason and revelation, nor between reason and faith, but rather between the truth/falsehood of alethic truth claims, and hence between the relative levels of harmony and dissonance in the knowledge relationship established between the knower and the object of knowledge. With these considerations in mind, we will now revisit the four objections to revelation outlined in the previous section.

(1) Historical contingency. Is the claim that Jesus Christ uniquely reveals the Triune God inherently irrational because the contingent truths of history can never constitute or embody eternal truth? Critical realism recognises that rationally warranted knowledge is dependent upon the establishment of viable causal explanations that are not bound to any *a priori* idealistic principles and which are able to penetrate beneath *a posteriori* surface descriptions of empirical data. Since we interrogate an object in order to enable its true nature to be revealed, and since the true nature of an object cannot be known in advance of the establishment of retroductive causal explanations, it follows that all *a priori* assumptions held by the questioner must themselves be open to question. Any failure to do so runs the risk of committing the epistemic fallacy by forcing the ontological reality of the object to conform to the questioner's *a priori* epistemic convictions. This is the chief failure of idealism, clearly visible in Hegel's forcing of reality into the procrustean framework of his dialectical logic. Similarly, empirical descriptions of the regular convergence of objects and events fall short of retroductive causal explanation, and consequently risk slipping into the epistemic fallacy by reducing reality to the level of surface appearances. This is the chief failure of empiricism, clearly visible in Hume's refusal to acknowledge causality on the grounds that, since we can observe only regular conjunctions of objects and events, 'causality' must be reduced to a human construction devoid of any purchase on reality. With regard to natural science, we can view astrology as a form of naive idealism, in which astrologers impose a set of *a priori* assumptions onto their observation of the movement of stars and planets. Similarly, we can view alchemy as a form of naïve empiricism, in which the recognition that the conjunction of different elements in the natural world can bring about changes in their structure (e.g. heat turns water into steam) leads to the search for conjunctions that will turn base metal into gold. The failure of both

astrology and alchemy is to seek to understand, respectively, the *causes* of the movement of stars and planets and the *reasons* why the conjunction of particular elements brings about changes in their structure. In sharp contrast, modern natural science has learned to *explain* the natural world by constructing theoretical models designed to reveal the underlying causal mechanism that configures particular objects and events, rather than relying on either idealistic principles or surface descriptions of empirical data.

The claim that immanent historical events cannot have any unique transcendental significance is predicated upon twin presuppositions: the idealist assumption that the immanent world is no more than a pale reflection of transcendent reality, and the empiricist assumption that history is no more than a contingent configuration of regular conjunctions of events. In rejecting these presuppositions, Christianity offers a retroductive causal explanation as to why there is rational warrant for asserting the eternal significance of particular contingent events, grounded in the *a posteriori* critical judgement that Jesus Christ uniquely reveals and embodies the true nature of the Triune God. Far from emanating from some impersonal absolute reality, or being the product of a demi-god shaping pre-existing matter using the Platonic Forms as a blueprint, the retroductive model generated by Trinitarian theologians infers that the created order is the result of an intentional act of free will on the part of the Holy Trinity unconstrained by any pre-existing matter or Platonic blueprint. As Irenaeus puts it, acting within the unity of the reciprocal loving relations of the persons of the Trinity, God the Father creates through his 'two hands': God the Son and God the Holy Spirit (Irenaeus 2004, pp. 487ff.). Because creation is ontologically grounded in the freely given grace and love of the Triune God, it is inherently real, good, meaningful and purposive. God intentionally creates an environment in which divine love can extend beyond the Godhead, and intentionally creates human beings in the divine image; that is to say, as persons able to enter into reciprocal loving relationships, both with one another and with their Creator. Hence the incarnation, the hypostatic union of God and humanity in Jesus Christ, constitutes the fulfilment of the divine purpose for creation.

> In the beginning was the Word [Greek λόγος, wisdom, reason, speech], and the Word was with God and the Word was God.... All things came into being through him.... And the Word became flesh and lived amongst us.... From his fullness we have all received, grace upon grace.... No one has ever seen God. It is God the only Son, who is close to the Father's heart, who has made him know.
>
> (John 1:1ff.)

Christianity thus offers a causal explanation as to why the immanent historical event of the life of Jesus Christ possesses unique transcendental significance, grounded not in any idealistic principle, nor in mere surface descriptions of Jesus' life, but in retroductive attempts to account for and explain his theological significance. This does not of course secure the truth of the Christian explanation of ultimate reality; it does, however, suggest that, given the insights of critical realism, it cannot be summarily dismissed as irrational. On the contrary, in rejecting the imposition of idealist and empiricist presuppositions it exposes the poverty of the assumption that contingent truths of history can never constitute or embody eternal truth. It is thus insufficient for idealist philosophers and empirical historians to reject Christian truth claims on the grounds of their inherently irrationality; instead, if they wish to debunk the Christian account of ultimate reality they must do so by generating alternative retroductive explanations of the ultimate order-of-things and demonstrate their greater comprehensiveness and superior explanatory power. However, our concern here is not with the respective virtues of Trinitarian, idealist and naturalist accounts of ultimate reality, but with the poverty of the claim that contingent historical events can never have eternal universal significance. Because this claim is self-evident only within a particular set of idealist and empiricist epistemic convictions that critical realism has subjected to fundamental criticism, it cannot as its stands be used, either to adjudicate between contested standards of rationality, or to dismiss Christian claims as irrational.

(2) Divine agency. Is it irrational to claim that divine agency can override the structural laws of nature and society? Bearing in mind the need to answer this question on the basis of explanatory causal theories, as outlined above, we can deal with this question in a more perfunctory manner. The crucial issue here is that of the ultimate ontological ground of reality. (i) Naturalism claims that the ontological foundation of reality is constituted by the impersonal structures of the natural order-of-things. The laws of physics are viewed as ontologically basic, so that the natural order-of-things constitutes the self-generating, self-sustaining and self-rationalising bedrock of reality that is neither dependent upon, nor reducible to, any more basic or fundamental reality. On this reading, even if higher strata of reality are irreducible to the basic laws of physics, they are nevertheless ultimately dependent upon them. (ii) Platonism identifies the ultimate ontological foundations of reality with the impersonal structures of the Platonic Forms: even if the demiurge responsible for the ordering of the world is seen as a quasi-personal agent, he is entirely dependent on the impersonal Forms, which he uses as a blueprint when imposing order upon pre-existent matter. (iii) Aristotelianism identifies the Unmoved Mover as the impersonal first cause of the web of causality that structures and organises the created order. (iv) Roy Bhaskar's ground-state is dispositionally personal, in that it possesses the potential to emerge as conscious being: this raises the question as to whether the *potentiality* for future personhood equates with the *actuality* of personhood in its potential state, and therefore whether the ontological ground of reality is an impersonal structure or a personal agent. Arguably all four, and certainly the first three, of these explanatory accounts of ultimate reality assert the ontological priority of impersonal structure over personal agency. In sharp contrast, the Triune God of Christianity is essentially personal. As we saw in the previous chapter, the Cappadocian Fathers resisted any notion that the persons of the Trinity were dependent for their existence upon any underlying impersonal substance, and insisted instead

that they owe their being and identity purely and exclusively to the reciprocal interpersonal bonds of love that constitute the ultimate ontological ground of the reality of the Triune God, and hence by extension of all created reality. In the incarnation the hypostatic union of divine and human agency established an indivisible bond between the creator and the created order that is intrinsically personal. Thus, according to Trinitarian Christianity the ontological basis of all reality is grounded in personal agency rather than in impersonal structure. Given the reality of epistemic relativism, the claim that personal divine agency cannot override the structural laws of nature and society is not self-evident; rather, it is an inference drawn from those explanatory causal theories that affirm the ontological priority of structure over agency. It follows that to affirm the intrinsic irrationality of any explanatory model that affirms the ontological priority of agency over structure is to do little more than assert (without reason) a preference for the ontological priority of structure over agency, and exclude (without reason) the voices of those who offer a (reasoned) defence of the ontological priority of agency over structure. On this reading there is nothing *inherently irrational* in claiming that divine agency takes precedence over the structural laws of nature and society. If this is the case, then it is inappropriate to speak of God 'overriding' or 'violating' the laws of nature, since to do so assumes the normativity and ontological priority of natural structures over divine agency.

(3) Mediated knowledge. Does the immediate apprehension of an object of knowledge enjoy epistemic privilege over mediate apprehension dependent upon testimonial evidence? The suggestion that epistemic immediacy enjoys epistemic privilege is rooted in a cluster of assumptions generated by the Enlightenment: the Cartesian attempt to dislocate the self from tradition via a hermeneutic of suspicion; the desire to establish a neutral observational space, or god's-eye 'view from nowhere', devoid of all subjective contamination; the attempt to bypass epistemic relativism by distinguishing between certain knowledge and uncertain opinion and belief; the search for indubitable epistemic foundations; the affirmation of an anthropology predicated on the notion of autonomous free-thinking individuals; the assertion of the innate possession of universal, culture-transcendent, reason. All of these constitute unobtainable ideals: it is impossible to dislocate ourselves from our received cultural traditions, and there can be no thoroughgoing hermeneutic of suspicion without a radical inconsistency between speech and action (empiricists who deny causality do not, as Bhaskar is fond of pointing out, generally leave buildings via third-floor windows); there is no neutral observational space, only the modernist claim to possess privileged access to a such a space, which in reality constitutes little more than a hegemonic assertion of cultural superiority; there is no possibility of bypassing epistemic relativism, so that all knowledge takes the form of more-or-less justifiable belief; there are no secure epistemic foundations; human beings are relational creatures, and rationality is not the primary mark of human identity (those with learning difficulties are not lesser human beings, and few university professors can claim to represent humanity at its idealised best); and it is far from clear that we possess an innate universal rationality independent of our culturally bound interaction with the world. At the empirical level we do not immediately

apprehend clear and distinct ideas generated by isolated packets of sense data; on the contrary, all empirical observation is mediated by received culture and is necessarily selective. Contra Locke, when we enter a room, we do not focus on atomistic pockets of sense data (clear and simple ideas) that we then construct into composite wholes (less clear and more complex ideas); rather we perceive the reality of holistic objects (cats, televisions, etc.) in a manner that is informed and mediated by our received cultural traditions. In identifying such objects we are necessarily selective in ways mediated by our established socio-cultural roles and expectations (a father entering a crèche may focus on his daughter, a school inspector may focus on the learning environment, a caretaker may focus on a leaking roof), since we cannot possibly focus indiscriminately on every object. Even the idealistic apprehension of clear and distinct logical principles is contingent, given the existence of a diversity of contrasting logical systems, all of which are probably the result of contingent *a posteriori* inference rather than secure *a priori* intuition. Further, there can be no epistemic certainty in the romantic possession of clear and distinct feelings and emotive responses to the world: the intensity of moral, aesthetic or spiritual sensibility cannot possibly be criteria for truth, unless we want to affirm the truth of racism, paedophilia and certain forms of malfunctioning religiosity. Critical realism's distinction between the domains of the empirical, the actual and the real rules out any appeal to the immediacy of directly experienced apprehension: instead, the individual's epistemic experiences must be located in the realm of the epistemic experiences of others, and accounted for at the level of explanatory theory. It is an unavoidable phenomenological fact that the vast majority of our knowledge of the world is mediated by third parties: my knowledge of the British monarchy, for example, is entirely mediated by third-party testimony since I have never – to the best of my knowledge – had a face-to face encounter with a member of the royal family. Thomas Kuhn reminds us that we learn science not by conducting experiments and theorising, as it were, 'from scratch', but by immersing ourselves in the scientific community and consulting scientific textbooks that mediate to us the rich history and tradition of scientific experimentation and theorisation (Kuhn 1970). We learn not by abstracting ourselves from learning communities in a futile attempt to establish unmediated contact with the object of knowledge, but by submerging ourselves in them in order to draw from the reservoir of received mediated knowledge. There is, then, no genuine alternative to an 'epistemology of mediation': the vast majority of our knowledge is dependent upon, and grounded, in mediated third-party testimonial evidence; and, that being the case, we cannot legitimately attribute epistemic privilege to the immediacy of direct apprehension.

(4) Testimonial authority. Is third-party epistemic testimony essentially authoritarian and hegemonic? This, broadly speaking, is a central claim of the postmodern tradition, which views all knowledge claims as brokered moves in an unavoidable power game. The claim is predicated on the assumption that it is impossible to make informed rational judgements between competing truth claims. Postmodernity arose in response to the totalitarian impulse of modernity, whose claim to possess secure knowledge effectively bypassed epistemic relativism. Critical realism steers a course between and beyond the polarity of

modernity and postmodernity. The equation of knowledge and power is dissipated once we recognise the contingent nature of all truth claims and affirm the possibility of judgemental rationality. This sheds light on the relatively common assumption that religious truth claims were intrinsically authoritarian: from the perspective of modernity, religious claims are intrinsically hegemonic because they lack any rational foundation (based on criteria predetermined by modernity itself) and therefore can only be asserted on authoritarian grounds. From the postmodern perspective, religious claims are intrinsically hegemonic simply because they are truth claims, rather than mere expressions of spiritual taste. Defenders of critical realism in the face of empiricist, idealist and postmodern opponents are only authoritarian if they seek to assert the truth of critical realism in a coercive manner without recourse to reasoned argument. Similarly, Christian theologians are only authoritarian if they seek to assert the truth of Christianity in a coercive manner without recourse to reasoned argument. If appeals to divine revelation are not intrinsically irrational, then they cannot be intrinsically authoritarian, though like all truth claims – however truth or false, and rationally or irrationally conceived – they can, but do not need to be, promulgated in an authoritarian manner. Critically realist epistemology recognises only one necessarily authoritarian factor, namely the intrinsic authority of the ontological reality of the object of knowledge itself. To reject or ignore the authority of the ontological object is to shape it into what it is not, and thereby to slip into the epistemic fallacy of forcing it onto the procrustean bed of our particular desires, prejudices and expectations.

Because our immediate epistemic encounters with ontological reality are severely limited, the vast majority of our knowledge of reality is mediated through third-person testimony. The modernist assumption that immediate encounters possess greater epistemic validity than mediated encounters is fundamentally misplaced: my immediate encounters with the natural order-of-things generate only superficial surface understanding, and any deeper and more substantial understanding is dependent upon knowledge mediated by natural scientists. Further, both immediate and mediated encounters require the scrutiny of judgemental rationality, since neither can avoid epistemic relativism. Critical realism thus supports and requires an epistemology of testimony (Lackey and Sosa 2006). The fact that Christianity is grounded in a retroductive account of God's self-revelation in the person of Jesus of Nazareth mediated through the testimony of the Christian Church does not warrant the conclusions, either that it is necessarily inherently irrational or that it necessarily possesses less epistemic warrant than claims to direct unmediated experience of God.

Revelation and Scripture

The rise of modern biblical criticism demonstrated that the Bible was the product of human authors bound to their own cultural contexts. This led to the common perception that the nature and status of the Bible as a vehicle of divine revelation constitutes an epistemic problem for Christianity. Though this is certainly the case, the exact nature of the problem needs to be carefully unpacked. Though Trinitarian Christians claim that the Bible is the inspired Word of God, they also affirm that it is simultaneously the product of human authors writing in culturally bound contexts. Without seeking to diminish the challenges posed by modern biblical criticism to classical Christianity, the weight of the epistemic problem lies in one particular response to this challenge: namely the rise of biblical fundamentalism. Christian fundamentalism is largely a modern phenomenon: faced with the reduction of the Bible to the status of one among many ancient religious documents, Christian fundamentalism sought to defend the status of the Bible as the inspired Word of God by appealing to a set of distinctively modern criteria. Specifically, it affirmed the Cartesian criterion of epistemic certainty and the positivistic criterion of a direct correspondence between language and empirical fact in presenting the Bible as an inerrant record of historical events. This was reinforced by a doctrine of plenary inspiration, according to which every word of the Bible was literally dictated to the biblical authors by the Holy Spirit. The impossibility of any direct empirical verification of these claims led to the authoritarian assertion of its truth, verifiable only indirectly via the affirmation of personal conviction. Christian fundamentalism seeks to affirm the infallibility and inerrancy of Christian Scripture.

> In its extreme form the dogma of the Infallibility of Scripture should mean that all parts of the Canon are directly and equally inspired by God, so that its every statement, whether concerning the mysteries of the divine Being, the processes of nature, or the facts of history, past or future, should be exactly and literally true.
>
> (Dodd 1978, pp. 20ff.)

The tendency of the liberal media to present Christian fundamentalism as prototypically representative of classical Christianity, rather than as a modern heretical aberration, led to a common assumption that Christianity is committed to an authoritarian, and fundamentally irrational, affirmation of the literal and inerrant truth of the Bible.

Investigation of the methods of scriptural exegesis employed in the early Church, including both the readings of Jewish Scripture embedded within the New Testament texts themselves and readings of the entire Christian Bible conducted subsequent to the formation of the New Testament canon, reveals the use of hermeneutical procedures that parallel contemporaneous Jewish and Hellenistic methods of exegesis, such as the differentiation between metaphorical, allegorical, typological and spiritual levels of meaning (Longenecker 1975). In the case of some New Testament writers, such as the author to the Epistle to the Hebrews and Saint Paul, the level of hermeneutical sophistication was remarkable when measured against other contemporary exegetes (Hughes 1979; Watson 2004). Exegetical methods continued to be refined and developed throughout the history of the Church (Kümmel 1973). Unlike modern biblical criticism, which tended to focus on the reconstruction of the historical origins of biblical texts, the primary hermeneutical concern of ancient and medieval exegetes was to address ontological issues regarding God's nature and actions in the world (Neill 1966).

Classical Christianity never questioned the human origins of the biblical texts, and had no notion of literal truth or historical inerrancy in the modern positivistic sense. At the same time, it insisted on affirming the divine inspiration of Scripture. In ancient Roman culture the Latin 'inspirare' (literally 'to inhale deeply') was used to refer to the process of stimulating persons by instilling a thought or idea in their minds. This understanding of inspiration has been carried over into contemporary culture, so that we might refer to a lecture as 'inspiring' because it breathes new insights into our minds and stimulates a response from us, or to an artist as being 'inspired' by a creative vision over which he does not have complete control. None of this subtracts from the fact that moments of intellectual and artistic inspiration are normally the fruits of hard graft grounded in an ongoing struggle, often through a process of trial and error, to master a particular medium and realise a particular vision. When Christian theologians speak of Scripture as 'inspired', they are generally referring to something akin to this. The biblical writers are inspired and driven to write by something 'beyond themselves', which either they or their readers identify as God's Holy Spirit. Though such inspiration may take the form of a specific 'peak experience' ('that particular lecture/act of worship was particularly inspiring'), more often than not it may simply be a consequence of their immersion within a particular community ('it is inspiring being part of the international community of critical realists/the catholic Christian church'). Such inspiration generates an urge to write: the Gospel writers recount the life of Jesus Christ within a theological framework; Paul writes letters to his fellow Christians in order to deepen their commitment to and understanding of the God revealed in the person of Jesus; the author of the Book of Revelation draws on the genre of apocalyptic literature in order to place the contemporary persecution of Christians in a cosmic perspective.

Christianity thus understands the Bible as a divinely inspired human construct. This leads it to oppose forms of 'biblicism' that view Christian Scripture as a source of unmediated access to God. In Christianity, mediation between God and creation takes place not within the biblical text, but through the ontological actuality of the hypostatic union of God and humankind in Jesus Christ. The Bible is a sacred book only insofar as it points beyond itself in proclaiming, testifying and mediating knowledge of God's concrete historical actions in the divine economy of creation and reconciliation. As such, the Bible plays a fundamentally different role in Christianity to that played by the Qur'an in Islam. The Qur'an constitutes God's full revelation to humankind mediated through the last of his chosen prophets, a replication in space and time of an eternal and uncreated original. The text of the Qur'an is recited to the Prophet Muhammad by the Angel Gabriel, and then recited by Muhammad to scribes who write it down. Thus Muhammad is not in any sense the author of the Qur'an but rather a passive conduit of God's message to humanity, a belief reinforced by the tradition of Muhammad's illiteracy. At the risk of oversimplification, the Qur'an plays a similar role in Islam to that of the incarnation in Christianity. The controversy surrounding Salman Rushdie's *Satanic Verses* was rooted in perceptions that the novel challenges the status of the Qur'an as unmediated revelation. Rushdie invokes the ancient tradition that Muhammad once dictated verses which permitted intercessory prayers to pre-Islamic Meccan goddesses, and latter withdrew them on the grounds that they were sent by the devil rather than by God. Rushdie also invokes the fictional figure of Salman the Persian, a scribe who discovers that Muhammad fails to notice changes he makes to the text of the Qur'an when transcribing the Prophet's dictation. Both invocations carry the implication that the text of the Qur'an may contain human as well as divine expressions – a position that was anathema to Muslims. The crucial point, in the present context, is that Christians recognise that the Bible, though divinely inspired, is a human product that offers indirect knowledge of God mediated through human agents, rather than direct unmediated knowledge flowing directly from God himself.

Explicit biblical teaching about the nature and status of Scripture is relatively sparse; the most detailed account is limited to the following:

> All scripture is inspired by God and is useful for teaching, for reproof, for correction, and for training in righteousness, so that everyone who belongs to God may be proficient, equipped for every good work.
>
> (2 Timothy 3:16f.)

Thus Scripture has a dual function: to reveal true teaching and correct false knowledge, and to teach and inspire Christians in the practice of their faith. In its revelatory role, the Bible makes known that which would not otherwise have been known: specifically the early Church's emergent retroductive account of the divine economy of creation and reconciliation, culminating in the hypostatic union of God and humankind in the historical person of Jesus. In appropriating this knowledge humanity, inspired by the Holy Spirit, must learn to read the Bible text as what it claims to be: a *theological* text. The unique understanding of the being and actions of God presented in Christian Scripture is not found in any other ancient philosophical or religious tradition. Insofar as the Bible makes available knowledge of God not available through any other channel, and insofar as Christians have good reasons for asserting the truth of this knowledge, then Christianity affirms the revelatory function of Scripture. It is important to recognise that this claim is not an *a priori* claim that functions to ground and justify Christian doctrine, but rather an *a posteriori* claim that seeks to make sense of Scripture in the light of its revelatory function. In its pedagogical role, the Bible transcends the mere impartation of knowledge, insofar as it testifies not to some abstract philosophical knowledge, but to salvific knowledge of existential import for all of humanity. As such, the Bible incorporates locutionary speech acts that identify actual states of affairs in the world, illocutionary speech acts that invite readers to participate for themselves in the emergent divine economy of salvation, and perlocutionary speech acts that seek to encourage or provoke readers' responses. The inspiration that empowered the biblical authors is the same inspiration that enables readers to respond to the Bible's truth claims by entering into a reciprocal relationship with God. Knowledge of God, that is to say, is inextricably bound up with knowledge of ourselves as relational beings.

> The unique character and authority of scripture as *revelation* is that it claims to be more than the provider of unique information, but also to be the bearer of saving knowledge, a vehicle of the word that 'is sharper than any two-edged sword, piercing to the division of soul and spirit, of joints and marrow, and discerning the intentions of the heart'.
>
> (Gunton 1995, p. 73, quoting Hebrews 4.12; emphasis in original)

Again, the Christian affirmation of the authority of the Bible as a vehicle of salvation is not imposed upon the text, but is rather inferred from the text in the light of the collective Christian experience of its intrinsic explanatory power. A student may embrace an 'unofficial' canon of established texts related of the philosophy of critical realism provided by her tutor as authoritative, but they can only become authoritative for her if her reading leads her to recognise and embrace the intrinsic authority of the texts themselves and thereby no longer rely on the authority of her tutor. Christians affirm the authority of the biblical text simply because it makes better sense of their lives and the world they indwell than any other text available to them. Here Gunton quotes Samuel Taylor Coleridge with approval:

> And need I say that I have *met* everywhere [in the Bible] more or less copious sources of truth, and power, and purifying impulses; – that I have *found* words for my inmost thoughts, songs for my joy, utterances for my hidden griefs…? In short whatever *finds* me, *bears witness for itself* that it has proceeded from a Holy Spirit.
>
> (p. 65; emphases added)

The Old Testament as mediated revelation

Christian Scripture reveals a personal God, rather than the impersonal or demi-personal deity of classical theism. For many this constitutes a stumbling-block. Thus Einstein rejects any notion of a personal God 'based on the authority of the Church' (Jammer 1999, p. 122). Belief in such a God is the product of human projection driven by fear of both life and death, demands a blind leap of faith, and generates mass indoctrination. 'I do not' he writes, 'believe in the God of theology who rewards good and punishes evil. My God created laws that take care of that. His universe is not ruled by wishful thinking, but by immutable laws' (p. 123). This charge of projectionism may be applied equally to Einstein's God: if the Judaeo-Christian God can be read as a projection of humanity writ large, so the God of classical theism can be read as a projection of the natural order-of-things write large. The expression of distaste directed towards the notion of a personal God appears to be closely linked to its implications for human freedom and dignity. According to Richard Dawkins, what makes his 'jaw drop is that people today should base their lives on such an appalling role model as Yahweh – and even worse, that they should bossily try to force the same evil monster (whether fact or fiction) on the rest of us' (Dawkins 2006, p. 248). Daniel Dennett expresses similar bemusement:

> [The] Old Testament Jehovah ... could take sides in battles, and be both jealous and wrathful.... Part of what makes Jehovah such a fascinating participant in stories of the Old Testament is His kinglike jealousy and pride, and His great appetite for praise and sacrifices.
>
> (Dennett 2006, pp. 206, 265)

Christian theologians tend to read such objections to a personal God as symptomatic of a culture committed to personal autonomy and the defence of the Protagorean claim that humankind must be the measure of all things. To many biblical theologians the interventions of Dawkins and Dennett in the field of Old Testament theology appear naïvely literalistic and positivistic, blind to the subtlety, artistry and complexity of the biblical text.

Walter Brueggemann suggests that close reading of the biblical text reveals a significantly different understanding of the nature of God. His *Theology of the Old Testament* seeks to combine fidelity to the biblical text with a concern to unpack its spiritual implications for contemporary Christianity (Brueggemann 1997). The '*Old Testament in its theological articulation is characteristically dialectical and dialogical, not transcendentalist*'; as such, it offers a mediated narrative of Israel's historical encounter with God, rather than direct unmediated access to the mind and will of God (p. 83; emphasis in original). The nature of the various texts collected in the Old Testament canon is such that they must be read intertextually and polyphonically, with due regard both to the differences and similarities between texts and to the various voices present within individual texts. Brueggemann employs as his hermeneutical key the metaphor of a courtroom trial, in which at various points within the texts both God and humanity are placed on trial and subjected to interrogation.

Brueggemann begins by outlining Israel's *core testimony* about God, with the caveat that, because human testimony is necessarily pluralistic and epistemically relative, it is not amenable to close systematisation. Hebraic Scripture testifies to an encounter with a creator God who makes a covenantal promise to Israel, issues commands regarding her moral, social and cultic relationships, delivers her from slavery and oppression, and leads her into an unknown future. Having established a covenant with his chosen people, God remains steadfast, merciful and faithful in his commitment to them: governing them as judge, king, warrior and father, and sustaining them as one who nurtures, evokes, values and enhances life. God, as testified to by Israel, is simultaneously sovereign and faithful: as sovereign he insists upon remaining God, and is properly jealous of any challenge to his glory and holiness; as faithful he clings in 'steadfast love' and solidarity to 'an enduring engagement with and involvement for Israel' (p. 297). The biblical text recognises that such a reciprocal covenantal relationship is not easy, since God's faithful solidarity cannot remain authentic if it ignores Israel's moral and spiritual failings as they challenge divine sovereignty:

> Yahweh is a Character and Agent who is evidenced in the life of Israel as an Actor marked by unlimited sovereignty and risky solidarity, in whom this sovereignty and solidarity often converge, but for whom, on occasion, sovereignty and solidarity are shown to be in an unsettled tension or in an acute imbalance.
>
> (Brueggemann 1997, p. 268)

Brueggemann then turns to Israel's *counter-testimony*, in which she subjects her core testimony of God's simultaneous sovereignty and faithfulness to cross-examination and testing. Crucially, this counter-testimony is not extraneous to the biblical text but rather embedded within it, so that the testing of God and of the Israelites' understanding of God 'constitutes a major dimension of Israel's life of faith' (p. 319). Typically, such counter-testimony takes the form of Israel's expression of grievance towards God in the face of perceived injustice, of which the injustice of exile (both in Egypt and Babylon) is the paradigmatic example. Brueggemann identifies three key features in this counter-testimony: God's apparent (1) hiddenness, (2) ambiguity and (3) negativity. (1) Out of her experience of the hiddenness and absence of God, Israel turns to the tradition of Wisdom Literature in an attempt to discern moral order and aesthetic meaning within creation itself, so that God becomes, indirectly, 'the hidden guarantor of an order that makes life in the world possible' (p. 336). (2) Out of her experience of the ambiguity and instability of God, Israel voices the complaint that, in his apparent willingness to deceive, entice and allure, God sometimes appears as 'an unprincipled bully who will coerce, manipulate, and exploit in order to have Yahweh's own way' (p. 362). (3) Out of her experience of the negativity and injustice of God, especially of his apparent willingness to allow the righteous to suffer, to curse human beings and to pursue violence, Israel looks beyond an unconvincing theodicy that makes human beings responsible for their anguish and instead, in a litany of complaint, accuses God of failing to keep his promises and thereby calls him to account. If the core testimony of Israel affirms Israel's experience of God's sovereignty and fidelity, her counter-testimony questions the truth of this core testimony in the face of her experience of God's hiddenness, ambiguity and negativity. Ironically, the very complaints and accusations directed against Israel's God by Dawkins and Dennett are already present in the Old Testament texts themselves. The outcome of Israel's counter-testimony is that, depending on specific experiences and contexts, the people of Israel follow a spiritual path that oscillates between '*self-abandoning praise*' and '*self-regarding complaint*' (p. 401; emphases in original). Brueggemann holds that the tension between these two responses 'belongs to the very character and substance of Old Testament faith, a tension that precludes and resists resolution' (p. 400).

Brueggemann then turns to Israel's *unsolicited testimony*. Just as, in a court of law, a witness may offer testimony not directly elicited by defence and prosecution counsels, so Israel attempts to speak of God beyond the boundaries imposed by her core testimony and counter-testimony. Israel does so because she delights in speaking of her God, wants to understand her God better, and desires to penetrate more deeply into her God's divine mystery. She is justified in this because she is aware that, since her relationship with God is reciprocal, God grants her the freedom to do so: '*Yahweh is committed to Yahweh's partners in freedom and in passion*' (p. 410; emphasis in original). This unsolicited testimony extends beyond Israel's covenantal relationship with God and concerns itself with the gentile nations and with creation itself. Israel's unsolicited affirmation of God's concern for the whole of humanity does not generate an abstract generic definition of personhood; rather, she understands humanity in the light of her established covenantal partnership with God, as that an authentic human being is one who is '*in relation to Yahweh*' and 'lives in an intense mutuality with Yahweh' (p. 453; emphasis in original). Israel understands humanity in terms of the dual relationships of divine sovereignty and human obedience, and divine fidelity and human freedom. The tension between these two polarities means that humanity 'is invited, expected, and insistently urged to engage in a genuine interaction that is variously self-asserting and self-abandoning, yielding and initiative-taking' (p. 458). This results in a spiritual '*dialectic of assertion and abandonment in the human person*', which operates as '*a counterpart to the unsettled interiority of Yahweh's sovereignty and fidelity*' (p. 459; emphases in original). This dialectic assumes a normative pattern: (1) God calls persons to hear, obey, discern and trust him; (2) faced with God's hiddenness, ambiguity and negativity, they act freely and responsibly, through complaint and petition, to demand God's fidelity; (3) as their experience of God's hiddenness, ambiguity and negativity gives way to a renewed understanding of God's sovereignty and fidelity, they turn to thanksgiving, praise and self-abandonment. This drama of rehabilitation is driven by hope for 'full knowledge of Yahweh, full communion with Yahweh … full enjoyment of an abundant earth … [and] full confidence in Yahweh at death' (p. 484). Israel predicates her ongoing spiritual pilgrimage on the mediated presence of God embodied in Israel's religious institutions and traditions: Torah, kingship, prophecy, cult and wisdom. These modes of mediation are simultaneously divine gifts designed to make God accessible to Israel, and human enterprises that are in constant danger of corruption.

Brueggemann suggests that the Old Testament embodies a dynamic dialectical theology in which human beings engage in a reciprocal relationship with God (ontological realism) in a manner that acknowledges human fallibility (epistemic relativism) and demands spiritual discernment (judgemental rationality). He wonders whether, when read in the light of the New Testament and Christian doctrine, in a manner that remains respectful of Judaism and rejects any notion of Christian supersessionism, Old Testament theology might offer a viable challenge to the hegemony of the 'metanarrative of military consumerism' that dominates contemporary Western civilisation (pp. 718ff.). Military consumerism grounds reality in the freedom of autonomous individuals to seek their own well-being through an economy of getting and using through force and violence, regardless of the fact that this results in the maintenance of a disproportion of wealth, power and access. Israel's testimony invites the court of humanity to attend to an alternative meta-narrative, one informed by undomesticated holiness, originary generosity, indefatigable possibility, open-ended interaction and genuine neighbourliness.

> Israel's testimony yields a world as deeply opposed to military consumerism as it is to any other alternative metanarrative that lacks the markings of the central Character.... Only in the presence of the richer, more dense metanarrative of Yahwism can the inadequacy of the dominant metanarrative be observed. And where the metanarrative of Yahwism is not fully and courageously voiced, the dominant metanarrative appears as the only available one.
>
> (Brueggemann 1997, p. 720)

11 Theological science

For Christians, the Bible is the divinely inspired human product of the early Church's attempts to make theological sense of divine action, and human responses to divine action, culminating in the person and work of Jesus Christ. As such, it functions as a mediatory witness to historical events whose authority is not imposed from outside, but is intrinsic to the biblical texts themselves: Christians experience, understand and judge these particular accounts of Jesus as more powerful, comprehensive and truthful than any other account available to them. From the start, Christian theologians recognised that apprehension of the alethic truth claims embodied within Scripture required a theological hermeneutic capable of penetrating beneath the surface of the biblical texts and engaging with the ontological reality of God's actions in history to which the texts bear witness. As a result, the Christian tradition developed retroductive models of its understanding of God, in the form of creedal statements which it recognised as the most powerful, comprehensive and alethically truthful explanatory accounts of the nature and actions of God currently available to it. At the same time, in recognising its creeds as the fallible products of human endeavour, it accepted the need to subject them to ongoing iterative refinement. The history of 'dogmatic' or 'systematic' theology is the history of the ongoing attempts of Christian theologians, working within the paradigmatic Trinitarian framework grounded in the Bible and articulated by the Ecumenical Councils, to deepen both its understanding of God and its understanding of the intellectual warrant for its theological assertions (Webster *et al.* 2007). In this chapter we will explore one instance of this ongoing history. As previously noted, the work of the Scottish theologian Thomas Torrance constitutes the first major exploration of the interface between Christian doctrine and critical realism. We will focus on his account of the epistemic basis and intellectual warrant of Christian truth claims rather than the material substance of his theology, concentrating especially on his magnum opus, *Theological Science*, delivered as lectures in 1959, though not published until 1969 (Torrance 1969). In comparing and contrasting the respective epistemologies of theological science and the natural sciences, Torrance sought to under-labour for the Christian community by reaffirming the inherent intelligibility of its classical Trinitarian doctrines and thereby recalling it to its proper object of worship.

Theological epistemology

Torrance's account of theological epistemology rejects both deductive and inductive reasoning. (1) Deductive arguments – such as the 'ontological argument' that God is by definition perfect, that it is more perfect to exist than not to exist, and that therefore God must exist – collapse because the coherence of linguistic statements alone does not guarantee their ontological purchase upon reality. Conceptual understanding must always give way to ontological reality: if we fail to allow this to happen, then we commit the epistemic fallacy of forcing reality into the procrustean bed of our own *a priori* assumptions and prejudices. As Einstein recognized, Newton's commitment to Aristotelian logic in general, and the notion of absolute time and space in particular, had to be abandoned in order to make way for the theory of relativity, and of relative time and space, demanded by a deepening understanding of the ontology of nature. (2) Inductive arguments – such as the 'cosmological argument' that the universe constitutes a network of causes, that the notion of an infinite regress of causes lacks explanatory power, and that therefore there must be a 'first (divine) cause' – collapse not merely because generalisations from regular occurring patterns of sense experience cannot predict irregularities, but, and more fundamentally, because surface descriptions of empirical reality cannot produce deep causal explanations as to why such patterns occur in the ways that they do. Thus, for example, the fact that we have seen only white swans does not rule out the existence of black swans: the colour of swans is best explained by reference to their DNA rather than observation of the regular appearance of swans of a particular colour.

Torrance argues that from the start we are unavoidably caught up in a web of knowledge relationships. Since we know only as *human beings*, there is a necessary personal coefficient to our knowledge of the world we indwell and in which we participate. This rules out both forms of objectivism that seek to bracket out the knowing subject and forms of subjectivism that seek to bracket out the object of knowledge. Because we participate in knowledge relationships prior to our reflective understanding of them, we always know more than we can actually tell. It follows that the necessary basis of all understanding is tacit belief grounded, not on indubitable foundations, but in intuitive apprehension. Tacit-intuitive faith in 'a transcendent reality, independent of our knowledge of it and accessible to all men, is the common determinant of the scientific enterprise' (Torrance 1984, p. 160). Where naturalists identify this transcendent reality with the natural order-of-things, theists identify it with God. In both cases the realistic pursuit of truth is ultimately dependent upon the tacit-intuitive acknowledgement of a 'concept of reality as that which has the independence and power to reveal itself in an indefinite range of unexpected manifestations' (p. 159).

The path from tacit-intuitive knowledge to explicit-reflective understanding follows the path from abductive, through retroductive to iterative reasoning. Abduction is 'fundamentally innovatory and creative, generating new ideas and insights in response to "surprising facts" ', and is normally sparked by 'a proleptic conception, an anticipatory glimpse, a tenuous and subtle outreach of the understanding with a forward thrust in cognition of something quite new' (McGrath 2009a, p. 44; Torrance 1984, p. 114). The abductive process is a common feature of the progress of natural science: Archimedes steps into his bath, an apple falls on Newton's head, and an abductive-intuitive seed is sown in the mind. The abductive process of making sense of the person of Jesus is clearly visible in the Gospels: 'They were all amazed, and they kept on asking one another, "What is this? A new teaching – with authority"!'; 'When his family heard it, they went to restrain him, for people were saying, "He has gone out of his mind"; 'But when the Pharisees heard it, they said, "It is only by Beelzebul, the ruler of the demons, that this fellow casts out the demons"' (Mark 1:27, 3:21; Matthew 12:24). One such abductive intuition gave birth to Christianity: 'He asked them, "But who do you say that I am?" Peter answered him, 'You are the Messiah"' (Mark 8:30).

Abductive intuition paves the way for retroductive reasoning, which takes the form of explanatory conceptualising, theorising, hypothesising and modelling. This is essentially a creative process that requires the proactive engagement of the mind as it strives to identify basic ontological relations and ideas.

> Once we have reached these basic relations or ideas, we connect them by theorems which enable us to deduce other ideas and relations, in an attempt to produce a theoretical and logically coherent structure through which we may seek to bring to light the inherent and coherent structure of nature.
>
> (Torrance 1984, p. 111f.)

Crucially, although the retroductive construction of explanatory models and theorems is a fallible human activity, it does not necessarily lead to forms of constructivism which reduce all such models to the status of fictions or conventions. Torrance insists that the veracity of any given construction is dependent upon our placing the 'concepts that creatively arise in our minds under the compulsion of the objective structures of nature', so that 'while basic scientific ideas and relations are intuitively reached they are objectively determined and controlled' (p. 111). Unlike deductive and inductive reasoning, retroductive reasoning from inference to the best explanation generates the concept of ontological stratification, according to which the universe comprises 'a sequence of rising levels, each higher one controlling the boundaries of the one below it and embodying thereby the joint meaning of the particulars situated on the lower level' (p. 159). The meaning of the whole is discerned in the higher strata of reality, and is 'not reducible to the laws by which the ultimate particulars of the universe are controlled' (p. 159). This means that 'as we move up the hierarchy of levels of reality, from the most tangible to the most intangible, we penetrate to things that are *increasingly real and full of meaning'* (p. 159; emphasis in original). Contra Richard Rorty, it is *not* self-deceptive to think 'that we possess a deep, hidden, metaphysically significant nature which makes us "irreducibly" different from inkwells or atoms' (Rorty 1980, p. 373).

Torrance quotes Polanyi in support of his claim that ontological stratification has fundamental theological significance: 'Such, I believe, is the true transition from science to the humanities and also from our knowing the laws of nature to our knowing the person of God' (Torrance 1984, p. 160; cf. Polanyi 1974, pp. 127ff.). Torrance invites us to read Christian doctrines as retroductive explanatory models of God's interaction with humankind. Following his lead, we can view the kerygma (Greek κήρυγμα, proclamation, preaching) of the first Christians as marking the first stages of a retroductive process that led to the production of the Ecumenical Creeds that constitute the paradigmatic explanatory framework through which Trinitarian Christians seek to iteratively explore and test their faith. Under the impact of Christ's death and resurrection, Peter's original leap of abductive imagination crystallised into a set of interlinked retroductive hypotheses regarding the person of Jesus: 'He is the reflection of God's glory and the exact imprint of God's very being'; 'In the beginning was the Word, and the Word was with God, and the Word was God . . . the Word became flesh and lived among us'; 'God was in Christ, reconciling the world to himself' (Hebrews 1:3; John 1:1, 14; 2 Corinthians 5:19). The theologians of the early Church, in the context of the encounter with Hellenistic philosophy, subjected the Christian hypothesis to ongoing iterative modification – understood as 'the successive and incremental revision of how we see or understand something in the light of insights disclosed through the process of the engagement itself' – and thereby generated the basic creedal statements of faith centred on the doctrines of the Trinity and incarnation (McGrath 2009a, p. 32). According to Charles Cochrane, such theological iteration and accompanying spiritual praxis enabled Christianity to assert its intellectual and moral superiority over a moribund Hellenistic culture (Cochrane 1944). Whether this historical judgement is true, and whether it supports claims to the veracity of Christianity, remain open questions. Our concern here is not directly with the truth of Christianity, but with the nature, status and warrant of its epistemic procedures. Torrance's argument is that, with exceptions imposed by the unique nature of the primary object of Christian knowledge, Christian theologians from the start tended to follow the abductive-retroductive-iterative process defended by critical realists, and as such, and prior to the explicit identification and articulation of their tacit critically realistic epistemic assumptions, already know more epistemically than they could actually tell.

The possibility and actuality of knowledge of God

Torrance's *Theological Science* opens with some preliminary observations about the actuality and possibility of knowledge in general, and of knowledge of God in particular (Torrance 1969, pp. 1ff.). At the general level, Torrance understands knowledge as a relationship between the knower and the object of knowledge: if either is absent then there can be no actual knowledge relationship and hence no actual knowledge. In all branches of knowledge we proceed from our actual knowledge relationships 'and seek to move forward by clarifying and testing what we already know and by seeking to deepen and enlarge its content' (p. 2). It follows that

> It is part of our situation that we are inevitably and inseparably *inside* the knowledge relation, from the start to the end, and so cannot step outside of ourselves to an indifferent standpoint from which to view and adjust the relations of thought and being.
>
> (Torrance 1969, p. 1; emphasis in original, quoting Brown 1953, p. 170)

It is only from within the context of an established knowledge relationship that we can ask how the object of knowledge can be known: 'genuine critical questions as to the *possibility* of knowledge cannot be raised *in abstracto* but only *in concreto*, not *a priori* but only *a posteriori*' (Torrance 1969, p. 1; emphases in original).

Turning to the particular question of knowledge of God, Torrance argues that:

> Christian theology arises out of the actual knowledge of God given in and with concrete happening in space and time. It is knowledge of the God who actively meets us and gives Himself to be known in Jesus Christ – in Israel, in history, on earth.
>
> (Torrance 1969, p. 26)

Operating from within the concrete *a posteriori* actuality of this knowledge relationship, Christian theologians affirm that knowledge of the Trinitarian God is possible because God elects to be known in this way.

> In Jesus Christ God has condescended to reveal Himself to us within our creaturely existence and contingency, and has assumed our humanity to meet us as man to man and to make Himself known to us within the conditions and limitations our earthly life, within our visible, tangible, temporal flesh.
>
> (p. 46)

> In Jesus Christ God has come to reconcile man to Himself so that man may be delivered from his self-enclosure and be restored to true objectivity in God and true subjectivity in himself.
>
> (p. 48)

> In Jesus Christ God has condescended not only to objectify Himself for man and to bestow His truth upon him, but also to provide from the side of man, and from within man, full, adequate and perfect reception of that truth.
>
> (p. 50)

It is, of course, possible that the knowledge relationship claimed by Christians is false, since the object of knowledge, the Trinitarian God who elects to enter into an interpersonal relationship with human beings, may not actually exist. However, those who doubt the veracity of Christian truth claims cannot legitimately demand a perspective outside of this knowledge relationship (whether putative or real) from which to judge them. Any such demand for a secure foundation of knowledge, including theological knowledge, outside of, or prior to, an actual knowledge relationship ignores the reality of epistemic relativism and bypasses the fact that knowledge necessarily proceeds from that which is already known, or is believed to be known. This does not make Christianity immune from rational critique, but it does disallow critiques based on Christian theology's perceived lack of a foundational base external to the actuality of its established (whether putative or real) knowledge relationship with God. Or, to put matters another way, Trinitarian theology is not dependent for its epistemic warrant upon the veracity and cogency of prior philosophical arguments for the existence of the God of classical Theism. Instead, the dispute between those who wish to affirm Christianity and those who wish to reject it must focus on the relative explanatory power of competing (concrete, *a posteriori*) retroductive explanatory models, rather than on the search for external (abstract, *a priori*) foundations on which to construct such models.

Because knowledge is relational, the nature and possibility of knowledge is co-determined by the nature of the knower and the nature of the object of knowledge. As previously noted, Torrance draws upon Polanyi in affirming a necessary personal coefficient in all human knowledge relationships: 'the personal factor inevitably enters into scientific knowledge for the very fact of our knowing [as persons] enters into what we know' (Torrance, 1969, p. 93; cf. Polanyi 1958). The fact that we understand the impersonal natural order as *persons* means that the knowledge relationships established by the natural sciences are never purely objective: this has profound and positive implications, above all with respect to ecological challenges arising from previous attempts to harness the powers of nature in impersonal, value-free, instrumental ways. When we enter into knowledge relationships with other persons the personal coefficient of knowledge works both ways, so that authentic interpersonal relationships are predicated upon the need to engage, in our own subjectivity, with the subjectivity of other persons. Failure to acknowledge the subjectivity of other people, and thereby to treat them merely as impersonal objects, has fundamental, and almost wholly negative, ethical consequences. Christian theology claims that the Trinitarian God, as a personal being in whose image human persons are created, chooses to meet us in his own subjectivity. Hence, to enter into a knowledge relationship with God is to enter into an intersubjective and interpersonal relationship. Whenever we enter into a genuinely interpersonal relationship, whether with another human being or with God, we risk the possibility that it may transform our lives: this is what Christians understand by 'faith' – not an epistemic last resort, in the form of a blind leap beyond reason, but the courage to risk the transformation of one's being and identity through engagement with the personhood of God.

This stress on intersubjectivity necessarily entails the objectivity of the other person in her personal subjectivity. To engage with other persons without respecting and acknowledging their distinctive objective identity is to subsume them within one's own frame of reference, and thereby ignore their personal subjectivity and reduce an 'I–Thou' relationship to an 'I–It' relationship. Authentic knowledge relationships are predicated on the primacy of the object of knowledge: the knower must relate to other persons as they actually are, and as they choose to relate to us, not as she might wish, prefer, imagine or assume them to be.

> God is *Person*, and when He objectifies Himself for our knowledge He does not cease to be *Subject*, to be Himself. He does not give Himself to us as a mere object subjected to our knowing, but as Subject who maintains Himself in implacable objectivity over against us, objecting to any attempt on our part to subject Him to our own knowing.
>
> (Torrance 1969, p. 38; emphases in original)

Torrance identifies three basic dimensions of realistic knowledge of God. (1) It is *rational*, insofar as rationality is understood as 'our ability to relate our thought and our action appropriate to objective intelligible realities' and to 'recognize and assent to what is beyond it' (pp. 11ff.). Such rationality must be distinguished from forms of neo-Kantianism in which the object of knowledge is subsumed and categorised within our own cognitive and conceptual frameworks. 'Scientific procedure will not allow us to go beyond the boundary set by the object, for that would presume that by the inherent powers of our own "autonomous reason" we can gain mastery over it' (p. 26). (2) It is *objective* insofar as, like all genuinely scientific knowledge, it is constituted at the formal level by 'a conscious relation to an object which we recognize to be distinct from ourselves but toward which we direct our thought as something intelligible and ascertainable' (p. 13). (3) It is *conceptual*, both in its acts of cognition and acts of expression:

> Right from the start it involves a structured understanding in formed acts of cognition, while the whole movement of theological thought consists in developing and clarifying the conceptual structure of this knowledge by constant reference to the object and by advancing in the cognitive modes of rationality set up between us and God as He communicates Himself to us.
>
> (Torrance 1969, pp. 13ff.)

Theological science and the development of natural science

Torrance now turns to the question of the relationship between theology and the natural sciences. Theology is not concerned with God *in abstracto*, but 'only as He has revealed Himself to man as Creator and Redeemer within the creaturely objectivities of the world' (Torrance 1969, p. 56). Because of this, theology must 'maintain faithful and responsible relations toward worldly objectivities in accordance with their creaturely nature' (p. 56). The theologian must be committed to an objective, realistic, scientific understanding of the created order, since this is the context in which God has placed human beings and elected to enter into relationship with them. In relating and responding to God, humanity cannot avoid simultaneously relating and responding to the divinely created order-of-things, including both the natural world and other persons. Despite the modern assumption of a fundamental tension between theology and natural science, 'theology can still claim to have mothered throughout long centuries the basic beliefs and impulses which have given rise especially to modern empirical science' (p. 57; cf. Foster 1973; Hooykaas 1973; Jaki 1980; Hodgson 2005). Specifically, theology is committed to the ultimate and essential intelligibility of God's creation, and its God-given freedom to function contingently apart from any divinely imposed necessity.

This is neither to deny that natural science has its origin in ancient Greece, nor that its progress has at times been improperly hindered by the Christian Church. Torrance traces such hindrance back beyond modern disputes with the likes of Galileo and Darwin to the dominance of Augustine's Christian Platonism in the early stages of Western Christendom, which perpetuated a negative attitude towards the physical world and envisaged salvation as an escape from the created order. Augustinianism effectively reduced the immanent created order to the role of providing sacramental intimations of a greater transcendent realm, rather than that of providing the material context in which God elects to dwell with, and relate to, human beings. This Augustinian–Platonic 'turn from the world' threatened the orthodox sense of the essential goodness and intrinsic intelligibility of creation. Despite this, the Christian commitment to the value and lucidity of creation was still nourished by the Church in the Middle Ages. As a consequence Thomas Aquinas, spurred by the rediscovery of Aristotle following Christian contact with Muslim scholars, was able to develop a medieval scholastic theology that recovered a sense of the order and intelligibility of nature. Aristotle's notion that the natural order is impregnated with causal mechanisms allowed Aquinas to read an eternal pattern of order off the face of nature. However, this positive move had two negative effects: first, it helped reduce the personal God of Christianity to Aristotle's impersonal 'Unmoved Mover', thereby opening the door to the sub-Christian Deism that was to dominate the philosophical theology of the Enlightenment; second, the Aristotelian interpretation of nature 'in the light of final and primary causes left little room for the element of real contingency in nature', upon which modern *a posteriori* experimental science depends (Torrance 1969, p. 60).

It was the Protestant Reformers, under the influence of both their reading of the Bible and appropriation of late medieval nominalism, who recovered a sense of the contingency of nature and integrated it with the Thomist sense of its inherent order and intelligibility. In doing so they effectively displaced 'the Stoic-Latin view of God as *deus sive natura*' in favour of 'the essentially Biblical view of God, the Creator of all things, and the active Redeemer of His people' (p. 59). They learned directly from William of Ockham, whose nominalism effectively freed particular events in the world from any notion of an eternal pattern of causal *necessity* (whether derived from Platonic Forms or Aristotelian causality), and thereby liberated God from his Deistic role as the necessary source and ground of such a closed causal nexus.

> Only when nature was liberated from medieval rationalism and disenchanted from its secret 'divinity', and only when it was realized that the order of nature, while intelligible to us theologically as a divine creation, precisely because it is a creaturely order, can only be known through observation and interpretation of the creaturely processes themselves, could the more or less static science of the ancient and medieval world give way to the great movement of modern science.
>
> (Torrance 1969, p. 61)

Modern empirical science emerged on the back of two basic assumptions: 'an ultimate orderliness behind the flux of nature', and an 'element of contingency in creaturely existence' (p. 61). If Augustine's Christian reading of Plato served to occlude the orderliness of nature, and Aquinas' Christian reading of Aristotle served to occlude the contingency of nature, then the Protestant recovery of the orthodox Christian doctrine of Creation, untainted by Platonic and Aristotelian elements, offered a model of God's creation as simultaneously ordered-yet-contingent.

Alongside the distinction between God and creation, the Reformers also asserted a distinction between grace and nature. The Deistic God of the Enlightenment was envisaged as essentially united with the natural order as its ultimate source and first cause. This in turn opened the door to a natural theology in which human beings could reason upwards from nature to God, and thereby, through the exercise of autonomous reason, engineer their own salvation. If the reduction of God to an impersonal first cause necessarily linked to the natural causal chain allowed for the quasi-divinisation of nature, it did so at the cost of reducing the divine-in-nature to the level of a mechanistic emanation from an impersonal and deterministic source; and if Deism allowed human beings to be attributed a share in the divine nature, it did so at the cost of reducing human beings to the level of neo-Platonic possessors of the divine spark of a potentially capricious deity. The Protestant principle of the absolute priority of divine grace, as revealed in the biblical witness to the life of Jesus of Nazareth, envisaged God creating, sustaining and pouring his grace out onto a world utterly distinct from its Creator. This gave creation a higher status as the object of divine love and attention, and human beings a higher status, as those – created only 'a little lower than God' and 'crowned with glory and honour' – invited into personal relationships with their creator, than that envisaged by deistic philosophy (Psalm 8:5). The distinction between grace and nature recalled theology to a proper sense of the 'otherness' of God and hence to the need for theological objectivity, and directed natural science towards its proper object and reminded it of the central importance of objectivity in its investigations of the natural order. This disallowed any theological veto over the scientific investigation of nature, since the natural order is a stratum of reality that God created to exist and function freely and independently of the divine will; and, vice versa, disallowed any natural scientific veto over the theological and/or metaphysical investigations of a transcendent or divine stratum of reality existing above and beyond the natural order-of-things. Thus both theological science and natural science share the same aims: to achieve a proper understanding of their respective objects of knowledge and strata of reality by entering into appropriate knowledge relationships with them. Such relationships must be concrete and *a posteriori* rather than abstract and *a priori*: the former allows the objective reality of the object to shape the mind of the knower, whereas the latter generates the epistemic fallacy of allowing the mind of the knower to shape the object of knowledge. The only fundamental difference between natural and theological science, albeit a significant one, is that the distinctive nature of their primary objects of knowledge, respectively the impersonal order of nature and the personal Trinitarian God, make different demands on the knower and hence require the development of different epistemic tools – tools that require progressive refinement in the light of the emergence of progressively deeper understandings of the objects they are designed to illuminate.

The nature of scientific activity

Torrance insists that any account of either theological or natural science cannot be abstracted from the actual processes of scientific enquiry in specific fields of study (Torrance 1969, pp. 106ff.). Nevertheless, it is possible to identify the basic contours of a generic scientific attitude applicable to all fields of learning and discovery. This consists of the vigorous application of our basic rationality as we seek to act towards objects in ways appropriate to their natures, the careful and systematic ordering of our thinking on the presumption of the inherent intelligibility of reality, and sensitivity and openness to the manifold and stratified order-of-things. This generic scientific attitude excludes all forms of secular 'scientism' and religious fundamentalism.

The traditional distinction between *scientia generalis* and *scientia specialis* retains a certain value. The notion of general science correctly identifies the fact that different spheres of scientific exploration share certain fundamental principles in common: a concern to discern the objective nature of particular objects of knowledge; an awareness of the nature of the knower and of modes of knowing appropriate to human nature; a commitment to formal scientific procedures; and a recognition of the nature and structure of thinking. However, general science has no independent existence apart from the concrete sciences, and as such cannot be abstracted from them and reified as an autonomous *scientia universalis*. In particular, general science must not be confused with philosophy: science is a means of discovering knowledge, while philosophy is a form of thinking that, despite its ability to help clarify our understanding of reality and of the nature of knowledge, cannot add anything positive or substantial to the raft of human knowledge. The fact that there is no *necessary* fit between the logical structures of philosophical thought and the logical structures of scientific theories is exemplified by the fact that the emergence of modern nuclear physics required the abandonment of Euclidean logic. Ultimately there are only particular sciences, which bear the responsibility of modifying general science in ways appropriate to the distinctive nature of their specific objects of knowledge. It follows that each particular science has its own particular aims, methods and modes of rationality generated *a posteriori* through concrete interaction with its allocated stratum of reality.

Theological science shares with the other sciences a common set of assumptions ordered around the fundamental principle of objectivity, as well as an obligation to adapt these assumptions in the light of the unique nature of its particular object. Christian theological science is concerned with human knowledge of God revealed in Jesus Christ, and with the explication of modes of knowing appropriate to the divine nature. It follows that 'theology must have its particular principles relevant to this unique object or field of knowledge where divine self-disclosure is involved' (p. 113). Theology remains scientific to the extent that its domain-specific procedures conform to the generic canons of general science embraced by all genuinely scientific endeavours. Torrance identifies five key markers of the scientific nature of theology.

1 Theological science is marked by a devotion to its proper object and a total reverence for objectivity.
2 Theological science is committed to the rigorous, disciplined, methodical organisation of knowledge.
3 Theological science entails a search for elemental forms through which to reduce the multiplicity of theological knowledge to a basic order in response to the claims of objective reality. The application of the principle of Ockham's razor to the construction of retroductive explanatory theories does not allow for the *identification* of theoretical constructs with ontic structures, though they must be determined by and point towards the ontic order inherent in the object of knowledge: 'it remains the *sine qua non* of scientific advance that [our theoretical constructs] penetrate through the mass of our conceptions and correlations to primary and elemental forms ... [and thereby] attain a profounder and clearer knowledge of nature' (pp. 118ff.).
4 Theological science approaches scientific inquiry via a 'form of questioning in which we allow what we already know or hold to be knowledge to be called in question by the object' (p. 120). This is not the same as Socratic questioning, which aims to clarify what we already know by sifting out false knowledge and recollecting true knowledge. Torrance holds that Socratic questioning is not scientific because its central point of reference is the mind of the knower rather than the object under investigation, whereas scientific questioning seeks to be responsive to the object of knowledge, so that the scientist is humbled before the object and allows its reality to shape her emergent questions.

> To achieve their end scientific questions must be ruthless and unrelenting, probing down into the deepest depths of our knowing in order to uncover and cut away all that hinders us from behaving in terms of the nature of the object, and in order to allow ourselves to be 'told' by the object what we cannot tell ourselves about it, and so genuinely to learn what is beyond what we already know or think we know.
>
> (Torrance 1969, p. 121)

Such scientific questioning must not be confused with scepticism and doubt, which – insofar as it affirms the knower's prior assumptions regarding the putative nature of an object of knowledge – is actually a form of self-certainty; rather, scientific questioning places faith in the inherent intelligibility of any given object of knowledge, despite the fact that such intelligibility may currently be beyond the capacity of human beings to grasp. Progress in science is possible because we refuse thoroughgoing scepticism in favour of a primal faith in the basic intelligibility of both creation and Creator.

5 Theological science requires a 'problematic' form of thinking in which we interrogate the object of knowledge in order to verify the truth of our claims about it. Different strata of reality require different verification procedures. If I read Torrance correctly here, this means that our judgements about other persons and our assessments of their characters and motivations are not amenable to overly rationalistic thought or experimental testing, but instead demand more holistic and intuitive levels of discernment. To suggest that such discernment is merely subjective is to artificially limit our knowledge of reality to the level of demonstrable certainty provided by the epistemic tools available to us. The fact that our understanding of other persons may appear less tangible than our understanding of chemical reactions measured in a laboratory does not mean that interpersonal understanding is any less truthful or real. We can and do identify the Holocaust to be a great evil, and are entirely justified in doing so, even though we may not have access to any strict formal procedures for verifying such a claim beyond that of intuitive moral discernment. With regard to our intuitive discernment of human character, among our best guides are the great novelists whose insight into human nature can only be properly expressed through the medium of fictional narrative, which enables the author to *show* that which could not otherwise be said. Since God reveals himself on an intersubjective level, the verification of theological statements is necessarily bounded by forms of spiritual intuition analogous to our intuitive discernment of other persons.

These five markers of the scientific nature of theology must be read in the light of the distinctive nature of the object of theological knowledge. The personal Triune God is revealed both as the sovereign lord of his creation who calls into question all human self-assertion, and as the divine servant who graciously under-labours for creation and washes the feet of his creatures.

The nature of theological truth

Torrance distinguishes between the ontological truth of being and the epistemic truth of knowing. Truth consists in the conformity of things in their inherent intelligibility to the eternal Word of God, so that the ultimate truth of creation is evident only in the light of its Creator. The primary form of theological truth is not abstract ideas, concepts and statements, but interpersonal knowledge relationships: God's truth is that 'which we can meet and know in our concrete existence through personal encounter and rational cognition' (Torrance 1969, p. 142). The ontological truth of God is final and ultimate; truth statements about God are necessarily contingent, and are possible only because God as ultimate truth meets us in Jesus Christ within our historical contingency. God's truth transcends the 'necessary' truths of human reason and the contingent truths of history: in Christ eternal truth becomes temporal without ceasing to be eternal. Christ is simultaneously the way, the truth and the life: grace is God's turning to us in Christ and communicating with us in a manner appropriate to our creaturely contingency.

> In Him God turns in Grace toward us and makes Himself open to us, summoning us to be open towards Him and to keep faith and truth with Him in Jesus, so that we may be true as God is true, and learn to do the truth as He does the truth.
>
> (Torrance 1969, p. 143)

The primary interpersonal and intersubjective truth of our knowledge relation with God can be articulated in secondary theological propositions. There is a circular relationship between theological propositions and theological judgements: theological propositions function to declare theological judgements, while theological judgements assess theological propositions. The interconnection between propositions and judgements presupposes a communal knowledge relationship. Theological propositions and judgements are not individual expressions of, or logical deductions from, generic human experience and thought. Rather, they are communal responses to

> a Word that is heard, a Truth that is communicated, or an Act that is done to us, a propositional question or rational communication that is directed to us from God himself in His Son, the Word made flesh, requiring recognition, response, decision for their articulation on our part.
>
> (Torrance 1969, pp. 163ff.)

Crucially, theological statements 'arise out of real dialogue initiated by God and not out of some irrational monologue we carry on with ourselves' (p. 164).

Theological propositions are existence statements that proffer truth claims about the nature of reality. There can be no proof of the truth of such statements independent of the knowledge relationships in which they originate. It follows that we can only convince others of their truth 'if we can get them to see or hear the reality they refer to as we see or hear it ... [and] share our *intuition* of the object given' (p. 165; emphasis in original). Torrance uses 'intuition' here to refer to 'our apprehension of a reality in its objectivity and unity, as a whole' (p. 165 n. 3). As such, it refers not to the psychology of the knowing subject, but to the establishment of a knowledge relationship through which the subject comes to 'intuit an object in a way appropriate to its nature' (p. 166). Existence statements are never complete because their meaning does not lie in them, but in the reality they seek to denote: in the act of intuition the mind is always open to a reality not fully comprehended and irreducible to abstract concepts. Existence statements and coherence statements are mutually dependent, so that a set of incoherent and mutually exclusive existence statements excludes the possibility of affirming the truth of the entire set; nevertheless, coherence statements cannot provide an additional source of knowledge to supplement the existence statements generated by concrete knowledge relationships – in this respect, existence statements are basic and primary. Theological existence statements are forced upon the theologian by the pressure of the objective reality of God: they emerge *a posteriori* out of divinely instigated knowledge relationships between God and humanity, 'and have a reference to the historical Jesus Christ as the objective reality beyond them, but in and through Jesus Christ they refer to God Himself as the ultimate objectivity' (p. 177). The internal coherence of existence statements is one, albeit highly significant, indicator of their truth or falsehood: theological statements are human constructs that strive to refer to objective realities in a coherent manner.

The logic of theology

Like all scientific statements, theological propositions possess their own specific logic by virtue of the distinctive nature of their object of knowledge. All scientific enterprises face the problem that 'being always breaks through the limits of our statements and outruns their logical forms' (p. 204). Theology faces the added difficulty that, because God by his very nature cannot be conceived of in the same way that we conceive of objects in the created order, 'the way in which statements refer to God is rather different from the way in which they refer to anything else' (p. 205).

To speak of 'the logic of God' is not to attempt to enter into the mind of God, but to understand God as he elects to reveal Himself to us: thus 'the logic of God' is revealed in the person of Jesus Christ, the incarnation of the eternal Logos of God. The actuality of the incarnation carries with it its own particular logic: (1) the logic of *grace*, in which God gives himself to us as the object of our knowledge without ceasing to be the divine subject, so that we know God only as God knows us; (2) the logic of *incarnation*, in which the hypostatic union between God and humanity in Christ provides the normative pattern of all theological statements; (3) the logic of *love*, by which the divine and human natures of Christ are united in one person, establishes love as the proper mode of interpersonal relations and the ontological ground of all reality, both created and uncreated; (4) the logic of *history*, grounded in the fact that since God engages ontologically with creation in the specific history of Jesus Christ, he is known only in concrete, communal, historical contexts, and not in abstract, individual mystical experiences that transcend particular spatio-temporal coordinates.

Because theological propositions are dependent upon and derive from historical events, they are a product of the Church by virtue of her tradition-specific testimonial relationship with the historical Jesus.

> If the Church as the Body of Christ constitutes the sphere where actual knowledge of the Truth is to be found, within that sphere the understanding of the Truth is articulated in a body of truths that drive from and ultimately cohere in the Person of Christ who is Himself the one Truth of God.
>
> (Torrance 1969, p. 212)

Put in more prosaic terms, if the Church had not acted as the custodian of the distinctively Christian retroductive explanation of ultimate reality, there would be no knowledge of the Trinitarian God.

To speak of 'the logic of man' is to seek to unpack the logic of human attempts to formulate accounts of the Trinitarian God. In seeking to enter into a loving relationship with us, God does not override our humanity through any deterministic predestination, but instead gives us the freedom to be ourselves, to overcome our self-imposed alienation from self, others and nature, and to enter freely into a proper relationship with our Creator. This includes the freedom to articulate theological propositions that employ human forms of thought and speech. The task of the theologian is to strive to bring human forms of speech into conformity with the realities they intend to refer to. Here the theologian faces a dual temptation: to confuse human expression with divine reality, or to allow human expression to subsume and colonise divine reality. Thus to speak of divine love is not to project human love onto God, but to identify the true nature of divine love as revealed in the person, teaching and actions of Jesus Christ, and to use this divine love as the measure with which to judge human relationships. The failure to allow the ontological reality of God to override epistemically relative human theological constructs leads Roman Catholicism to forms of ecclesiastical idealism and the advocacy of papal infallibility, and the Protestant Churches to forms of anthropological idealism and the advocacy of biblical infallibility. Insofar as theological propositions are unavoidably fallible human constructs, they always fall short of the realities they seek to describe. Insofar as theological propositions are able to establish a partial ontological purchase, they offer truthful – though necessarily provisional and incomplete – accounts of divine reality. The theological pursuit of deeper truth entails the logic of question and answer through which divine reality is progressively revealed as our abductively grounded retroductive explanations are iteratively refined in the ongoing attempt to establish for our theological statements a stronger purchase upon reality and a greater internal coherence.

It follows that theological statements must be open to critical re-examination and reconstruction, and that they must never be used instrumentally to enforce a premature epistemic closure. The responsibility of the theologian is to strive to bring her statements into conformity with reality in order to ensure that they retain their function as disclosure models. This requires the repeated testing of existence statements in the light of the established knowledge relationship between God and the Church: a relationship rooted in the ontological actuality of the incarnation, attested in Scripture, transmitted in tradition, proclaimed in preaching, and sustained in worship and prayer. It also requires the repeated analysis of coherence statements to ensure the coherence of the inner logic of existence statements, while constantly guarding against the idealist trap of allowing coherence statements to override existence statements.

Torrance identifies similarities and differences between theological science and other scientific endeavours. He views theology as a positive and independent science operating on its own ground and developing its own distinctive modes of inquiry. It differs from other sciences only because of the uniqueness of its primary object of investigation. Theological science is possible only because God elects to reveal himself to humanity, though in responding to this revelation it remains a fallible human endeavour. As such, it faces the same boundaries as all other scientific activities, seeking to rigorously test its truth claims in the light of the reality it seeks to comprehend more fully and understand more deeply. At the same time, because theology cannot bypass the claims of God on humanity and the responses of humanity to God, and since this dialectic of claim-and-response concerns ultimate reality and the ultimate good of humanity, it is distinct from other sciences. Because the proper object of theological investigation is God, theology cannot trespass into the realm of other scientific endeavours with their particular objects of knowledge, and must instead respect them as independent branches of knowledge. As such it can only speak about other independent sciences if they themselves transgress their appointed spheres and trespass into the domain of theology. Thus, for example, given the stratified nature of the created order, reductive attempts to dismiss Christian truth claims on the basis of genetic determinism or biological behaviourism constitute an unwarranted and illegitimate privileging of one particular domain of scientific investigation over another. If the truth claims of Christian theology are to be dismissed, they must be dismissed *as theological claims*, and not as pseudo-scientific claims transgressing the domain of the natural sciences.

12 Radical orthodoxy

Beyond secular reason

Colin Gunton's account of divine revelation and Thomas Torrance's account of theological science – which informed our discussion of judgemental rationality vis-à-vis Trinitarian theology in the previous two chapters – both set out to establish the inherent rationality of Christian theology in a manner that does justice to its intrinsic integrity. They argue that, taken on its own terms, and in comparison with other rational enterprises, Christian theology legitimately lays claim to a capacity for judgemental rationality that enables it to penetrate beyond cultural relativism and establish epistemic purchase on ontological realities without recourse to a blind leap of irrational faith. As such, any refutation of Christian truth claims cannot take the short route of dismissing them as inherently irrational, but must rather take the longer route of a sustained intellectual critique grounded in the presentation of an alternative retroductive account of the ultimate order-of-things capable of demonstrating its greater inclusivity and explanatory power.

In this chapter we explore an alternative 'defence' of the integrity of Christian truth claims, as presented by a prolific group of contemporary theologians operating under the umbrella of 'Radical Orthodoxy' (Milbank 1990a, 1990b, 1997, 2003; Ward 1995, 1996, 2000; Pickstock 1998; Milbank *et al.* 1999; Milbank and Pickstock 2001; cf. Hanby 2003; Miner 2004). Since the early 1990s Radical Orthodoxy has emerged as a distinctive and controversial theological school, rooted in (though not exclusive to) the Anglican tradition, and of sufficient stature to have generated a wealth of secondary commentary (Hemming 2000; Smith 2004; Ruether and Grau 2006; Shakespeare 2007). Unlike Gunton and Torrance, most theologians who identify themselves with Radical Orthodoxy follow John Milbank in rejecting all attempts to demonstrate the compatibility of Christian theology with the canons of occidental reason generated by the Enlightenment – including those, driven by a desire to protect the integrity and distinctiveness of Christian theological discourse, that are willing to critique and reformulate modern accounts of reason and rationality – by defending the startling and altogether more radical claim that human intellectual enterprises that are not grounded in the Christian account of reality are themselves inherently irrational.

Milbank's *Theology and Social Theory* attempts 'to disclose the possibility of a sceptical demolition of modern, secular social theory' from the perspective of Christian orthodoxy (Milbank 1990a, p. 1). He argues that 'the most important governing assumptions of such theory are bound up with the modification or the rejection of orthodox Christian positions' (p. 1). As such, occidental reason cannot legitimately claim to be grounded in objective universal human reason, but rather, and in common with all intellectual enterprises, is irretrievably bound to local constraints and considerations. Consequently, and contra Gunton and Torrance, Christian theology has nothing to gain by striving to demonstrate a level of compatibility (however qualified) with the legacy of the Enlightenment and its claim to access universally warranted reason. Rather, if the 'Christian perspective is persuasive, then this should be a persuasion intrinsic to the Christian *logos* itself, not the apologetic mediation of a universal human reason' (p. 1; emphasis in original). The claims of Radical Orthodoxy are introduced here in order to highlight the richness and diversity of orthodox Christian responses to the epistemic challenges of modernity, and – primarily for reasons of space – no attempt will be made to adjudicate between the perspectives of Gunton and Torrance on the one hand, and Milbank and his colleagues on the other. Given the orthodox Christian commitment to ontological realism and epistemic relativism, Milbank's distinctive approach to the employment of judgemental rationality constitutes a significant contribution to our exploration of the relationship between Christianity and critical realism.

Radical orthodoxy and secular paganism

Milbank rejects the claim that secular philosophy is grounded in universally valid canons of reason. In accord, at least in part, with the critical realist notion of epistemic relativism, he follows Foucault in identifying an irrevocable link between knowledge and power. Genealogical investigation of what he terms the 'secular episteme' reveals a social, economic, political and intellectual history marked by a series of contingent developments unbounded by universal reason. The secular realm is thus a contingent human creation, and it is only the dominance of its plausibility structures in the post-Enlightenment era that allows it to claim to be the result of inevitable intellectual progress. Social science embraces a set of assumptions about truth and values regarding the human condition that are profoundly questionable and which arose in direct opposition to Christian theology. As such, secular discourse 'is actually *constituted* in its secularity by "heresy" in relation to orthodox Christianity', and, insofar as its ontological assumptions and humanistic values valorise a naturalistic world in which human beings occupy the summit of the evolutionary chain, constitutes a form of neo-paganism that worships the created order rather than the Creator (p. 3; emphasis in original). Ironically, though liberal theologians continue to seek to justify Christian faith by embracing 'secularization and the autonomy of secular reason', social theory itself, especially in its postmodern turn, 'increasingly finds secularization paradoxical, and implies that the mythic-religious can never be left behind' (p. 3). Despite this, mainstream secular theory continues to claim to be able to understand the world without reference to God or transcendence, and seeks to exercise a colonial economy of power over theology. It strives to discredit theology in the public sphere on the ground that it does not conform to the standards of secular universal reason, and in seeking to relegate religious belief to the private sphere it simultaneously immunises itself from theological critique and buttresses a capitalist economy predicated on the ultimate power and authority of financial exchange. Faith becomes an optional extra, which individuals are free to consume at will, provided they do so as consenting adults behind closed doors. However, since secularism is a product of contingent judgements rather than universal reason, it is itself a faith system that claims a privileged status which it seeks to protect through a totalitarian economy of power and colonial domination.

Milbank goes on to argue that the secular is complicit in a political ontology of violence, which he defines as 'a reading of the world which assumes the priority of force and tells how this force is best managed and confined by counter-force' (p. 4). Positivism, in seeking to free humanity from all religious superstition and metaphysical constraint, established a humanistic quasi-religion rooted in the worship of personal autonomy. However, an ontology predicated on individual freedom is inherently violent because it cannot possibly avoid the clash of human interests, and hence is left with no option other than to seek to counter the violence inherent in its structures with various forms of counter-violence. Thus, for example, the blanket evocation of tolerance to manage this clash of interests is radically underdetermined and inevitably violent: since toleration of homosexuality necessarily requires intolerance of homophobia, the defender of human autonomy has no option but to invoke Popper's 'paradox of tolerance' and 'claim, in the name of tolerance, the right not to tolerate the intolerant' (Popper 1966, p. 265). Though Hegel came close to deconstructing secularism in his attempt to locate the Christian logos within his idealist system, the Marxist response to Hegel embraced an intrinsically violent romanticised 'gnostic plot about a historically necessary fall and reconstruction of being' (Milbank 1990a, p. 4). Similarly, Nazism, in utilising the technological fruits of empirical science in pursuit of its romantic vision of the will-to-power of the Aryan race in the face of its supposed contamination by sub-Aryan and sub-human elements, 'was nothing but an unhindered attempt to raise man as God, to unleash and perfect the power of human freedom' (Milbank 2003, p. 179; Shakespeare 2007, p. 13).

Secularism, according to Milbank, is just another pagan religion, worshipping at the altar of various objects and forces immanent within the created order. The pre-Christian paganism of ancient Greece and Rome sought to mitigate the power of the brute forces of chance and fortune through sacrifices designed to appease and thereby control the gods. Similarly, the post-Christian paganism of the secular age seeks to mitigate the power of those universal forces implicit in nature, society and human agency that, following the departure of the Triune God, now constitute the ontological bedrock of reality. In its modern guise it does so through the exercise of universal reason utilised in support of a variety of secular utopias: technology tames nature; monetarist theory directs the market forces of capitalism; the affirmation of human rights curtails the abuses of autonomous self-interest; revolutionary planning steers the progress of history. Secular paganism also possesses a postmodern guise. True, postmodernity recognised the link between the evocation of universal reason and totalitarian violence, and sought to manage such violence by deconstructing universal reason; however, the resulting valorisation of an economy of unconstrained freedom had the effect of uncritically sustaining the liberal-capitalist status quo, thereby doing violence to the majority who have little share in the goods and produce of capitalism. Thus the logic of violence-contra-violence endemic to the project of universal reason is sustained even with the eclipse of reason itself, since postmodernity continues to hold fast to the primary human value underlying the Enlightenment's affirmation of universal reason, namely the ontological and moral priority of human autonomy and self-regulation. Like Greek and Roman paganism, the redux paganism of secularism lacks any viable alternative to the ontological priority of those impersonal rational structures and personal agencies over which it seeks to establish a modicum of control and restraint through a further exercise of power: in the absence of the Christian ontology of love, power resists power and violence counters violence.

The 'secular *episteme* is a post-Christian paganism, something in the last analysis only to be defined, negatively, as a refusal of Christianity and the invention of an "Anti-Christianity"' (Milbank 1990a, p. 280; emphasis in original). Ironically, the seeds of this redux paganism lie in Christianity itself, specifically in the failure of Christians to hold fast to their distinctive vision of the Triune God and preserve Augustine's distinction between the earthly city of Rome and the heavenly city of God. The late medieval scholastic theologians Duns Scotus and William of Ockham shared the belief that the medieval synthesis of faith and reason threatened to make the sovereignty of God vulnerable to the colonising tendencies of human reason. Their nominalist philosophy offered a flattened vision of reality as a series of univocal objects interacting with one another without any substantial metaphysical grounding. This in turn generated a vision of a deity differing from the world in degree rather than in kind. Given the univocal nature of human and divine volition within this nominalist framework, the protection of divine sovereignty required the introduction of a fundamental distinction between impoverished human will and the absolute omnipotence of the divine will, to the extent that the divine will utterly transcended human comprehension. This philosophical vision of a monistic deity of absolute power constituted an elemental departure from the theological vision of a Trinitarian God existing in reciprocal relationship, freely participating in creation, and lovingly revealing himself to his creatures.

> In the thought of the nominalists, following Duns Scotus, the Trinity loses its significance as a prime location for discussing will and understanding in God and the relationship of God to the world. No longer is the world participatorily enfolded within the divine expressive *logos*, but instead a bare divine unity starkly confronts the other distinct unities which he has ordained.
>
> (Milbank 1990a, p. 14; emphasis in original)

The impact of this turn from Trinitarianism to Deism and Unitarianism is clearly visible in the philosophy of Descartes. His systematic hermeneutic of suspicion, grounded in the exercise of universal reason, led, via a flirtation with the possibility of a deceptive 'malignant demon', to the assertion of the existence of a deity acting as the first cause and primary source of order in the universe. Having rejected the ontology of love proclaimed by Trinitarian faith, and faced with the possibility that reality is ontologically grounded in the arbitrary whim of an all-powerful and potentially malevolent deity, the project of universal reason embraced the task of taming and domesticating those primal forces of violence, whether divine, natural or socially constructed, which now appeared to constituted the ontological bedrock of reality following the departure of the personal Triune God of love. The ensuing rejection of an all-powerful deity, and subsequently of the modern primacy of human reason, left only the postmodern primacy of personal autonomy, and faced humanity with no alternative other than to seek to counter the ensuing clash of human interests through an increasingly unsustainable appeal to the innate goodness of humanity once freed from the twin hegemonies of God and reason.

According to Milbank there can be no compromise between Christianity and secularism because the Christian ontology of love and secular ontology of violence are mutually exclusive. We cannot properly know the world independently of God, and it cannot exist without divine sustenance: remove God from the equation and, despite the best rational efforts of humankind, the world ultimately becomes meaningless and collapses into non-being. The attempts of liberal and progressive theologians to forge a compromise with secularism are misplaced and futile. To place nature alongside God as a second divine reality, or ontological foundation, is to undermine the sovereignty of God, and hence the sovereignty of love and peace. Here Milbank focuses on attempts by liberation theologians to forge a partnership between Christianity and Marxism. As a Christian socialist, he embraces a vision of society driven not by market forces but by reciprocal giving and receiving according to ability and need. However, insofar as Marxism seeks to combat capitalism with revolution it simply perpetuates the cycle of violence unleashed by secularism. Since Christianity is committed to non-violence, a violent means can never justify a peaceful end. The only way to defeat violence is to act consistently and harmoniously with the will of the God of love and peace. In the Eucharist Christians participate communally in a reciprocal relationship with the Triune God: a liturgical act that is truly revolutionary, liberating and transformative because it is the only one capable of directly challenging secular economies of violence (Pickstock 1998). If non-coercive love enjoys ontological priority over the demi-reality of coercive violence, then love will inevitably triumph. Love does not need to enter into partnership with violence to aid its victory; on the contrary, it must actively resist the forces of violence by bringing the men and women of violence into direct confrontation with an alternative economy:

> You have heard that it was said, 'An eye for an eye and a tooth for a tooth.' But I say to you, Do not resist an evildoer. But if anyone strikes you on the

> right cheek, turn the other also ... and if anyone forces you to go one mile, go also the second mile.... You have heard that it was said, 'You shall love your neighbour and hate your enemy.' But I say to you, Love your enemies and pray for those who persecute you, so that you may be children of your Father in heaven.
>
> (Matthew 5:38ff.)

In resisting compromise and even dialogue with secularism, Milbank affirms the exclusive truth of the Christian story. At the heart of the Christian faith is the sacrament of the Eucharist, in which Christians participate in the nature of God in a dynamic communion that is made possible by the hypostatic union between humanity and God in the incarnate Christ, a joining together which replicates the hypostatic union of the three persons of the Trinity. Christian exclusivism has nothing to do with a theocratic striving for power in the world: whenever the Church pursues this path, it rejects its ontological roots in favour of an economy of violence. Instead, the ontological priority of love grounds and justifies Christian exclusivism: the Church's role is to under-labour on behalf of the world by unapologetically proclaiming and witnessing to the truth, unpalatable to many, that the alpha and omega of creation is a divine economy of love in which God graciously invites humanity to participate.

Milbank addresses the question of the epistemic grounds of his ontological assertions by suggesting that Christianity prefigures the linguistic turn in modern philosophy. Historically, Christianity challenged the classical philosophical grounding of metaphysics in fixed substances: since God is personal, the basic ontological category is dynamic relationality rather than static substance. Persons relate to one another through language. Milbank follows Vico in suggesting that belief in fixed substances is a relic of paganism, insofar as it valorises the underlying structures of the created order rather than its Creator, and that in naming substance qua substance philosophy seeks to impose and enforce order on and in the world. Christianity, on the other hand, employs language as a tool of communication between different persons, human and divine; as such, and in line with much post-structural philosophy, the proper reference of language is not metaphysical objects but interpersonal relationships. Christianity is thus grounded not in a set of cognitive propositions but in a dynamic relationship with God expressed through narrative, principally the narrative enacted in the Eucharist. Truth is constituted by the dynamic relationship between the knower and the object of knowledge; where the object is personal, truth is constituted by the nature of interpersonal relationships. As a result, there can be no epistemic justification of a knowledge relationship that does not proceed from the relationship itself: there is no way of justifying a knowledge relationship by appealing to universal reason. We encounter God in the dynamic and creative relationship rooted in the Christian narrative and enacted in the Eucharist. The only ground for asserting the truth of the Christian story is to appeal to the attraction and lure of its intrinsic aesthetic brilliance. The intrinsic attraction of the Christian story addresses human desire. Resisting any rigid separation between agape and eros,

Radical Orthodoxy identifies human desire as central to the knowledge relationship. Humankind's desire for God, which flows from the fact that we are created in the image of God as relational creatures, is at the heart of Christian life. However, human desire has become entrapped and corrupted by sin: this is nowhere more evident than in capitalism's stimulation of misplaced desire and cultivation of cravings that cannot possibly be satisfied by consumer goods. 'If our desire is not oriented to the transcendent and beautiful, then we will be impoverished and at the mercy of those who manipulate desire for worldly ends' (Shakespeare 2007, p. 145). Participation in the Christian narrative and entry into a reciprocal relationship with God results in the transformation of desire, as human craving finally learns to focus on its proper object (Milbank and Pickstock 2001, pp. 108ff.). The only justification for the Christian story is intrinsic to the story itself: by participating in the story, we enter into a gracious reciprocal relationship with God grounded in a dialect of divine beauty and human desire. Human attempts at self-justification through the exercise of universal reason, enwrapped in economies of violence and seeking to make God accountable to creation, subvert the actual ontological order of things and in doing so renounce any legitimate role in the quest for the epistemic justification of Christianity.

The hegemony of occidental reason

In the rest of this chapter I will work through the implications of Milbank's position by bringing his work into dialogue with Doug Porpora's critically realist discussion of inter-faith dialogue, as presented in *Transcendence: Critical Realism and God* (Porpora 2004). 'Can believers in different religions sit down together and argue over their rival truth claims in a manner that has any prospect of persuading anyone of anything?' (p. 109). Porpora's response to his own question is broadly sceptical, and in discussing the issue he limits himself to offering 'a propaedeutic to a propaedeutic on inter-religious dialogue' (p. 109). He introduces an important distinction between 'dialogue' and 'debate': the former being concerned to establish mutual understanding and tolerance, the latter being concerned to address contested questions of truth. Within the literature of inter-faith dialogue there is 'a strong reluctance to debate the rival truth claims of different religions'; instead, 'the function of dialogue, we are told, is to listen, to understand, and to build community' (p. 112). Inter-faith dialogue is at an embryonic stage: much of the literature constitutes an internal debate within Christianity about the problems and possibilities surrounding dialogue, and there is relatively little substantial communication between different religious traditions, and that which does takes place is primarily concerned to establish mutual understanding rather than adjudicate between conflicting truth claims. Porpora is clear that debate in pursuit of truth is an intellectual possibility, since religious traditions make truth claims (ontological realism) that are disputed (epistemic relativism) and open to critical assessment (judgemental rationality). However, he questions whether, at the present time, such debate is morally, socially and politically justifiable.

> After introducing myself to the literature on inter-religious dialogue, I am no longer certain that such dialogue can or should be done now. The reason, however, is political or social rather than philosophical. Our different religious communities do not seem ready to conduct such a dialogue dispassionately or harmoniously. That being the case, we probably should not engage in it.
>
> (Porpora 2004, p. 125)

We will consider Porpora's moral objections in the next section, and concentrate here on his affirmation of the philosophical possibility of inter-faith debate.

As we have seen, Milbank's default position appears to preclude both the value and viability of inter-faith debate. He shows little interest in establishing a deeper and more empathetic understanding of the faith of secularists, and shows no expectation of learning anything of value from the phenomenon of secularism. On the contrary, he insists in advance that secularism is rooted in an incoherent ontology of violence, wrongly claims privileged access to a universal reason, is historically naïve and insufficiently self-critical, and as a redux paganism is entirely antithetical to Christianity. Since there can be no compromise with such an outlook, no conciliation between the secular ontology of violence and Christian ontology of peace, the only valid conversation is one designed to convert secularists to the exclusive alethic truth of Christianity and persuade them of the greater aesthetic beauty and attractiveness of the Christian story. If judgemental rationality vis-à-vis Trinitarian theology may be shown to work in such an apparently extreme case as Milbank's Radical Orthodoxy, then it is almost certainly applicable to other, less extreme cases, such as the defence of the intrinsic rationality of Christian theology set out by Gunton and Torrance.

We begin with a procedural observation before moving to the substance of the argument of this section. Throughout this chapter the term 'inter-faith' is used in an inclusive sense. As Porpora points out, since it could 'turn out that no truth claim made by any religion is even partially correct ... atheism in all its forms must likewise be an interlocutor invited to what is called inter-religious dialogue' (p. 110). This claim needs to be rephrased positively: there are many worldviews that reject any notion of transcendence, theistic or otherwise, and to define them purely in terms of their opposition to theism is to do them a disservice by absenting the positive substance of their alternative worldviews. As a working hypothesis, I will assume that different worldviews – transcendent, immanent, theistic, naturalistic and so forth – offer conflicting accounts of the ultimate nature of reality. Critical realism recognises that explanatory accounts of 'higher' strata of reality are non-reducible to the 'lower' strata from which they emerged. It follows that a complete explanation of a particular stratum, even if this were achievable, could not possibly provide an ultimate explanation of all the various strata viewed as a totality. Given that all the strata of which we are currently aware appear inherently rational and intrinsically open to judicious explanation, it is reasonable to suppose that the totality of reality, the totality of different strata in our universe in their complex interrelationships – and, by extension the totality of other universes, realities and modes of being currently unknown to us – is amenable to rational explanation. If the biological domain is dependent upon but irreducible to the chemical domain and the chemical domain is dependent upon but irreducible to the physical domain, is there anything that the physical domain is dependent upon and irreducible to? Here the naturalist will argue along the lines that the physical world is somehow self-generating, self-sustaining and self-rationalising; the theist will argue that the physical world is dependent upon a Creator God; the proponent of the philosophy of meta-Reality will argue that the physical world emanates from a more primal ground-state. Wittgenstein argues, with reference to the justification of obeying a rule, that 'If I have exhausted the justifications I have reached bedrock, and my spade is turned. Then I am inclined to say: "This is simply what I do"' (Wittgenstein 1968, §1.217). Transferring these epistemic concerns to the ontological domain, we can say that attempts to arrive at *the* ultimate spade-turning explanation of the totality of reality in the transitive domain are true only insofar as they correctly identify the ultimate bedrock of reality, and explain the ultimate order, meaning and purpose of the totality of reality in the transitive domain. Any notion of inter-faith dialogue that excludes non-religious and non-transcendent accounts of the ultimate order-of-things must be rejected in favour of an inclusive approach, both because such accounts are potentially true, and because an exclusive focus on religious accounts can inadvertently disguise the misplaced anti-religious assumption that a non-religious secular worldview constitutes a self-evidently true default position that, unlike religious positions, does not need defending.

Turning then to the substance of the present section, we begin with the observation that, despite his rejection of secular universal rationality and apparent preference for aesthetic preference over cognitive choice, Milbank does not support any notion of *epistemic* incommensurability between Christianity and secularism. He rejects secularism not because he finds it incomprehensible but because he claims to comprehend it all too well and, as a result, has no option but to affirm its *ontological* incommensurability with Christianity. Milbank is clearly not a relativist, and as such rejects any form of thoroughgoing judgemental relativism; however, this raises the question of the rational grounds on which he rejects secularism, given his denunciation of universal secular reason. Bhaskar argues that (1) the notion of two entirely incommensurate and untranslatable languages, and (2) the notion of total and uniform mutual comprehension constitute unrealisable limiting cases (Bhaskar 1986, pp. 70ff.; Collier 1994, pp. 89ff.). Negatively, the exercise of judgemental rationality between conflicting theories is impossible unless there is at least a minimal degree of (1) convergence and meaning filiation and (2) divergence and meaning variance. Positively, the exercise of judgemental rationality between conflicting theories is possible because there is disagreement in the transitive realm about the nature of a common object in the transitive realm. Thus Milbank is engaged in an argument in the intransitive realm about the question of the ultimate nature of reality in the intransitive realm. Despite Milbank's rhetoric of opposition, which comes to a head in his rejection of secular reason's claim to access universal reason, the opposition is clearly not absolute since Milbank claims to understand his secularist opponents when declining to enter into dialogue with them. But if, as Bhaskar argues, '[c]ommunication is impossible unless some descriptive and practical presuppositions are shared in common', we are bound to ask what it is that Milbank holds in common with secularism, given that it is *not* a shared commitment to universal secular reason.

Milbank's rejection of universal secular reason appears to be related to modern foundationalism: the claim to be able to identify 'foundational or basic beliefs which guarantee their own truth' and 'are accessible to any rational person, irrespective of their historical or cultural contexts' (McGrath 2002, p. 21). Foundationalism has its immediate roots in Descartes' pursuit of epistemic certainty, his 'quest for some fixed point, some stable rock upon which we can secure our lives against the vicissitudes that constantly threaten us' (Bernstein 1983, p. 18; cf. Descartes 1970, pp. 61ff.). Whether these foundations are found in empirical observation, idealistic conceptualisation or romantic intuition, they have in common a desire to enable human beings to assert a level of control over their identities and destinies in the face of an intimidating and menacing universe. In doing so, they respond to the apparent violence of the universe by embracing the epistemic fallacy – itself no stranger to epistemic violence – that promises a means of ordering the universe according to the will and needs of humanity without regard to the nature of reality itself. In attempting to master the totalitarian drive of modernism, the postmodern celebration of the epistemic freedom to create reality according to one's personal desires and preferences simply replicates the act of hubris that gave birth to modern foundationalism. Whether in its modern foundational form, or postmodern anti-foundational form, universal secular reason is in error insofar as it affirms a form of universal reason that seeks to allow humankind authority over the world. If this reading is correct, then Milbank does not oppose reason per se, merely the reification of a particular account of reason that wrongly claims the status of universality and which is predicated on the pro-active assertion of the human will.

Non-foundational or contextual epistemology recognises that all knowledge inevitably emerges from within the spatio-temporal specific contingencies of culture. Because we do not have access to a god's-eye view-from-nowhere, there 'is no "vantage point" from which we can view and evaluate our beliefs, except on the basis of our already existing beliefs' (McGrath 2002, p. 35; cf. Nagel 1986). As Thomas Kuhn points out, we learn science not by assuming an objective or abstract perspective, but by immersing ourselves in the traditions of the scientific community (Kuhn 1970). Such immersion requires our personal engagement with that which we seek to understand, so that knowledge is constituted by the relationship between the knower and the object of knowledge. Since we necessarily carry our prejudgements and pre-understanding with us in any investigation, we must be willing to allow the object to challenge and shape our understanding if we are to avoid imposing our pre-given assumptions onto the object. This means that human rationality (contra Kant) and the inherent rationality of the object of knowledge (contra positivism) are never either transparent or given: we have to interrogate reality, and allow reality to interrogate us, in order to establish both knowledge itself and the appropriate epistemic procedures through which to continue the epistemic task.

In his discussion of contextual epistemology, Alister McGrath draws attention to Otto Neurath's well-known nautical analogy:

> We are like sailors who on the open seas must reconstruct their ship but are never able to start afresh from the bottom. Where a beam is taken away a new one must at once be put there, and for this reason the rest of the ships is used as support. In this way, by using the old beams and drift-wood, the ship can be shaped entirely anew, but only by gradual reconstruction.
>
> (Neurath 1973, p. 198; cf. McGrath 2002, pp. 34ff.)

It may be instructive to extend this analogy and apply it to inter-faith debate and dialogue. Let us envisage (with apologies for the very British nature of the analogy) two ships, HMS *Christian* and HMS *Secular*, encountering one another on the high seas. The crew members decide, somewhat against precedence, to debate the relative virtues of their respective ships. It is apparent from the start that, despite their ships' very different designs, communication between the respective crews, though difficult, is not impossible. It is *difficult* because they have conflicting understandings as to what counts as a good ship, the best way to reconstruct a ship while under sail, and – most fundamentally of all – what the purpose of a ship is and why a crew should bother to sail one in the first place. At the same time, however, it is *possible* because both crews share a common humanity, a common language, a common profession, and sail a common ocean. After a cautious start, in which the respective crews are anxious to get to know one another and learn about each other's ships, the debate proper gets underway: Which ship is best? What is the best way to sail? What is the point of sailing? Problems begin to emerge when the crew of HMS *Secular* appeals to the crew of a third ship, HMS *Universal Reason*, to act as adjudicator in the debate, on the grounds that its crew members possess privilege knowledge about shipping that is unbiased, neutral and universally valid. This problem is intensified when the crew makes the additional claim that further debate is impossible unless the crew of HMS *Christian* recognises the absolute authority of HMS *Universal Reason*. This throws the crew of HMS *Christian* into turmoil: some perceive HMS *Universal Reason* as a threat and want to abandon the debate; others advocate modernising HMS *Christian* using HMS *Universal Reason* as a model; others want to continue the debate by challenging the authority of HMS *Universal Reason*. This latter group is somewhat nonplussed, because the crews of HMS *Universal Reason* and HMS *Secular* appear to share an identical understanding of shipping that is very different from their own. Further, though HMS *Christian* has a somewhat traditional design, it has sailed the high seas relatively successfully for many centuries; HMS *Secular*, on the other hand, enjoys a distinctively modern design and is a relative newcomer to the ocean. Despite this, the crew of HMS

Secular claims that the crew of HMS *Christian* knows very little about sailing, and encourages crew members to scupper their ship and seek a berth on HMS *Secular*. As a result, the crew of HMS *Christian* begins to suspect that the crew of HMS *Secular*, mesmerised by the 'law of fashion', is behaving irrationally. Thus the debate throws up not merely ontological questions about the ultimate nature of ships and the craft of sailing, but also epistemic questions about the rational status of nautical knowledge. In this situation, nothing can possibly be resolved by an appeal to the authority of HMS *Universal Reason*, since her authority is fundamentally disputed. In this situation the only option available is to extend the terms of the debate, so that it embraces not merely disputed ontological questions regarding the nature of ships and sailing, but also contested epistemic questions regarding the nature and status of nautical knowledge. In philosophical terms: inter-faith debate must address not only contested ontological truth claims about the intransitive domain, but also contested epistemic claims in the transitive domain regarding appropriate rational procedures and truth criteria. Given the actuality of epistemic relativism and the fact that all understandings of rationality are necessarily contingent and open to public dispute, it is profoundly irrational for a community to claim that a particular position is irrational simply because it does not conform to their own preferred understanding of rationality.

This is not to suggest that different traditions cannot ease their way towards a shared understanding of epistemic procedures and criteria for rational judgement. With respect to epistemic procedures, we have already had cause to note the Jesuit theologian Bernard Lonergan's appeal to four 'transcendental precepts': 'The basic form of alienation is man's disregard of the transcendental precepts, Be attentive, Be intelligent, Be reasonable, Be responsible ... the basic form of ideology is a doctrine that justifies such alienation' (Lonergan 1973, p. 55). I used these precepts as the basis for my critique of Sean Creaven's attack on Christianity in *Against the Spiritual Turn*, suggesting that his position was fundamentally flawed due to his failure to be properly attentive, intelligent, reasonable and responsible with regard to the relevant evidence and issues when constructing his argument (Creaven 2010;Wright 2011). I did so in the hope and expectation that, though these minimal procedural precepts had been articulated by a Christian theologian, Creaven would be able to accept their application to academic debate in general, whether religious or secular. However, the veracity of my argument did not depend upon Creaven's acceptance of them: we could have engaged in rational debate even if Creaven had chosen to reject any or all of them, or if he had chosen to challenge my own particular understanding of precisely what it means to be attentive, intelligent, reasonable and responsible about Christianity. This brings us to the crucial point of the argument of this section: the possibility of informed inter-faith debate is *not* dependent upon any prior agreement between the various parties regarding the nature and extent of human reason and the identification and application of epistemic procedures.

With respect to criteria for rational judgement between conflicting theories, Bhaskar has consistently argued that one theory trumps another theory by virtue of its greater and more comprehensive explanatory power, so that we are justified in preferring one theory over another whenever it is able to explain, within its own particular frame of reference, more of the phenomena under investigation than any other theory. There is, I think, something of a consensus that Bhaskar's position here would benefit from further elaboration (Groff 2004, p. 21). Building on the work of Basil Mitchell, I have suggested elsewhere five criteria for adjudicating between conflicting theories: congruence with life experience, internal coherence, fertility, simplicity and illuminatory depth (Wright 2007, pp. 219ff.; cf. Mitchell 1973, pp. 50ff.). In retrospect, and cautious of the dangers of forcing reality into the procrustean bed of our *a priori* assumptions, I would now prefer to regard them as contingent 'generic indicators' rather than fixed 'formal criteria'. If informed debate between conflicting theories is not dependent upon prior agreement about epistemic procedures, then it cannot be dependent upon any established criteria for adjudicating between true and false truth claims.

My suggestion is that for all Milbank's antipathy towards secularism, his thoroughgoing critique of its epistemic and ontological assumptions, and his refusal to enter into debate with any intention to do other than to persuade liberal secularists of the error of their ways, he is nevertheless actively engaged in a debate with secularism, and that this engagement, insofar as it addresses both ontological and epistemic issues, is fundamentally rational. There is, I suggest, much work still to be done on the issues raised in this section; however, as matters stand, Porpora's claim that judgemental rationality about religion is philosophically possible is entirely justifiable, even in the extreme case of Milbank's trenchant rejection of universal reason. Though I suspect Porpora may judge differently, I suggest that this is so even among those whose motivations are partisan and oriented towards the conversion of interlocutors.

This argument requires one further qualification. Milbank's resistance to the claims of universal reason is such that he clearly finds an affinity with those standing towards the postmodern relativist end of the spectrum of theologians and philosophers concerned with the employment of judgemental rationality. Gunton, Torrance and Milbank are all representative of the tradition of Christian theologians committed to Trinitarian orthodoxy and deeply suspicious of the incursions of the secular Enlightenment into the field of theology. However, whereas Gunton and Torrance, if forced to choose, would almost certainly express a preference for modernity over postmodernity on the grounds that Christianity has always made cognitive truth claims, Milbank demonstrates a clear preference for postmodernity in electing to make judgements between conflicting theories on the basis of aesthetic preference rather than cognitive choice. I am not convinced that the difference between the two is as clear-cut as might be assumed: first, because the distinction been postmodern aesthetic narrative and modern cognitive proposition is a false duality; second, because rational choice need not be reduced to a narrowly conceived rationalism. Kierkegaard, in his *Concluding Unscientific Postscript*, challenges the reduction of thinking to rationalism where personal interests are involved. Aesthetic choice does not preclude reason, and appears far closer to the actual procedures individuals adopt in their interpersonal exchanges. Critical judgement has a personal dimension that embraces the existential commitments of the whole person, and as such is perhaps closer to the Aristotelian notions of wisdom (Greek σοφία, sophia) and practical wisdom (Greek φρόνησις, phronesis) than modern rationalism (Dunne 1997).

The hegemony of liberal polity

The weight of Porpora's suspicion of inter-faith debate rests on practical issues of morality, specifically the potential threat it offers to both individual flourishing and socio-political well-being. In the light of his engagement with inter-faith literature, and of his own personal experience of inter-faith dialogue as a Roman Catholic raised a Presbyterian and married to a Jewess, he expresses uncertainty as to whether critical realists can and should encourage inter-faith debate at this particular point in time. 'I now reflect that as important as an inter-religious search for truth may be, the world is not ready for it' (Porpora 2004, p. 115). In developing his argument, Porpora makes two important preliminary remarks. (1) Inter-faith *debate* cannot avoid contested questions of truth in the way inter-faith *dialogue* often manages to do. Those non-realist theological 'fictionalists' who bypass questions of truth by insisting that religious language is merely self-referencing and enjoys no substantial purchase upon ontological reality are supported by a political groundswell owing to their supposed ability to promote tolerance in an even-handed and charitable manner. However, though fictionalism is 'undeniably tolerant … it extracts too high a price from all religions' because it 'withdraws from all religions any claim of saying anything true or consequential' (p. 112). But what kind of tolerance is it that refuses to acknowledge the truth claims of religious traditions and dismisses them as inconsequential prior to any informed debate? Despite possessing the veneer of openness and tolerance, fictionalism is fundamentally disrespectful of the beliefs and truth claims of others: 'I tolerate you [only] because you cease to make any claims on me, particularly the kind of claims that truth makes' (p. 112). (2) Much inter-faith dialogue, as opposed to inter-faith debate, tends to avoid contested truth claims by seeking to identify moral principles common to all religious traditions. Here Porpora references, with approval, the adoption of a global inter-faith ethic at the 1993 meeting of the Council for a Parliament of the World's Religions (p. 119; cf. Küng 1991, 1995). However, he goes on to point out that such agreement cannot disguise fundamental ontological differences between the world's religions, and that it is impossible to ignore the link between moral values and ontological commitments. Thus, inter-faith debate cannot avoid, whether by fictionalising religious narratives or reducing theology to morality, contending with the very real tensions generated by contested religious truth claims. Religious doctrines are typically exclusivist, so that to affirm the ontological claims of one tradition is necessarily to negate the ontological claims of other conflicting traditions. This means that interface dialogue cannot easily avoid the impulse for the different parties to proselytise and seek converts. In the case of Christianity in particular, its colonial past means that its desire to debate will always carry the undertones of 'hegemony and imperialist oppression' (Porpora 2004, p. 115). Similarly, in post 9/11 America, 'Christians and Jews both persistently associate Islam with extremism and terrorism' (p. 117). If genuine debate cannot avoid raising such tensions, in doing so it must contend with high levels of ignorance and prejudice on the part of ordinary religious adherents, of whom 'a great many remain parochial, uninformed and even positively ethnocentric' (p. 117). Porpora argues that in the face of such ignorance we should do all we can to encourage *dialogue for mutual understanding* between faith traditions. However, he also argues that this entails ensuring that *debates about truth* are excluded from the agenda. Since few religious adherents are able to respond appropriately to challenges to their faith, such debates need to be restricted to the professional theologians. Porpora concludes: 'Our different religious communities do not seem ready to conduct such a dialogue dispassionately or harmoniously. That being the case, we probably should not engage in it' (p. 125).

Milbank's reading of inter-faith debate is almost diametrically opposed to Porpora's. Milbank points out that calls for inter-faith dialogue tend to issue from liberal political economies. The liberal commitment to a secularised 'universal reason' generates the belief that practical reason can provide a neutral space within which such dialogue can take place. However, this ignores the fact that liberal values have a complex genealogical relationship with classical Greco-Roman civilisation and the Judaeo-Christian tradition, and are inextricably bound to pragmatic reconfigurations of power following the disintegration of Christendom. In particular, the liberal values of freedom, tolerance and dialogue are embedded in the globally dominant secular discourse of Western culture. This constitutes the paradox that drives Milbank's critique of inter-faith dialogue in its liberal form: 'The moment of contemporary recognition of other cultures and religions ... is itself ... none other than the moment of total obliteration of other cultures by Western norms and categories, with their freight of Christian influence' (Milbank 1990b, p. 175).

The very notion of inter-religious dialogue assumes that religious traditions are variants on a common genus:

> [I]t is clear that other religions were taken by Christian thinkers [in the modern era] to be a species of the genus 'religion,' because these thinkers systematically subsumed alien cultural phenomena under categories which comprise Western notions of what constitutes religious thought and practice.
>
> (Milbank 1990b, p. 176)

The notion of religion as a 'genus' is a product of modernity: pre-modern dialogue between faith traditions made no such assumption. Milbank contends that it is impossible to arrive at a viable definition of 'religion' that does not ignore essential features of particular traditions: thus Hick's pluralist theology of religion encompasses Christianity within its orbit only by reducing ontological claims about the incarnation to the status of non-realistic myth. Liberalism claims that dialogue constitutes a privileged route to truth, and that it is only feasible provided that all parties embrace a set of common assumptions that make possible 'a sympathetic comprehension of the perspective of the other' (p. 177). Liberal inter-faith dialogue tends to require participants to accept that their various traditions are all expressions of a common genus, from which they emerged and to which they will return. Each religious grouping is free to embrace its own distinctive tradition, but only on condition that it tolerates and affirms the value of alternative traditions.

> One can only regard dialogue partners as equal, independently of one's valuation of what they say, if one is already treating them, and the culture they represent, as valuable mainly in terms of their abstract possession of an autonomous freedom of spiritual outlook and an open commitment to the truth. In other words, if one takes them as liberal, Western subjects, images of oneself.
>
> (Milbank 1990b, pp. 177ff.)

Such a strategy avoids the concrete socio-political contexts and specific truth claims made by particular religious traditions by reducing religion to the level of cultic practice: on this reading the essence of religion is to worship and contemplate God or the transcendent, a practice conducted by different religious traditions *despite* their being rooted in particular socio-political contexts from which they proclaim contingent and ultimately disposable truth claims regarding the precise nature of the transcendent. The liberal call to sympathetic dialogue thus turns out to be a mode of betrayal of the specific self-understandings of different religious tradition, and as such constitutes an imperialism grounded in the arrogance of Western locality. Milbank offers three reasons why this distinctively liberal understanding of dialogue cannot provide a viable foundation for inter-faith dialogue.

(1) In reducing the distinctive particularity of specific religious traditions to the level of cultic practice and isolating it from the generic domains of ethics, aesthetics and politics, liberalism effectively privatises religion and makes inter-faith dialogue a servant of modern secular politics, in which the public sphere is owned and policed by a liberal hegemony and inter-faith dialogue functions to affirm the freedom of individuals and communities to practise their faith in the privacy of their own cultic space, on condition that they tolerate and respect the freedom of other individuals and traditions to do likewise.

(2) In identifying dialogue as a privileged mode of access to truth, liberalism asserts the ideological view – opposed by socialists and radical feminists – 'that the sole principle of justice is the according to everyone the rights of free action and expression, whatever their natural-social status' (p. 182). According to Milbank, such an ideology is morally unacceptable. The freedom of the oppressed to give voice to their oppression does not constitute justice; on

the contrary, is serves merely to perpetuate an unjust status quo. True justice lies in the *obliteration of oppression*, so that *the oppressed are no longer free* to give voice to their oppression. Further, the liberal freedom to speak regardless of natural or social status properly rules out any form of racism or sexism that claims a particular race or gender possesses exclusive access to truth. However, liberalism illegitimately applies this line of reasoning to religion. According to Rosemary Ruether, the 'idea that Christianity, or even the Biblical faiths, have a monopoly on religious truth is an outrageous and absurd religious *chauvinism*' (p. 183, quoting Ruether 1987, p. 141; emphasis added). This suggests that 'religion' possesses the same ontological status as 'race' and 'gender', that it should be treated as a commodity that should be freely circulated and globally available, that different religious traditions are variants on a common genus, and that consequently Christianity 'is only permitted to supply innocuous cultural variants in this process, just as women and men, blacks and whites, are supposed to experience a single human reality in diverse ways' (Milbank 1990b, p. 183). However, if religions are not culturally contingent expressions of a common genus, but rather discrete and ontologically incommensurable accounts of ultimate reality, then the argument cannot apply and Christian exclusivism cannot be intrinsically chauvinistic or 'religionist'. Christianity makes truth claims about the ultimate nature of reality, in the same way that natural science makes truth claims about the natural order. Natural science claims access to universal truths, and is not simply one expression of a common genus that may also include astrology and alchemy. There is nothing inherently chauvinist or 'scientist' about this, since natural science is not simply one possible relative way of looking at the world, but an account of the natural order-of-things that has a legitimate claim to possess greater explanatory power than both astrology and alchemy. In the same way, Christianity claims that its account of ultimate reality possesses greater explanatory power than other religious and secular alternatives, and presents its claims in the public domain for examination and debate. Liberalism's critique of Christian exclusivism seeks to absent Christian truth claims from public debate on ideological grounds, since it knows, in advance of any debate, that in a plural economy religious truth claims cannot be, or must not be allowed to be, exclusive; and it seeks to do so aggressively, charging Christians who hold fast to their belief in the greater explanatory power of Christianity on rational grounds with chauvinistic and religionist attitudes, and thereby associating them with racists and sexists. The fundamental difference between Christians, racists and sexists is that – regardless of whether or not some Christians exhibit sexist and racist attitudes and behaviour – Christianity offers *reasons* for its truth claims that are open to critical assessment, whereas racists and sexists are unable to do so. Christianity would only constitute a form of chauvinism if it were devoid of all rational support; however, the claim that all religions, including Christian, lack epistemic warrant is itself a chauvinist product of the hegemony of the liberal Enlightenment. To discriminate against a person or group of persons on the grounds of their race or sex is not comparable to rejecting a particular religious or secular worldview, or theology or philosophy. If this were not the case, then critical realism would be inherently chauvinistic and unable to reject idealism, positivism and pragmatism without being tainted with the same brush as racists and sexists – a position that is transparently absurd.

(3) If religious identity includes an unavoidable socio-political and doctrinal dimension, it cannot be reduced to the level of culturally bound cultic observance. This being the case, justice cannot possibly be achieved through consensus, because consensus can only be achieved by doing violence to concrete religious identities. 'Only by significantly altering its traditional attitude to sacred law, and thereby its entire received character as a social product, can Islam, for example, bring its treatment of women into line with modern Western, let alone feminist, assumptions' (p. 184). Milbank observes, in passing, that traditional Islamic attitudes to women are not necessarily further from feminist goals than some modern Western stances. Though Milbank does not elaborate upon this, we might venture the suggestion that the freedom of Western women to dress as they choose may possibly in some cases serve to turn them into objects for male consumption, whereas the freedom of Muslim women to wear a veil may possibly in some cases serve to withdraw them from that particular capitalist market: 'Look at my eyes, not at my body – I am a person, not a commodity.' Milbank's central point, however, is that dialogue designed to promote inter-religious agreement 'nearly always betokens the triumph of Western attitudes and a general dilution of the force of traditional religious belief' (p. 184). Whenever Islam asserts itself as a political body it generally tests the boundaries of liberal tolerance. Western toleration of religion proceeds by limiting religion to the private sphere and asserting a secular hegemony over the public sphere of law, politics, knowledge and education. The liberal call to inter-faith dialogue is an invitation, not to a neutral meeting ground, but to a 'place where the other religions and even Christianity itself to some degree, have been most engulfed by the dominance of secular norms' (p. 184).

Milbank claims that the Western celebration of pluralism does little to serve the cause of justice.

> [Y]oking the good causes of socialism, feminism, anti-racism and ecologism to the concerns of pluralism, actually tends to curb and confine them, because the discourse of pluralism exerts a rhetorical drag in a so-called liberal direction, which assumes the propriety of the West-inspired nation-state and the West-inspired capitalist economy.
>
> (Milbank 1990b, p. 175)

Secularism's attempts to ground freedom and justice in a supposedly tradition-transcending universal reason functions merely to buttress Western imperialism. Liberalism's ideological commitment to the notion of a common human nature, dislocated from and transcendent of the concrete particularities of cultural identity, serves to impose an economy of the 'One' onto the plurality of the 'Many'. Where the liberal insists that human beings possess a common natural identity prior to the historical accidents of religious identity, most religious adherents insist that their natural and religious identities are inseparable. Liberalism's failure here lies in its lack of a satisfactory account of the relationship between the One and the Many. Because the hegemony of the liberal tradition-transcending One inevitably does violence to the embodied particularity of the Many, its consensual account of justice is necessarily ontologically violent.

This leads Milbank to his most startling claim: the only viable account of the relationship between the One and the Many is that provided by the Christian ontology of reciprocal love. The bond of love that ontologically unites the persons of the Trinity, and the Trinity and creation through the hypostatic union of God and humanity in the incarnation, and which is therefore the only ontologically viable means of uniting humanity, requires and necessitates the mutual recognition and affirmation of the identity of the 'Other as Other'. This means that inter-faith dialogue must proceed on the basis of an unmediated recognition of difference and a desire to convince the other of the truth of one's own position: anything short of this will constitute a descent into violence. If a critical realist enters into debate with a logical positivist, she does violence both to herself and to her interlocutor if she dishonestly seeks to occlude the differences between their respective positions and fails to argue vigorously that critical realism is a more powerful philosophy than logical positivism. Better the discontented philosopher than the contented sheep. If it is true, as Milbank contends, that secularism is ontologically violent and Christianity ontologically peaceable, then the liberal goals of freedom, justice and social harmony can only be achieved through Christianity. Not to argue this case forcefully, but instead to search for common ground between the two traditions, is to seek an impossible compromise between love and violence, and to do so by adopting the ground rules established in advance by an intrinsically violent ontology.

If Milbank is correct, then Porpora's suggestion that society is not yet ready for inter-faith dialogue constitutes an affirmation of a status quo in which all religious traditions suffer under the hegemony of a secular imperialism that seeks, in the name of freedom and tolerance, the violent conversion of all peoples, nations and faiths to the ideology of Western liberalism. To wait until 'the time is right', as Porpora suggests, is to wait until such hegemony has obtained absolute power. If Western liberalism is indeed fatally flawed in this way, and if it continues to insist that there is no alternative other than to worship at the altar of a consumer capitalism predicated on and buttressed by an incoherent liberal rhetoric, then non-violent revolution rather than appeasement is the only viable option. If we are to achieve freedom and justice, then we must build on foundations of stone not sand, and this means getting our ontology right. This requires the ultimate victory of Christianity, since only Christianity proclaims love as its ontological foundation. A Christianity true to its ontological foundations can never be coercive or imperialistic; rather, it must under-labour for God's creation, recognising, affirming and serving the unique value and particular identities of all human beings as they live, move and have their being within the eternal and unconditional embrace of divine love, and striving to proclaim and persuade in thought, word and deed the exclusive truth that creation is ontologically grounded in the perfect reciprocal love of the one Triune God, Father, Son and Holy Spirit. This being the case, a Christian epistemology cannot be grounded on the rationalistic presuppositions of the hegemonic tradition of universal reason, and, according to Milbank, is entirely justified in refusing this option. The cogency of the argument of this chapter is not linked in any direct way with any assessment of Milbank's radical position that might ensue. To attempt to refute or defend Milbank necessarily requires the application of some form of judgemental rationality; this being the case, any such refutation contains within itself an acknowledgement of the rational challenge laid down by a form of orthodox Trinitarian theology that rejects out of hand Western occidental reason without a necessary reduction of Christian faith to a blind irrational leap into the dark – Milbank's position may be radical, but it is neither blind nor intrinsically irrational.

Part V

Judgemental rationality and the historical Jesus

13 Towards a critically realist historiography

Both the philosophical tradition of natural theology that culminated in eighteenth-century Deism, and the Abrahamic tradition of revealed theology that gave birth to Judaism, Christianity and Islam, claim to discern God in the structures of the natural, aesthetic and moral order-of-things. However, according to the latter, God is more clearly discernible in historical events: divine agency operating within creation reveals God more fully than the imprint of the Creator visible in the structures of the created order. The claim that God was incarnate in Jesus of Nazareth means that Christianity is inextricably grounded in historical events that are open to historical scrutiny. According to orthodox Christianity, the life of Jesus was not merely symbolic of an eternal ideal; rather, it constituted, embodied and actualised a unique spatio-temporal event that ontologically established and defined, once and for all, God's relationship with creation. Consequently history matters to Christians, and does so deeply and profoundly. This partly explains, though perhaps does not excuse, the positivist insistence by fundamentalist Christians that the Bible constitutes throughout a literal record of past events. From the outset orthodox Christianity recognised the diversity of literary forms contained in the Bible: myths and legends, poems and songs, parables and folklore, prophetic pronouncements and wisdom sayings take their place alongside historical narratives. The first Christians were also aware of the complexities of historical reportage, and recognised that it involved seeking to explain the causes of events rather than merely constructing a factual chronology. Thus they adopted four Gospels into the New Testament canon because of the perceived value of their theological explanations of historical events, fully aware in doing so of the problems involved in harmonising their different accounts of the life of Jesus. To say that history matters to Christians is not to suggest that Christian faith is reliant on the results of historical scholarship. The writing of history is, after all, a fallible process, and any work of historical scholarship is open to revision. It is, however, to suggest that the results of historical investigation, grounded in the judgemental rationality of historians, constitute one important thread in the multi-stranded cord through which Christianity strives to affirm its coherence and integrity and clarify its self-understanding.

Critical realism has had a relatively limited impact upon historiography, as compared to its influence in other fields such as natural science, sociology and theology. Gregor McLennan's critically realist study of Marxism and historical method has not to date generated sustained debate among historians, and surveys of historical methodology and the philosophy of history engage with critical realism, at best, only tangentially (McLennan 1981; Bentley 1999; Burns and Rayment-Pickard 2000; Lemon 2003; Day 2008). Historical study of the New Testament is an exception to this pattern: two scholars, Ben Meyer and Tom Wright, have sought to develop a critically realistic historiography applicable to the historical investigation of the life of Jesus and origins of Christianity (Meyer 1989, 1995; Wright 1992). Their efforts have generated an extensive methodological debate, and stimulated a raft of substantial historical studies utilising the insights of critically realistic historiography (Meyer 1979; Sanders 1985; Borg 1993, 1998; Wright 1996, 2000, 2003; Newman 1999; Denton 2004; Stewart 2008). Neither scholar appears to have a direct relationship with the school of philosophical critical realism inaugurated by Roy Bhaskar; instead, both draw inspiration from the parallel tradition of Christian critical realism: Wright draws directly from Thomas Torrance, and Meyer from Bernard Lonergan.

In this chapter we will draw from the work of Meyer and Wright to establish the contours of a critically realistic historiography, using the broad traditions of positivist and idealist historiography as counterpoints in order to tease out its distinctive features. In Chapter 14 we will explore and critique the impact of positivist and idealist historiography on the modern 'Quest for the historical Jesus'. In Chapter 15 we will present the fruits of Wright's attempt to develop a retroductive historical explanation of the life of Jesus of Nazareth and its relationship to the emergence of the early Church.

Positivist historiography

Margaret Thatcher's assertion that history can be reduced to 'factual information … in a clear chronological framework' was condemned as positivistic by her critics (Thatcher 1993, p. 595; Ross 2000, p. 73). Just as positivism reduces reality to regular conjunctions of objects or sense data devoid of any metaphysical or theological underpinning, so positivist historiography reduces history to a sequence of factual events strung together in chronological order. The principle of verification, of testing truth claims against sense experience, enables the positivist to distinguish between true and false statements, and thereby discard the meaningless assertions of moral, aesthetic, metaphysical and theological discourse. In the case of historical claims, where direct empirical verification is no longer possible, verification must take place indirectly, by weighing the balance of probability. In evaluating a historical source, the positivistic historian must first distinguish between value judgements and historical 'facts'. Value judgements do not constitute objective historical facts, but rather subjective responses to such facts, in the form of aesthetic, moral, metaphysical or theological interpretations overlain on the events, either by the historical actors themselves or by later historians. The pursuit of historical objectivity requires the historian to strip away all such subjective interpretations and thereby expose the objective historical facts. Though this does not preclude the historian describing the beliefs and values held by historical agents, it does disqualify her from evaluating them and addressing the question of their truth. For the positivist historian, claims that in Germany under Nazi rule 'the country was depopulated', 'millions of people died' and 'millions of people were killed' all constitute legitimate factual statements. However, the claims that 'millions of people were *massacred*', and that the deaths 'were part of a single organised campaign of *brutal* killing', stray into the illegitimate realm of subjective assessment, because the terms 'massacre' and 'brutal' possess a perlocutionary force that transcend the purely factual (Bhaskar 1998, p. 59; Wright 2007, p. 169). Having stripped away all value judgements, the positivist historian must now sift through factual statements in order to separate the true from the false. According to David Hume's 'calculus of probability', the likelihood of a historical event having occurred is directly proportional to (1) its conformity to the regular conjunction of events that constitute our ordinary everyday experience of the world, and (2) the trustworthiness of the historical testimony. By this measure it is probable that Alfred Cobban's account of the execution of Marie Antoinette is accurate, (1) because executions are common during violent revolutions, and (2) because Professor Cobban is a respected historian. By the same measure, it is unlikely that the Gospel accounts of Jesus' resurrection are historically accurate, (1) because people rising from the dead do not form part of our ordinary everyday experience, and (2) because the claims of religious enthusiasts are frequently untrustworthy.

The suggestion that positivist historiography reduces history to factual events organised in chronological sequence requires qualification. Auguste Comte warned against imposing a 'systematic empiricism' on 'historical observation' in a manner that is 'radically opposed to the true spirit of positivist philosophy' (Burns and Rayment-Pickard 2000, p. 113). Historical facts remain idle, sterile and uncertain if separated from 'the intellectual attitudes and speculative sign-posts indispensable to their scientific exploration' (p. 114). A truly scientific history must move beyond the mere chronicling of events by ensuring that observations are related to historical laws, which are discovered by identifying regular sequential patterns in historical data, and are not to be confused with the positing of causal explanations. Comte recognises the 'danger that continuous use of scientific theories may distort real observation, reading into it the confirmation of speculative prejudices that have no real basis' (p. 114). However, any such danger is obviated by the fact that the 'chief characteristic of positive theories is the continuous and systematic subjection of imagination to observation' (p. 114). It was Carl Hempel who did most to provide Hume's 'calculus of probability' with the scientific grounding envisaged by Comte. Positivistic social science employs a deductive-nomological procedure, seeking to deduce general laws of human behaviour from close observation of the regular conjunction of events: poverty correlates positively with poor health and low life expectancy; voting patterns reflect economic interests; capitalism breeds inequality; power tends to corrupt, etc. Once such laws have been established, the historian can utilise them when evaluating historical testimony. The ensuing positivist science of history – dubbed by Popper a form of 'historicism' – aims at historical prediction, 'attainable by discovering the "rhythms" or the "patterns", the "laws" or the "trends" which underlie the evolution of history' (Popper 1957, p. 3). The dialectical nature of the relationship between historical judgements and historical laws enables the historian to refine both over time. Since human behaviour is less uniform than nature, historical law tends to lack the predictive accuracy of natural law, and as such is elliptic, elastic and more probabilistic than deterministic.

The self-imposed embargo on evaluative judgements did not prevent positivist historians from presenting history as a story of human progress; indeed, the inexorable passage from religious through metaphysical to positivistic accounts of reality is central to Comte's historical vision. True, Hume's general cynicism regarding human nature led him to play down the possibility of historical progress. However, the views of John Stuart Mill and Henry Buckle are more representative of positivistic historiography: the former held that historical development was inevitable, with human stupidity and genius serving merely to temporarily impede or accelerate progress; the latter held that increasing mastery of the laws of nature and concomitant control of the natural world served to generate an environment ever more conducive to human progress (Burns and Rayment-Pickard 2000, pp. 102ff.). Thus the positivistic affirmation of fact and rejection of value is not followed through consistently: in affirming human progress positivist historians make value judgements.

Despite the ongoing influence of positivism upon the historian's craft, there is an emergent consensus that:

> There is not, nor can there be, any such thing as a bare chronicle of events without a point of view. The great Enlightenment dream of simply recording 'what actually happened' is just that: a dream. The dreamer is once more the positivist, who, looking at history, believes that it is possible to have instant and unadulterated access to 'events'.
>
> (Wright 1992, p. 82)

There are no such things as 'bare and unvarnished facts': our understanding of the past requires us to make selections from a vast range of evidence, and any attempt to set out the totality of 'facts' without selection will inevitably fail owing to 'the sheer overwhelming volume of information' (p. 83). Even a video recording, which at first glance may appear to offer direct access to empirical data from the past, inevitably involves a process of selection: the camera records a certain set of images taken from a particular perspective, the film editor selects extracts from the recorded material and places them in particular sequences, and different viewers focus on different aspects of the finished film. Tom Wright suggests that selection processes are ingrained in the fabric of our everyday existence: 'We do this all the time ... selecting tiny fragments of our lives and arranging them into narratives, anecdotes, family legends, and so on' (p. 83).

The 'legacy of positivism often seduces us into imagining that a "fact" is a "purely objective" thing, unalloyed by the process of knowing on anybody's part' (p. 83). In reality there are no pure facts: our discernment of objective reality cannot bypass our subjectivity; our grasp of ontology cannot circumnavigate epistemic relativity and judgemental rationality. This unity of fact and value runs deep: it is not simply that on an epistemic level we cannot engage with objective facts apart from our subjective values; rather, at the ontological level reality itself is inherently value-laden and meaningful (Collier 1999).

Our understanding of reality cannot be limited to the *empirical domain* of our observation of regular patterns in empirical events: if we are convinced by our experience that all swans are white, we have never visited Dawlish in Devon and seen the black swans introduced to the town from New Zealand by John Nash. Neither can our understanding be limited to the *actual domain* of the sum total of objects, events and experiences: even if extensive enquiries indicate that no human being has ever reported seeing a swan that is a colour other than black or white, we cannot rule out on purely empiricist grounds the possibility of the existence of green swans inhabiting an entirely unexplored part of the world, or the possibility of the future emergence of orange swans. A rich explanation of reality needs to pass beyond both the *empirical domain* of our own experiences, and the *actual domain* of the totality of experiences, objects and events, and to penetrate the *real domain* of the underlying causal structures and generative mechanisms that sustain and generate different states of affairs in the world (Bhaskar 1997, pp. 56ff.). In doing so, we can *explain* the colour of swans with reference to their biological make-up and DNA, just as we can *explain* that rubies are red owing to the presence of the 'impure' element chromium in the mineral corundum. Such explanation transcends the limitations of a statistically 'significant' set of reported observations. In the same way, we can *explain* the past, not by generating an 'objective' chronology of historical facts, but by accounting for the interaction of a variety of causal mechanisms, in the form of social structures and individual agencies, that were responsible for the configuring of particular historical events.

The insistence on congruence between the interpreter's experience of reality and the claims of historical testimony generates hermeneutical closure, driven by the presumption that historical testimony has validity only insofar as it conforms to the naturalistic worldview of the positivist historian. As Thomas Torrance notes, the positivists' strategy of protecting objective fact from subjective distortion by appealing to their own understanding of human nature tends to 'assimilate historical investigation and observation to the modes of thought that obtain in the natural sciences', in particular to 'a form of natural science conceived after a Newtonian or pre-quantum model' (Torrance 1969, pp. 314ff.). The deductive-nomological identification of laws of both nature and human nature generates a deterministic worldview that affirms necessity over contingency. As a consequence, 'many historians, operating with the concept of a closed continuum of cause and effect as a primary canon of credibility, develop a positivist and often a deterministic view of history' (p. 315). This ignores the relativity of the historian's perspective, undermines any possibility of historical testimony opening up new and potentially more truthful ways of understanding reality, and denies the historian 'the experience of being pulled up short by the text' because 'its meaning is not compatible with what we had expected' (Gadamer 1979, p. 237). Positivist historians, insofar as they embrace their own understanding of reality as the only valid criterion for judging the past, tend to lack the intellectual virtue of humility, which demands recognition of the epistemic contingency of all interpretive frameworks.

> A person trying to understand a text [should be] prepared for it to tell him something. That is why a hermeneutically trained mind must be, from the start, sensitive to the text's quality of newness. But this kind of sensitivity involves neither 'neutrality' in the matter of the object nor the extinction of one's self, but the conscious assimilation of one's own fore-meanings and prejudices. The important thing is to be aware of one's own bias, so that the text may be present itself in all its newness and thus be able to assert its own truth against one's own fore-meanings.
>
> (Gadamer 1979, p. 238)

Ernst Cassirer notes, with reference to modernity, that 'it is customary to consider a major shortcoming of this epoch that it lacked understanding of the historically distant and foreign' (Cassirer 1951, p. x). As a result, 'in naive overconfidence it set its own standards as the absolute, and only valid and possible, norm for the evaluation of historical events' (p. x). The Renaissance's rediscovery of a classical antiquity no longer filtered through the lens of medieval Christianity generated a new historical consciousness. As heir to this tradition modernity was not indifferent to history per se, merely mistrustful of the Christian account of history. If the past could not be looked to as a source of new knowledge, or of knowledge capable of resisting and challenging prevailing modernist worldviews, it could nevertheless be 'useful' and 'beloved' in offering a superficial, if inescapable, feeling of congeniality towards classical antiquity. The Graeco-Roman tradition had sought to oversee the triumph of reason over myth, most visibly in the writings of Lucretius and Cicero, only to see that achievement 'undermined' by Christianity. With the dawning of the Enlightenment history turns full-circle as, once again, reason claims to assert itself over religious 'myth'. History, on this reading, has nothing new to tell us: its value lies purely in its ability to confirm that which we already think we know. Ultimately, positivistic historiography perpetuates a version of the epistemic fallacy.

Idealist historiography

If positivism cultivates a naturalistic worldview in which reality is ultimately reducible to regular occurrences in the material world, then idealism affirms the primacy of the non-materialist realm of ideas. For Plato the realm of the Forms constitutes an eternal reality that the ephemeral world of matter reflects only partially. Cartesian dualism had a crucial impact upon modern forms of the realist/idealist debate, forcing protagonists to operate with a set of polar opposites: mind/matter, self/world, subjectivity/objectivity, scepticism/certainty, value/fact, etc. Where positivism reduces history events to a chronology of factual events, idealism views such events as reflections of eternal values. If we dislocate the self from the world, and look for the meaning of history in our established idealistic values rather than in the contingency of past events, then we 'cannot escape the grave error of reading back into the past what we want to think in the present, so of reducing history in the end only to the mode of encountering and understanding ourselves' (Torrance, 1969, p. 316). Kant's phenomenalism, in denying the possibility of penetrating beyond the phenomenal realm of things-as-they-appear-to-us to the noumenal realm of things-as-they-are-in-themselves, established a modern idealistic tradition that may be traced from Hegel's Absolute Idealism through Romanticism to post-structuralism. Common to all three is a belief in the active role of the mind in constructing accounts of reality. In limiting knowledge to the realm of phenomenal appearances Kant effectively deprived it of any underlying ontological referent, thereby 'giving rise to a romantic idealism where the human spirit could range at will, uncontrolled by scientific evidence or knowledge' (Torrance 1980b, p. 27).

(1) Absolute Idealism. Hegel sought to overcome Cartesian dualism by representing reality as an organic unity, in which mind and matter evolve together in dialectical relationship. He affirms the ontological being-in-becoming of Spirit (Geist/Mind), via a deterministic process of the progressive realisation of self-consciousness and freedom. Consequently his *Philosophy of History* regards history as an ongoing series of increasingly sufficient manifestations of Spirit: 'Universal history … is the exhibition of Spirit in the process of working out the knowledge of that which it is potentially' (Hegel 1956, pp. 17ff.; Lemon 2003, p. 215). As the progressive realisation of Spirit, history is driven by a necessary teleology: 'this is the only aim that sees itself realised and fulfilled; the only pole of repose amid the ceaseless change of events and conditions, and the sole efficient principle that pervades them' (Hegel 1956, pp. 19ff.; Lemon 2003, p. 215). Since Spirit manifests itself above all in nation-states and national cultures, history is primarily the study of the progressive development of political structures from primitivism through to maturity, though Hegel does acknowledge the role of world-historical individuals, those 'Heroes' who stand out from the crowd in their unconscious enactment of Spirit at particular stages in its development. Crucially, the philosophical principles underlying the historical process can be established prior to historical investigation: 'Hegel accepts the need for a pre-existing analytical framework with which to approach "the facts" … [derived] from established principles gained from Philosophy' (Lemon 2003, p. 216). This does not rule out the need to study history empirically, since these ideal principles need to be related to specific historical events. Hegel here cites the case of Johannes Kepler, who, he suggests, 'must have been familiar *a priori* with ellipses, with cubes and squares, and with the ideas of their relation' prior to discovering the laws of planetary motion via empirical observation (Hegel

1956, p. 64; Lemon 2003, p. 216). Thus history must necessarily conform to a pre-established idealistic framework: the historian, cognisant in advance of the necessary thrust of historical development, has the task of filling in the specific details – a kind of historical 'painting by numbers'.

(2) Romantic Idealism. The roots of Romantic historiography may be traced back to Rousseau's critique of the Enlightenment as the triumph of corrosive culture over virtuous nature. He finds an inductive basis for the general law that human culture generates moral degradation in historical studies of the rise and fall of ancient civilisations: 'history teaches that people first lived close to nature in a pristine ignorance that enabled their "natural" virtues of straightforwardness and honesty to flourish' (Lemon 2003, p. 174). Where Hegel prioritised reason, the Romantics highlighted emotion: the freedom inherent in the immediacy of sensibility is contrasted with the constraints intrinsic to the reification of sense. This led them to tell a story of *the loss and future recovery of a golden age of primal innocence*, in which human beings once lived, and will once again live, by instinct rather than reflection. The task of the historian is to emplot historical events within this primal mythical story. The process of 'romantic emplotment regards the story as a triumph, the hero (which may be an individual, a class, a nation, or an idea) transcending the formidable difficulties that beset their past, transforming themselves and their world' (Day 2008, p. 175). The myth assumes various literary forms: romance, comedy, satire, etc. However, it is tragedy – in which the hero faces suffering and even death *en route* to redemption and transformation – that provides the prototypical form of Romantic meta-narrative of the future utopian recovery of an occluded idyllic past. Hayden White's study of Romantic historiography draws attention to the ways in which this Romantic myth is imposed upon history by emploting historical events within its narrative framework (White 1973). In his *Addresses to the German Nation* Fichte responded to Napoleon's occupation of Berlin by telling a story of the past greatness of the German race and contemporary infiltration and contamination by the Jews, and calling for a 'People's War' (*Volkskrieg*) designed to re-establish national unity and purify the Arian race (Fichte 2009). Such Romantic emplotment has been regularly repeated: in Nazi Germany (the future triumph of the purified Aryan race), in the hippy culture of the 1960s (the stardust and golden 'Woodstock Generation' returning to the Garden of Eden), in the New Age movement (the spiritual reconnection with divinised nature), and in the racist rhetoric of the right-wing British National Party (the re-establishment of a mythical 'all-white' Britain). That the moral and material substance of this Romantic primal narrative is free-floating is unsurprising, since the roots of Romantic emplotment are situated in the subjective grievances and hopes of particular individuals and groups. Crucial, for our present purposes, is that once again we encounter the imposition of a predetermined historical vision onto historical data.

(3) Linguistic Idealism. Postmodernity constitutes a radicalisation of the Romantic strand of modern thought, rather than a negation of modernity per se. The Enlightenment's dream of freedom was undermined by the emergence of an impersonal, rationalistic, bureaucratic and deterministic social order that severely restricted personal autonomy. The Romantic reaction sought to recover a

personal, affective, relational and free-flowing community in which sense gave way to sensibility and the human spirit was freed to pursue its own destiny. However, the fruit of this newfound freedom was tyranny and totalitarianism: Rousseau's exercise of unconstrained primal emotivism led directly to the gates of Auschwitz. The belief that reason could free humanity from the tyranny of God, and that sensibility could liberate humanity from the tyranny of reason, led to the stark realisation that human sensibility itself could be equally as tyrannical, if not more so. The postmodern reaction was to shift the grounds of the debate: the problem was not the failure of God, or human sense, or human sensibility to lead us to paradise, but that humanity elected to seek paradise in the first place – it was the quest itself that was misplaced. The theological, rationalistic and emotive thirst for paradisiacal security, which generated a series of totalitarian meta-narratives designed to order and control reality, must now be abandoned. Hence Jean-François Lyotard's call to 'wage a war on totality', Michel Foucault's call 'to abandon all those discourses that once led us to the sovereignty of consciousness', and Julia Kristeva's insistence that selfhood is transient and negotiable (Lyotard 1984, p. 82; Foucault 1989, p. 202; Kristeva 1994). If there is no God above, no nature beneath, and no stable self within, we are freed to live as we desire, freed to engage in a directionless nomadic wandering, freed to participate in the playful cycle of an ongoing construction, deconstruction and reconstruction of imaginary realities, freed to embrace 'the fundamental principle of understanding what it means to be human in the modern (and postmodern) era' – freedom itself (Schwöbel 1995, p. 57). If language can no longer connect with God, describe nature, or express our inner selves, we are free to use words as playthings, constructing an abundance of language games in order to satisfy our desires and exercise our imaginations. What then of history? It becomes cyclical: a never-ending round of eternal recurrences lacking in any teleological direction, as valorised by Nietzsche and Schopenhauer. The writing of history becomes a form of disengaged rhetoric, in which historical 'facts' constituting the raw material that the postmodern historian is free to utilise at will by employing a range of discursive strategies to emplot a variety of different narratives. As Hayden White implies, in idealist historiography historical facts cease to be the positive substance of history and instead become the footnotes supporting and justifying our given values and ideals (White 1973).

Idealist historiography, whether in its Absolute, Romantic or post-structural guise, tends to locate the meaning of historical events in idealised value systems that transcend the events themselves, and as a consequence is predisposed to emplot historical data within pre-established extra-historical narratives. There is no sense in which the significance of historical events might lie in the events themselves: instead, *a priori* assumptions regarding the ultimate meaning of history are read back into historical testimony. As such, idealist historiography engenders the process of 'inventing "history" by a backwards projection of ideology' (Wright 1992, p. 85). Allowing idealistic eisegesis to replace realistic exegesis constitutes a cardinal error in the interpretation of historical testimony. The poverty of idealist historiography is transparent: to emplot historical events arbitrarily within a pre-established narrative framework of our own making is to impose our own established values onto the past, thereby reducing history to the role of exemplifying and valorising our primary beliefs and values in a manner that risks doing violence to the past. The past does not exist to be airbrushed into images that merely confirm our prejudices and preferences. Bernard Williams suggests that despite its prevailing distrust of the hegemony of *truth*, contemporary society retains a deep-rooted concern for *truthfulness*, in the form of 'a pervasive suspiciousness, a readiness against being fooled, an eagerness to see through appearances to the real structures and motives that lie behind them' (Williams 2002, p. 1). If the historian subjects historical data to a hermeneutic of suspicion in order to establish the true nature of past events, then she has a responsibility to acknowledge, scrutinise and account for her own ideological and methodological commitments insofar as they impinge upon this process.

One final observation: despite their apparent polarity, there is not actually a great divide between positivist and idealist historiography. At first glance, the suggestion that idealist historiography feeds off positivist historiography may appear difficult to justify. However, it is the unverifiable nature of the principle of verification that exposes positivism's covert idealism: ultimately, the positivist historian orders and shapes the past by imposing the idealist criterion of probabilistic verification onto historical data. Despite the claim to bracket subjectivity and deal only with objective facts, it is the positivist historian's experience of regular conjunctions of events, assessed against normal everyday experience, which controls her understanding of the past. Thus, Hume's notion of the constancy and uniformity of human nature is imposed upon history rather than derived from it. As Robert Burns points out, Hume acknowledges that 'the ancient Greeks differed in many aspects of their behaviour – incest, infanticide, homosexuality, suicide – from modern Frenchmen' (Burns and Rayment-Pickard 2000, p. 34; cf. Hume 1902, pp. 333ff.). However, he fails to resolve the ensuing dichotomy between the *a posteriori* empirical results of his own historical investigations, which reveal 'that the characters of men are to a certain degree inconstant and irregular', and his *a priori* claim that the task of history is 'to discover the constant and universal principles of human nature' (Hume 1902, pp. 88, 83). If positivist historiographers ultimately emplot historical data within a preconceived interpretative framework, then precisely the same accusation may be set at the feet of idealist historiographers. Though idealist historiography starts from a different premise than that embraced by positivist historians, it arrives at the same outcome: the insertion of historical data into a preconceived procrustean framework. Idealism's turn from fact to idea trains the spotlight on the subjectivity of the historian and opens the door to forms of anti-realism, constructivism and relativism. Wright suggests that the idealist assertion of the primacy of mind over external reality ultimately casts the historian into a morass of fantasy (Wright 1992, pp. 33ff.). If we accept that historical understanding is necessarily coloured by the perspective of the historian, yet at the same time continue to hold fast to the unobtainable ideal of a neutral viewpoint, then our knowledge of the past will inevitably become irredeemably subjective.

Critical realist historiography: ontological realism

We turn now to a critical realist alternative to positivist and idealist historiography, in which history 'is neither "bare facts" nor "subjective interpretations", but rather the *meaningful narrative of events and intentions*' (Wright 1992, p. 82; emphasis in original). Positivist historiography scours historical testimony, stripping away value judgements and false testimony to reveal factual events arranged in chronological order. John Locke provides the basic epistemic model here: the mind receives atomistic packets of sense data (historical 'facts') which it arranges into complex ideas (chronological order). This has ontological implications: reality is reducible to atomistic facts, since complex wholes exist only nominally, as constructions of the historian. However, on a more holistic view, when we see (say) a cat, we encounter a substantial ontological reality directly, rather than construct the idea of 'cat' from isolated packets of sense data (black colour, soft feel, purring sound, etc.). Further, we discern the creature in a specific spatio-temporal context: it is inextricably related to its environment, and possesses both a past history and future potential. Thus the cat is located in a web of meaning that we articulate by telling stories. Wright argues that stories constitute a basic mode of human life: 'It is not the case that we perform random acts and then try to make sense of them; when people do that we say that they are drunk, or mad' (1992, p. 38). The stories through which we live out our lives are not reducible to either bare fact or abstract ideal. Rather, our identities are rooted in and constituted by our life stories: a husband working in the garden may be simultaneously gardening, preparing for winter, pleasing his wife by taking exercise, and responding spiritually to creation and its divine creator. Here the husband's primary intentionality to garden is intrinsically related to the annual cycle of domestic activity, the narrative history of his life and marriage, and his eternal destiny (MacIntyre 1985, p. 207).

Wright identifies stories as a key element in our worldviews, alongside symbols, praxis, and basic or ultimate questions and answers (Wright 1992, pp. 38ff.). As the aetiological myths of primal societies show, the location of stories on the map of human knowledge is more fundamental than explicitly formulated beliefs. We tell stories because this is how we perceive and relate to the world. The notion that stories are a basic constituent of human life does not carry the implication that they can be reduced to fictitious constructions dislocated from reality. We tell stories about ourselves and our place in the ultimate order of things because we believe them to be true. Stories may offer truthful accounts of past events (e.g. accounts of Jesus telling parables), or act as truthful indicators of the actual nature of reality (the parables themselves). The fact that to tell stories requires the rhetorical construction of plots and designation of characters, and that the same historical realities are capable of generating different stories, need not imply any thoroughgoing relativism. Though there are many different ways of constructing narrative representations of reality, we can nevertheless judge some to be more truthful than others without reducing them to verifiable facts or privileging one form of representation over another. The stories we tell confirm, subvert or modify alternative stories: the Christian story constitutes a partly confirmatory and partly subversive retelling of the Jewish story; the Islamic story constitutes an emplotment on the stories of Judaism and Christianity within an alternative Muslim narrative; the post-Enlightenment naturalist worldview rejects the theological stories of the three Abrahamic faiths and denies them any purchase on reality. Contra much liberal praxis, stories that embody universal truth claims cannot be reduced to privatised, fictional and non-realistic worldviews: a non-liberal worldview that contradicts the liberal one needs to be contradicted ('that is not how things actually are') rather than patronised ('of course you are free to believe that, provided it remains within the boundaries of your own private faith community'), and failure to do so allows power structures victory over rational debate.

Where positivism seeks to bracket the subjectivity of the historian in order to reveal the objectivity of the past, idealism gives subjectivity priority, insofar as the historian is deemed to be in possession of a pre-existing narrative into which historical data can be emplotted. Critical realist historiography rejects such a dualistic distinction between the passive reception of facts and active construction of meaning. Contra Descartes, we are not isolated creatures, dislocated from the external world and reliant on knowledge located in the inner receptacle of the mind: as Wittgenstein observes, 'there is a kind of general disease of thinking which always looks for (and finds) what would be called a mental state from which our acts spring as from a reservoir' (Wittgenstein 1969, p. 143). Rather, since we indwell the world and participate directly in reality, knowledge is constituted, not by our inner mental states, but by the ongoing *relationship* between the subjective knower and the object of knowledge. There is thus an unavoidable personal coefficient of knowledge. 'A person who is trying to understand a text is always performing an act of projection': the questions we formulate when investigating a past event are dependent upon expectations and assumptions ingrained in our given worldviews and belief systems (Gadamer 1979, p. 236). Without prior prejudices, it would be impossible for the historian to interrogate the past: this is so even for positivist historians who, in seeking the purity of an objective description of the past, embrace 'the fundamental prejudice of the enlightenment ... the prejudice against prejudice itself, which deprives tradition of its power' (pp. 239ff.). The pursuit of historical knowledge proceeds via abductive intuition, the retroductive construction and testing of new explanatory narratives, and their subsequent iterative consolidation and refinement. The subjectivity of the historian is mirrored in the subjectivity of participants in past events: the proper objects of historical attention are not objective facts, but 'human subjects and communities, with their experiences, deeds, views, words, and their truth and falsity, in inter-action with each other and with nature' (Torrance 1969, p. 313). The concerns of the past are intermeshed with the concerns of the present: the Gospel writers articulate truth claims regarding the eternal significance of Jesus of Nazareth that are replicated by contemporary Christians, and that either confirm or challenge the worldviews and belief systems of contemporary historians. To explore a past event that directly impinges upon one's own worldview, whether positively or negatively, with studied indifference is to do a disservice both to oneself and to the event itself; all the more so once we recognise 'that the full story of the "inside" of an event may only be unfolded gradually, in the light of subsequent and consequent events' (Wright 1992, p. 115). Thus, it was 'only in the years after the Second World War, as the truth of the "Final Solution" came to be known, that one could really understand what had been going on all along in Germany in the 1930s' (p. 115). If history involves the narration of meaningful events rather than the identification of bare facts, it cannot bypass the intersection of the subjective horizons of meaning of both historical participants and historical investigators. Such subjectivity does not diminish the intellectual status of history; on the contrary, it 'binds the historian to all other disciplines', since all knowledge, in every intellectual domain, necessarily involves a personal coefficient (p. 113). There can be no unmediated access to ontology: the pursuit of truth necessarily proceeds via the epistemically relative subjectivity of both historical actors and the historians of their actions.

The crucial point, however, is that past events constitute substantial ontological realities that actually occurred regardless of whether the historian is aware of them or not. Though the subjectivity of the historian is a necessary and unavoidable factor in the pursuit of historical knowledge, this does not permit her to impose her own assumptions, whether positivist or idealist, upon the past. Rather, a critical realist historian should be concerned to allow the past itself to establish its own reality and shape and reshape the subjectivity of the historian. For ideological purposes, one historian may wish to maximise the horrors of the Holocaust, while another may wish to minimise and even deny them. Though their ideological commitments cannot be removed from the interpretive frame, both have a moral and intellectual responsibility to bring their minds into conformity with the historical actuality of the Holocaust itself, insofar as it can be accessed by the mediating data and testimony, rather than bring the past into conformity with their own ideological commitments.

Critical realist historiography: epistemic relativism

Critical realist historiography is committed to the pursuit of ontological truth, in the form of a holistic narration of intrinsically meaningful events rather than an atomistic description of chronologically ordered facts. It is also committed to epistemic relativism, recognising that historical data is open to a range of alternative narratives, few of which are likely to be absolutely wrong, and some of which may have legitimate claims to being more truthful than others. This immediately undermines the totalitarian claims of positivism and idealism.

> History, then, is real knowledge, of a particular sort. It is arrived at, like all knowledge, by the spiral of epistemology, in which the story-telling human community launches enquiries, forms provisional judgements about which stories are likely to be successful in answering these enquiries, and then tests these judgements by further interaction with the data.
>
> (Wright 1992, p. 109)

The narration of meaningful historical events must avoid surface descriptions of regular occurrences of events and shun the instrumental emplotment of historical data within a preconceived idealist framework. Instead, it must seek to construct deep retroductive explanations of contingent events: Why did 'X' happen, when we might equally have expected 'Y' or 'Z'? Such deep explanation requires the historian to attend to the *aims*, *intentions* and *motivations* of participants in historical events, as well as the constraining and enabling social structures within which they act. Wright uses 'aim' to refer to the fundamental orientation of a person's life in relation to a particular mindset and worldview, 'intention' to refer to the application of the aim in a particular (in principle repeatable) context, and 'motivation' to refer to the actualisation of the aim in a specific (one-off) situation (pp. 109ff.). Thus, to explain my purchase of company shares, it is necessary to account for my *life aim* (in relation, say, to my capitalist worldview and desire to become rich), my *general intention* (say, to invest money for financial gain), and my specific *motivation* (to purchase those particular shares at that specific time). Similarly, to explain Jesus' stage-managed entry into Jerusalem riding a colt a few days before his arrest, trial and crucifixion it is necessary to account for his *life aim* (in relation to his subversive reading of the first-century Jewish worldview and desire to inaugurate the Kingdom of God), his *general intention* (to take his message to the Jewish heartland of Jerusalem), and his *specific motivation* (to proclaim his Messianic status by enacting the prophecy of Zechariah 9:9f.).

According to Collingwood, to generate such deep explanations the historian must seek to penetrate the 'inside' of events, thereby re-enacting the past and reconstructing the thinking of historical participants (Collingwood 1946, pp. 213ff.). This requires empathy, intuition and imagination as well as close attention to the whole range of historical evidence, directed towards a noetic engagement with the inner logic of the historical events under scrutiny and designed to generate a powerful retroductive explanation of the past. Torrance embraces Collingwood's basic position:

> It is certainly true that it is the historian's task to discover the intrinsic logic of historical events by interpreting them as embodying intention or as expressive of purpose, and that in this way he really is concerned with an 'inner' and intelligible aspect of historical phenomena and not simply with a bare chronicle of *res gestae*.
>
> (Torrance 1969, p. 316)

However, he expresses concern that 'Collingwood took too idealistic a view of history and too materialistic a view of natural science', and thereby tended to reduce history to the expression of mental experience within a closed materialist frame (p. 316). With regard to 'mental experience' Wright argues that, both in everyday life and historical investigation, it is entirely possible to explain a person's aims, intentions and motivations without the necessity of resorting to Freud, Jung or depth psychology (Wright 1992, p. 112). With regard to a 'closed materialistic frame', Wright suggests that a rich historical explanation needs to place 'the complex network of human aims, intentions and motivations, operating within and at the edges of the worldviews of different communities and the mindsets of different individuals', alongside relevant social structures, and the interplay of contingency and regularity in the natural environment (p. 113). 'The task of the historian is thus to address the question "Why?" at all possible levels, down to its roots in the way the people under investigation perceived the world as a whole' (p. 112). It is possible to go a stage further here, and suggest that it is not sufficient simply to identify the fact that Jesus' aims, intentions and motivations were intrinsically related to a set of theological beliefs and commitments: given that theology is an entirely legitimate intellectual exercise, any retroductive account of the life of Jesus cannot justifiably avoid the question of the truth of Jesus' theological claims in general, and his belief that God is an active agent in the historical process in particular. Just as economic history requires economic insight, and political history needs political acumen, so it follows that theological history demands theological discernment.

Critical realist historiography: judgemental rationality

We understand the past retroductively, by generating and refining historical narratives in an ongoing quest to produce increasingly powerful and comprehensive explanations of past events. Since we have no direct access to the ontological actuality of the past, it follows that such narratives necessarily lack secure foundations. Consequently, we must both accept epistemic relativism and embrace judgemental rationality. Just as, given 'a range of possible meanings [of a poem] the critic is justified in selecting that meaning which, in the light of the poem as a whole (and any other relevant evidence), makes the best sense', so the historian is justified in judging 'between competing [historical] theories on the basis of their intrinsic merits as explanations of [historical] reality' (Mitchell 1973, p. 51; Lopez and Potter 2001, p. 9). The fact that at 'the foundation of well-founded belief lies belief that is not founded' does not imply judgemental relativism (Wittgenstein 1975, §253). Given the contingent nature of even the most powerful retroductive explanations, historians must continually strive for iterative clarification, since our 'epistemic responsibilities do seem to require that we be open to changing any *one* of our beliefs if given sufficient reason to do so' (Marshall 2000, p. 145; emphasis in original).

The historian seeks to assess the 'culmination of probabilities, independent of each other, arising out of the nature and circumstances of the case which is under review' (Newman 1979, p. 230). However, the historian cannot appeal to any fixed criteria when judging between competing retroductive explanations of historical events. This is because the primacy of ontology over epistemology requires us to refine our epistemic tools in response to the ontological reality we seek to comprehend. Without such refinement there would be no methodological progress in the discipline of historiography. Historians do, of course, employ specific interpretive methods and procedures, and strive to provide reasons for their judgements; however, such tools and judgemental criteria are provisional measures, subject to development and refinement. Just as an astute political commentator will seek to 'read' a developing political event, so an astute historian will seek to 'read' history, and neither will attempt to do so by mechanistically applying a pre-established set of interpretive criteria. Thus the craft of the historian combines scientific rigour and intuitive wisdom. Here Willard Quine appeals to 'receptivity to data, skill in reasoning, and yearning for truth', while, as we have seen, Bernard Lonergan advocates attentiveness, intelligence, reasonableness and responsibility (Quine and Ullian 1978, p. 4; Lonergan 1973, p. 55). Though the historian does not have access to fixed 'formal criteria' when seeking to adjudicate between conflicting historical narratives, she can nevertheless draw on a series of contingent 'generic indicators': attentiveness, inclusivity, congruence, coherence, fertility, simplicity and depth (Wright 2007, pp. 219ff.).

(1) Attentiveness. Any powerful retroductive explanation will, above all, be attentive to the available evidence. In the case of the history of Jesus of Nazareth, this requires close attention to the primary sources: the Gospels, other canonical and non-canonical Christian texts, Gnostic manuscripts, and writings of Jewish and Greek historians, together with a raft of archaeological evidence. To make sense of these primary sources in historical context, close attention needs to be paid to evidence that sheds light on the cultural milieu of first-century Palestine. Specialist academic study requires that both primary and contextual evidence be read in their original languages (Greek, Hebrew, Aramaic, etc.). In addition, due attention must be given to the reception history of the interpretation of the primary evidence, from the first century through to the present day, and to the fruits of contemporary scholarship.

(2) Inclusivity. A historical account is likely to possess greater explanatory power if it is inclusive of all the available evidence. To ignore, exclude or circumnavigate a set of historical data begs the question of the adequacy of the explanatory account proffered by the historian. The principle of the hermeneutical circle, that the parts and whole of a raft of evidence reciprocally support and illuminate one another, is a crucial factor in critically realistic historiography. No cluster of data can be ignored or circumnavigated, and should only be excluded if there is a strong justification for doing so. Positivist historiography tends to emphasise the inclusion/exclusion of atomistic events without reference to the greater whole, while idealist historiography tends to emphasise the greater whole without reference to the individual parts.

(3) Congruence. Because historical actors and historians participate in the same ontological reality, one should anticipate a level of congruence between their experiences. However, this expectation needs qualification in the light of the epistemic relativism of the historical understanding of both parties: thus, for example, though miraculous events were not considered especially extraordinary by most first-century Hellenists and Palestinians, they are rejected as impossible by many modern historians. The resolution of such incongruence cannot proceed simply by rejecting one epistemic position in favour of the other. Rather, the critically realistic historian must adopt a stance of critical openness, directing a dual hermeneutic of trust and suspicion towards both the testimonial evidence

and the process of historical retrieval, and subjecting both to critical assessment. Because the primary task of historians is to explain the past rather than buttress their own beliefs and commitments, any resolution of incongruence that prioritises the experience of the historian over that of the historical actors needs to be justified, and all attempts to affirm the integrity of the historical data exhausted. This criterion may also be applied to the historical sources themselves: because the primary authors of historical testimony inhabit the same reality and identical, or at least similar, cultural milieux, one would expect to observe congruence between them. However, the emergence of new beliefs, practices and worldviews cannot be ruled out, and when there is evidence of such emergence taking place the task of historical explanation becomes crucial. Thus, for example, the simultaneous congruence and incongruence of Judaism and Christianity in the first century demands historical explanation. The quasi-positivistic strategy of identifying all statements attributed to Jesus that are incongruent with Judaism as creations of the first Christians does nothing to *explain* the distinctiveness of the Christian worldview. Instead the strategy simply presumes that Jesus' teaching had nothing original to say and was therefore entirely congruent with Judaism, and that the teaching of the first Christians was entirely original and therefore completely incongruent with Judaism. Such 'passing of the explanatory buck' creates more problems than it solves, and appears to reflect an ideologically driven desire to distinguish the moral teachings of the 'good' Jesus from the theological teachings of the 'corrupt' Paul, and in doing so prejudges any attempt to explain the simultaneous congruence and incongruence of Judaism and Christianity with reference to Jesus himself.

(4) Coherence. Historical knowledge concerns real events no longer accessible to sense experience. This means that the truth of our historical claims, the convergence between historical narrative and historical event, must be judged in terms of the greater coherence of one explanatory historical account over any other. In affirming this, the historian is in harmony with both the methodology of natural science, which is continually dealing with unobservable entities, and critical judgements in the field of the humanities. A historical account that is internally coherent will be more persuasive than one wracked with inconsistencies. Thus the suggestion that Christianity was the product of a deliberate fraud perpetuated by Jesus' disciples following his execution is inconsistent with the ethical stance adopted by the first Christians and their willingness to suffer martyrdom rather than recant their faith (as well as reflecting the popular postmodern fascination with speculative conspiracy theory – Did President Bush *really* order the attack on the World Trade Center?!), whereas the suggestion that the first disciples sincerely believed – rightly or wrongly – in the truth of their claims regarding Jesus is entirely consistent with their moral commitments and praxis.

(5) Fertility. A worldview is fertile if it is able, without compromise to its integrity, 'to permit the acceptance of alien beliefs when its holders meet with good reasons to hold them true' (Marshall 2000, p. 147). Thus the ability of Christianity to embrace evolutionary theory without detriment to its established theological commitments is an indicator of its fertility. Similarly, a historical

explanation proves itself fertile if it is able to accommodate new testimonial evidence without the need for any fundamental revision. Thus, for example, though the discoveries of the ancient libraries at Qumran and Nag Hammadi generated fundamental shifts in our understanding of the Jewish and Gnostic context of early Christianity, historical scholarship grounded in the hypothesis that the New Testament documents provide a broadly authentic account of the life of Jesus and origins of Christianity has been able to accommodate these shifts with relatively little need to revise its core explanatory narrative.

(6) Simplicity. The principle of Occam's razor, that the simplest explanation is normally the best, suggests that the more straightforward a historical explanation, the greater its ability to interpret the historical data at face value, then the more powerful it is likely to be. Towards the middle of the first century Christianity emerged as a distinctive worldview simultaneously rooted in the worldview of Judaism yet clearly distinct from it. This event demands a historical explanation: by far the simplest account, supported directly by both Christian and non-Christian contemporary texts, is that Christianity emerged in response to the life and teaching of Jesus of Nazareth. An alternative explanation, that Jesus was just another Jewish prophet and that Christianity was generated by Saint Paul, requires the creation of a complex and convoluted hypothesis regarding the intentions of Jesus and motivations of Paul that lacks direct evidential support. Generally speaking, when a simple hypothesis is available that conforms to the available evidence taken at face value, there is no need to generate a complex explanation that requires the historian to add a series of qualifications and riders to the available evidence.

(7) Depth. A robust and persuasive historical explanation is likely to possess hermeneutical depth and be able to 'trump' alternative explanations by offering a richer and profounder account of past events. This is something that is easier to experience than explicated: few who have encountered Bhaskar's dialectical philosophy doubt that it is altogether deeper than Hegel's, though some might struggle to explain precisely why this is the case. Similarly, few who have read Shakespeare in any depth doubt that his writing possesses deeper insight into the human condition than, say (and, once again, with all due respect), the novels of Geoffrey Archer, though some would struggle to explain exactly why this is so. In both cases, the best response to the uninitiated is simply to say 'Go, read, and experience at first hand'. Just as a good scientific theory 'calls attention to its own beauty and partly relies on it in claiming to represent empirical reality', so a good historical hypothesis calls attention to its own inherent depth, dynamism and beauty, and relies on it in claiming to successfully explain historical reality (Polanyi 1958, p. 133). If this all sounds a little nebulous to the ear of the positivist historian, we must remind ourselves that to interpret history is to interpret historical actors, and that our understanding of other people necessarily crosses the boundaries between science and art, reason and intuition, sense and sensibility.

14 The quest for the historical Jesus

Christians worship a vulnerable God, one who, for the salvation of humankind, came down from heaven, was incarnate in Jesus of Nazareth, and was executed by the Roman authorities in occupied Jerusalem sometime between 29 and 33 CE. The historical actuality of the life of Jesus Christ, God incarnate, fully divine and fully human, has always been fundamental to classical Christian faith. For the first Christian theologians the Docetic suggestion that Jesus merely appeared to be human and was not therefore subject to the realities and contingencies of history, together with the Ebionite suggestion that Jesus, though the greatest of prophets, was not actually divine, constituted the two paradigmatic distortions of classical incarnational theology. Colin Gunton has shown how these heresies have been replicated in the modern era (Gunton 1983). (1) Docetism reappears in the guise of Deism, the dominant theology of the Enlightenment, which denied any possibility of an immutable deity becoming a mutable human being, since that would allow the accidental truths of history to override the necessary truths of reason. Though the life of Jesus could be read idealistically, as a symbolic representation of an eternal ideal, it could not be seen to embody, let alone generate and actualise, eternal truth. Thus the eternal significance of Jesus is effectively dislocated from his historical actuality. (2) Ebionism reappears in the guise of naturalism, the emergent philosophy of the mainstream Enlightenment, which denied the existence of any transcendent realm above, beyond or ingrained within the natural order. On a naturalist reading, though Jesus was certainly a flesh-and-blood human being, he could not possibly be divine, or representative of any form of transcendent reality. Idealists sought to protect eternal transcendent truth by denying any necessary relationship between the eternal ideals of Christianity and the Jesus of history; naturalists sought to deny eternal transcendent truth by reducing Jesus to an ordinary human being. Thus, with the dawning of the Enlightenment, the ancient heresies of Docetism and Ebionism were reconstituted in distinctively modern guises: either the eternal significance of Jesus has no necessary connection with his historical person, or the historical person of Jesus has no eternal significance. Where once Paul's proclamation of Christ crucified constituted a 'stumbling-block' to ancient Jewish scholars who could not accept Jesus' divinity, and 'foolishness' to Hellenistic philosophers who could not accept God's mutability, so now assertions of Jesus' divinity constitute an affront to empiricist- and positivist-minded naturalists unable and unwilling to discern any eternal significance in the historical Jesus, while assertions of the eternal transcendental significance of the all-too-human Jesus constituted an affront to rationally minded idealists unable and unwilling to accept that eternal ideals can be embodied in and constituted by contingent historical events. Insofar as orthodox Christianity persisted in seeking to identify, in the hypostatic union of Christ's divine and human natures, the essential unity of contingent historical events and eternal transcendent truths, it rejected philosophical idealism and naturalistic reductionism in favour of a heady combination of historical and theological realism.

Just as Christians worship a vulnerable God, so their faith is vulnerable to the contingencies of history, and by extension to the processes and results of historical research. Any retroductive account of the ontological significance of Jesus of Nazareth provided by orthodox Christians will, inevitably, be concerned with the possibility of revealing a genuine convergence between the historical Jesus and the Christ of faith proclaimed by classical Christianity. The fact that the Christian historian approaches the historical data with a particular theological agenda does not imply that the writing of Christian history is intrinsically subjective. This is because all historians necessarily approach historical data with one or other set of prejudices and preconceptions; those who claim *not* to do so merely embrace the prejudices and preconceptions of a modernist worldview that claims to inhabit the neutral objective space established by (supposedly) universal reason. If all historical investigation is necessarily 'contaminated' by the subjectivity of the historian, then the alethic truth of their historical narratives must ultimately be judged against the objective historical realities testified to by historical documents that are themselves inevitably wrapped up in the subjectivity of *their* authors.

Positivistic historiography generated a story of the progressive erosion of Christian faith in the face of trenchant historical criticism. Though this is by no means the whole story, critical scrutiny of the Bible undoubtedly raised fundamental questions about the historical foundations of Trinitarian Christianity. This was perhaps most clearly evident in the emergence, at the beginning of the nineteenth century, of the so-called 'Quest for the Historical Jesus'. We will refer to this Quest as the 'Original Quest', to distinguish it from the 'New Quest' and 'Third Quest' which emerged in twentieth-century scholarship. The Original Quest was, in the main, conducted by scholars convinced of the essential discontinuity between the Christ of Christian faith and the Jesus of history, and as such constituted 'an explicitly anti-theological, anti-Christian, anti-dogmatic movement' (Wright 1996, p. 17). Its ideological commitments combined (1) a positivistic concern to strip away the supernatural 'Christ of faith' proclaimed by the Church, in order to reveal the unadulterated 'Jesus of history', and (2) an idealistic concern to account for the eternal significance, or insignificance, of both the Jesus of history and the Christ of faith. We will begin with the positivist concern to recover a picture of the Jesus of history supposedly obscured by the Christ of faith, and then consider idealist attempts to recast the orthodox understanding of the eternal significance of the historical Jesus.

The Original Quest: positivism

It was David Hume who most clearly articulated the two core positivist principles underlying the nineteenth-century Quest for the Historical Jesus: the inviolability of natural law, and the inability of reports of miracles to provide epistemic warrant for Christian belief (Hume 1902, pp. 109ff.; cf. Mackie 1982, pp. 13ff.). Given that all knowledge is grounded in empirical experience, and that some knowledge claims have greater epistemic warrant than others, Hume insisted that we hold our beliefs in proportion to the strength of the available evidence. Though he accepted the theoretical possibility of miracles, he offered two interlinked reasons for rejecting them. (1) He argued that miracles are *intrinsically improbable*. A reported historical incident is unlikely to have occurred if it does not conform to the 'constant and regular conjunction' of events that constitute our ordinary everyday experience of the world (Hume 1902, p. 111). Hume defines a miracle as a 'violation of the laws of nature', and argues that, since the laws of nature are codified accounts of regular conjunctions of events, miracles are highly improbable (p. 114). At times he goes even further: since 'a uniform experience amounts to a proof, there is here a direct and full *proof*, from the nature of the fact, against the existence of any miracle' (p. 115; emphasis in original). (2) He argued that historical accounts of miracles are *inherently untrustworthy*. Our beliefs about the past are reliant on the primary testimony of eyewitnesses and the secondary evidence provided by transmitters of such testimony. Though, in the normal course of events, we tend to accept the testimony of others at face value, we must be alert to the possibility that it may be mistaken or even intentionally deceptive. We have particular reason to exercise suspicion whenever witnesses contradict each other, are of a doubtful character, have a vested interest in persuading us to believe their stories, or deliver their testimony with undue hesitation or excessive assertiveness. The intrinsic improbability of miracles places the burden of proof on the testimonial evidence.

> When anyone tells me, that he saw a dead man restored to life, I immediately consider with myself, whether it be more probable, that this person should either deceive or be deceived, or that the fact, which he relates, should really have happened.
>
> (Hume 1902, p. 116)

The tendency of the uneducated mind to embrace the strange and marvellous inclines 'the generality of mankind to believe and report, with the greatest vehemence and assurance, all religious miracles' (p. 119). Reports of the miraculous 'abound among ignorant and barbarous nations', and those apparently civilised people who continue to believe in them are reliant on the uneducated testimony of their superstitious ancestors (p. 119). To shift the balance of probability in favour of miracles would require testimony of extraordinary power: as a general maxim, 'no testimony is sufficient to establish a miracle, unless the testimony be of such a kind, that its falsehood would be more miraculous, than the fact, which it endeavours to establish' (pp. 115ff.). Unsurprisingly, such testimony is not forthcoming: 'there is not to be found, in all history, any miracle attested by a sufficient number of men, of such unquestioned good-sense, education, and learning' to tilt the burden of proof in favour of the miraculous (p. 116).

It was Herman Reimarus' *Wolfenbüttel Fragments*, published posthumously by Lessing in 1778, that set in train the Original Quest for the Historical Jesus (Voysey 1879). Though Reimarus chose not to make his historical investigations public during his lifetime, their eventual publication stimulated a movement whose protagonists determined 'to destroy Christianity (as he knew it) at its root, by showing that it rested on historical distortion or fantasy' (Wright 1996, pp. 16ff.). Reimarus' commitment to a rationalistic natural religion led him to expunge from the Gospel narratives any reference to revelation, miracle and the supernatural. This process, according to Reimarus, revealed Jesus as a fanatical religious reformer, convinced he is the long-expected Jewish Messiah anointed by God to free the country from Roman occupation and establish a new religio-political order. However, his attempt to instigate a popular uprising failed, and he died on the Cross defeated and desolate. Driven by worldly ambition, the disciples stole the body of Jesus, announced his resurrection from the dead, proclaimed him as a heavenly saviour, and thereby sowed the seeds of a Catholic Church ruled by a priestly hierarchy (Voysey 1879, pp. 94ff.). In perpetuating this fraud they produced a range of documents, inconsistent in detail and supernatural in outlook, which eventually formed the canon of the New Testament. Reimarus regarded miracles as 'unnatural events, as improbable as they are incredible', and viewed the miracle stories contained in the Gospels as the products of an intentional fraud perpetuated by the leaders of the early Church (p. 70). Given the credulity of their poorly educated congregations, it was a simple task for them 'to invent as many miracles as they pleased, without fear of their writings being readily understood or refuted' (p. 73). What was invisible to lay members of the early Church is transparent to the modern historian: evidence of the fraud abounds in the Gospel texts.

David Friedrich Strauss' commitment to rationalism, Hegelianism, and a Darwinian inspired scientific materialism led him to reject the historicity of the miraculous events described in the Gospels on *a priori* grounds (Strauss 1873). However, his account of the origin of the miracle stories, as set out in *The Life of Jesus Critically Examined*, departed significantly from that of Reimarus (Strauss 1973). Rather than postulating deliberate fraud, Strauss posited the unintentional accretion of mythological elements during the period of the oral transmission of stories about Jesus. Myth, for Strauss, 'is the representation of an event or of an idea in a form which is historical, but, at the same time characterized by the rich pictorial and imaginative mode of thought and expression of the primitive ages' (p. 53). The biblical authors' 'ready disposition to derive all things down to the minutest details, as soon as they appear particularly important, immediately from God' reflects a primitive Semitic folk consciousness at odds with the philosophically sophisticated mindset of both Graeco-Roman civilisation and the modern Enlightenment (p. 78). Crucially, Strauss held that the attempts of modern scholars to strip away the Gospels' mythical accretions and reveal the actual Jesus of history expose an unbridgeable gulf between the Jesus of history and the Christ of faith (Strauss 1977).

According to Adolf von Harnack, historical science in the generations following Strauss learned 'to pass a more intelligent and benevolent judgement on [the Gospel] narratives, and accordingly even reports of the marvellous can now be counted amongst the materials of history and turned to good account' (Harnack 1904, p. 25). He offers four reasons for this shift in thinking. (1) Since Hellenistic culture had no conception of the laws of nature, it could not envisage miracles as violations of such laws; it did, however, enjoy a widespread assumption, not limited to the religious sphere or restricted to the uneducated, of the occurrence of 'marvellous' events (pp. 25ff.). (2) Because miracles were frequently attributed to living persons by contemporary eyewitnesses in the Hellenistic era, there is no need to explain the Gospel accounts as *retrospective* acts of intentional fraud or inadvertent myth making (p. 27). (3) Though the universal search for meaning and purpose transcending the 'blind and brutal course of nature' ultimately belongs 'to the realm of fantasy', such fantasy has a positive role to play in our attempts to transcend ourselves and our environment and as such should not be lightly dismissed as outmoded superstition (p. 27f.). (4) Since our knowledge of the laws of nature is incomplete, we should not rule out the possibility of offering naturalistic explanations of miraculous events; in particular, given our knowledge of the mind–body relationship, we cannot rule out the possibility of psychosomatic healing that does not contradict the laws of nature (pp. 28ff.). Thus, with the important proviso that violations of the laws of nature are impossible because 'what happens in space and time is subject to the general laws of motion', Harnack concludes that the modern historian may legitimately affirm that Jesus was recognised as a miracle worker by his immediate contemporaries, and discern a historical kernel within many of the miracles stories (p. 27).

The tradition from Reimarus through Strauss to Harnack focused on a single question: 'How can we make sense of the Gospel narratives, given the impossibility of miracles, supernatural occurrences and divine interventions in the natural order?' The solution was twofold: (1) to expunge all such references from the Gospels, leaving only incidents that conformed to the positivist principle of the 'constant and regular conjunction of events'; and (2) to dismiss the historical testimony as untrustworthy, the result of deliberate fraud, mythical accretion or eyewitness misinterpretation. The positivist presuppositions determine the results: given the inviolability of natural law, the Gospel narratives cannot provide epistemic warrant for Christian belief. Underlying this is a further unquestioned presupposition driven by the agenda of biblical scholars, namely that the function of miracle stories in the Gospel narratives is to provide epistemic warrant for Christianity, rather than reveal the ontological reality of God.

The Original Quest: idealism

Though the protagonists in the nineteenth-century Original Quest for the Historical Jesus were committed to the inviolability of nature and drew indirectly upon Hume's rejection of miracles, their core outlook was idealist rather than empiricist or positivist. Having expunged the miraculous from the Gospel narratives and placed the remaining 'facts' in chronological order, they turned to the question of the eternal significance of the events surrounding the life of Jesus and the emergence of the early Church. Strauss introduced a distinction, later taken up and developed by Rudolph Bultmann, between *Historie* and *Geschichte*: the former referred to the 'historicity' of the life of Jesus of Nazareth as uncovered by critical exegesis, the latter to the 'historicality' of the eternal significance of the Christ of faith (Jones 1991, pp. 36ff.). If *Historie* uncovered the brute facts of history, then *Geschichte* sought to expound the lasting eternal value and immediate existential significance of these facts. As we have already seen, idealism rejected any notion that historical events could possess any substantial transcendental value, since contingent historical events could never override necessary eternal truths. They could however act as immanent symbols, indicators and exemplifiers of eternal truths. Given these idealist presuppositions, the central task was to emplot the life of Jesus within one or other pre-established idealist narrative, thereby revealing the ultimate significance of the idealised Christ of faith overlaying the empirical Jesus of history.

As a child of the rationalist Enlightenment, rather than the tradition of modern idealism instigated by Hegel, Reimarus constituted an exception to this general rule. As a deist he accepted the existence of God as the disinterested first cause of a cosmos possessing a moral order accessible to universal human reason, and rejected the existence of a personal interventionist God willing to contemplate the eternal punishment of those who failed to believe in him and accessible only by the few through revelation and faith. Following Christian Wolff, he had no time whatsoever for religious practice and piety, which he regarded as crass fanatical superstition. Having rejected both the fanaticism of Jesus and the fraudulence of the disciples, he found nothing of value in the Gospels to emplot within his deistic worldview. As such he represented an extreme wing of Enlightenment thought. Diderot and Voltaire, despite their trenchant rejection of Christianity, were both willing to emplot Jesus' moral teachings, once stripped of their theological substance and emancipated from Christian misrepresentation, in the Enlightenment's central meta-narrative of moral progress. Both identified a 'contradiction between a moral teacher of some distinction and the corruption of tactically skilled communities', and chose to embrace the former and reject the latter (Buckley 1987, p. 40). The eternal significance of Jesus was not that he was God incarnate, but that he was a historic exemplifier of eternal moral norms that could be established independently through the exercise of universal reason.

In sharp contrast, Strauss was willing *both* to affirm the enlightened moral virtue of the historical Jesus *and* to accept that the mythological story of the Christ of faith overlain on the historical narrative is not completely devoid of value. Previously Giambattista Vico had rejected both the valorisation ('unsurpassable expressions of philosophical and human wisdom set down in matchless poetry') and denigration ('riddled with ignorance, superstition, and disgraceful praise of barbaric behaviour') of mythology (Lemon 2003, p. 128). The ancients were neither seeped in primal wisdom nor superstitious fools, but rather ordinary human beings constrained by the limitations of their cultural contexts. Similarly Nietzsche insisted that, despite its limitations, myth 'touches that which is beyond the grasp of a merely factual history' and provides 'a horizon which allows us to orientate our way in the flux of life' (Burns and Rayment-Pickard 2000, p. 137; cf. Nietzsche 1993, p. 117). According to Strauss, the Gospel writers sought (albeit unconsciously) to express the abstract Hegelian ideal of the triumph of Spirit over nature through the concrete elemental language of myth. Strauss' attempt to demythologise the Gospels and expose this substantial philosophical core drew directly upon Hegelian dialectic. Human beings are indeed miracle workers, insofar as the human spirit is able to assert its authority and freedom over deterministic nature (Neill 1966, p. 16). Thus Strauss emplots the Gospel accounts of the historical Jesus into the pre-established Hegelian narrative of the dialectical progress of Spirit towards self-realisation, and as a result reads into the Gospels an idealistic vision to which he is already committed. Ultimately, Strauss' reading of the Gospels reflects Feuerbach's reductive account of theological statements as human projections: religious myths are expressions of humankind's deepest aspirations, and in generating such mythology human beings effectively create a God after their own image (Feuerbach 1957).

Harnack sought to affirm the eternal value of the Jesus of history by locating him firmly within the romanticised worldview of nineteenth-century Liberal Protestantism. He had little time for the Christ of faith. The identification of Jesus as the incarnate God and second person of the Trinity was a theological construction of the early Church that distorted the pristine message of Jesus: a 'Hellenisation' of Christianity, brought about by its encounter with Greek philosophy, which generated theological and philosophical speculation that departed radically from the teachings of the historical Jesus. For Harnack Jesus was essentially a moral teacher: his parables illuminated eternal moral principles in a manner accessible to the uneducated, rather than proclaiming theological truths; his proclamation of the coming Kingdom of God had nothing to do with the eschatological culmination of cosmic history, and everything to do with the gradual emergence of an ethic of universal brotherhood and sisterhood; his parables were not vehicles of theological education, but tools of moral edification sustaining and enriching the nineteenth-century bourgeois myth of the inexorable evolutionary progress of humanity towards a utopian future. The historical Jesus thus becomes both the primary exemplifier of the moral vision of a liberal order grounded in Christian values stripped of the superfluous theological and metaphysical trappings of classical Christianity, and the primary justification of a secularised Christian culture in which the Church provides the nation-state with a moral buttress.

It was Albert Schweitzer who brought the nineteenth-century Quest for the Historical Jesus to a close. His detailed survey of the literature concluded that it had simply replaced the traditional picture of Jesus clothed in Christian dogma with a new picture of Jesus clothed in the hopes and aspirations of nineteenth-century liberalism (Schweitzer 1945). George Tyrrell's oft-quoted critique of Harnack may be applied to Schweitzer's reading of the Quest as a whole: 'Whatever Jesus was, He was in no sense a Liberal Protestant ... the Christ that Harnack sees, looking back through nineteen centuries of Catholic darkness, is only the reflection of a Liberal Protestant face, seen at the bottom of a deep well' (Tyrrell 1963, pp. 22, 49). For Schweitzer the key to interpreting the Gospels lay in locating them in their cultural milieu: Jesus shared with many contemporaries an eschatological belief in the immanent end of the world, and proclaimed a radical ethic geared to the coming crisis. However, the implications of Schweitzer's suggestion were not immediately followed up. For Karl Barth, undoubtedly the greatest of modern theologians and lifelong socialist, the outbreak of the First World War revealed the sterility of Liberal Protestantism. His liberal teachers, including Harnack himself, joined other liberal theologians in signing a manifesto supporting the war policy of Kaiser Wilhelm II. 'To me they seemed to be hopelessly compromised by what I regarded as their failure in the face of the ideology of war'; their lack of seriousness in the reciprocal relationship between theology and praxis indicates that 'their exegetical and dogmatic presuppositions could not be in order' (Busch 1976, p. 81). For Barth the key hermeneutical mistake of Liberal Protestantism's reading of the Gospels was to replace exegesis with eisegesis, reading extrinsic meaning into the text rather than drawing intrinsic meaning from it. The Gospels tell a story which, if taken seriously, invites us to emplot our stories within God's story rather than seek to domesticate God's story within our own. The task of theology is not to project a God constructed in our own image, but to retell the story, intrinsic to the Gospels, of God's proactive intervention in his creation, through which we may be transformed into the divine image. To read back our own assumptions about the ultimate meaning of history into the Gospel narrative is to fall into the trap of 'inventing "history" by a backwards projection of ideology' (Wright 1992, p. 85). Barth drew the conclusion that historical scholarship cannot possibly perform the task of establishing the intellectual foundations for Christian faith; any attempt to reconstruct the Jesus of history behind the Christ of faith constitutes a form of theological self-affirmation and self-justification at odds with the integrity of Gospel story itself.

The New Quest: positivism

The half century following the publication of Schweitzer's review of the Original Quest saw a marked decline in scholarly research into the historical Jesus. Schweitzer's exposure of the subjectivity of liberal scholarship, coupled with the unpalatable nature of his own vision of Jesus as apocalyptic visionary, left liberal scholarship uncertain about the future trajectory of historical scholarship. The rise of Form Criticism, with its focus on the ways in which oral tradition shapes historical data, led to scepticism about the possibility of reconstructing the life of Jesus and an increased interest in the light the Gospels shed on the first Christian communities (Bultmann 1972). At the same time Barth's suspicion of utilising biblical scholarship in an attempt to buttress faith led to a waning interest in historical study of the life of Jesus among orthodox scholars. Rudolf Bultmann sought to combine these two factors, arguing historically that it was impossible to recover any substantial knowledge of the life of Jesus, and theologically that it was the Christ of faith rather than the historical Jesus that possessed existential significance for Christians (Bultmann 1958, 1966; cf. Miegge 1960). In retrospect, it is clear that this division constituted yet another form of Docetism, since the eternal ideal represented by Bultmann's Christ of faith bore no necessary relationship to the Jesus of history. The status quo shifted significantly in 1953, when Ernst Käsemann – conscious of the appropriation of a supposedly non-Jewish Christ by Nazi ideology, alert to 'the dangers of idealism and docetism', and aware 'that if Jesus was not earthed in history then he might be pulled in any direction, might be made the hero of any theological or political programme' – directly challenged the presuppositions of Bultmann's followers (Wright 1996, p. 23). Käsemann's now infamous paper on 'The Problem of the Historical Jesus' proved to be a decisive turning point in twentieth-century biblical scholarship, inaugurating what quickly became known as the 'New Quest for the Historical Jesus' (Käsemann 1964, pp. 15ff.).

Wright argues that, despite its initial promise, this New Quest quickly became bedevilled with the same positivist and idealist assumptions and working practices that dominated the Original Quest. This is especially prevalent in the reconstruction of the historical Jesus promulgated by the North American 'Jesus Seminar'. Established in 1985 by Robert Funk and John Dominic Crossan, the Seminar contains an inner core of reputable New Testament scholars, alongside less well-known figures and non-specialists, drawn from a limited number of institutions (Funk and Hoover 1993; Funk 1997, 1999). The controversial nature of their conclusions, together with their concern to disseminate them as widely as possible, has made the Seminar something of a *cause célèbre* across North America.

> [T]he initial flyer advertising the seminar spoke in classic positivist terms of 'the quest for fact and history, for honesty and candour, for the truth and its consequences'. It invited as participants those who preferred 'facts rather than fancies', 'history rather than histrionics', and 'science rather than superstition'.
>
> (Wright 1964, p. 31)

Seminar members follow the established practice of positivist historiography in focusing on the atomistic parts of the historical data, attempting to distinguishing between authentic and inauthentic sayings and events, and only then seeking to weave the remaining authentic data into a coherent whole. The central, and most remarkable, feature of the work of the Seminar is the practice of using different coloured beads to vote for whether they consider an event or saying attributed to Jesus to be authentic (red), probably authentic (pink), probably inauthentic (grey) or definitely inauthentic (black). The Seminar appears to be driven by a hermeneutic of suspicion: the burden of proof lies with those who wish to affirm the historicity of a saying or event. Judgements regarding authenticity are made via the application of pre-established criteria rather than the assembly of holistic retroductive explanations. Thus, for example, the criterion of dissimilarity assumes that a saying is more likely to be authentic if it stands apart from teachings of Judaism and the early Church. Seminar members tend to hold the view that the actions of Jesus must conform to a naturalist worldview: though many accept that Jesus practised healing without recourse to ancient medicine or magic, they tend to insist that any successful miracle must have a psychosomatic explanation. The consensus view of the Seminar is that Jesus was a sapiental wisdom teacher rather than an apocalyptic prophet. As an itinerant Hellenistic-Jewish sage, he used parable and aphorism ironically in order to encourage his hearers to break the bounds of Jewish theology and the established social order, adopt an unconventional 'outsider' social role, and strive to bring healing and harmony to God's creation.

Given the Seminar's inbuilt scepticism – less than one-fifth of the sayings of Jesus in the canonical Gospels and Gnostic Gospel of Thomas are deemed 'authentic' or 'probably authentic' – and radical conclusions, it comes as no surprise to find them subject to sustained criticism on both methodological and substantive grounds. Just as observation of the constant and regular conjunction of events is no substitute for retroductive explanation, so the constant conjunction of scholarly consensus, as indicated by the regular appearance of voting balls of a certain colour, is no substitute for the development and testing of coherent retroductive hypotheses. Thus, if the sizeable minority of Seminar members were to exercise proper caution in voting a saying 'probably authentic' on the basis of a common well-argued retroductive model, while a small majority were to vote it 'definitely inauthentic' for a variety of different and entirely incompatible reasons, then the reported consensus of the seminar would show it to be 'probably inauthentic'.

It is not only the democratic procedure that is academically flawed. The criteria for identifying authentic sayings and events also lack intellectual integrity. One influential criterion is that short, pithy statements are more likely to be authentic than long, complex narratives, on the ground that the former are more disposed to memorisation and accurate oral transmission than the latter. There are strong arguments suggesting that the criterion misrepresents the nature of the transmission of oral tradition, and that an extended parable is equally disposed to memorisation, if not more so, by virtue of the fact that its narrative structure functions, in much the same way as an extended joke, as an *aide-mémoire*. Even if we accept that short, pithy statements are more disposed to accurate oral transmission, this does not warrant the conclusion that Jesus was primarily a teacher of condensed nuggets of esoteric wisdom. Such a conclusion is grounded in the epistemic fallacy of confusing our epistemic access to an object with the ontology of the object itself: even if is the case that short, pithy statements provide the best available access to Jesus' teaching, it does not follow that his teaching consisted primarily of short, pithy sayings. The criterion of dissimilarity, which, as we have already noted, asserts that sayings are more likely to be authentic if they are not found in the teachings of Judaism or the early Church, is similarly flawed. It generates the radical conclusion that Jesus was not an eschatological prophet, on the assumption that since both John the Baptist and the early Church embraced eschatological beliefs about the in-breaking of God's kingdom, any eschatological statements attributed to Jesus in the Gospel tradition must be rejected as either a residue from the teaching of John or a creation of the early Church. The normative historical procedure of seeking to locate and understand historical persons within their historical milieu supports the common-sense view that Jesus is highly likely to have inhabited the same culture of eschatological expectation as Judaism and the early Church. However, the Seminar's criterion dictates that this is highly unlikely to have been the case, and as a result generates a Jesus more akin to a Greek Hellenistic wisdom teacher than a Jewish Palestinian prophet. This dislocation of Jesus from his immediate cultural context generates 'an eccentric Jesus who learned nothing from his own culture and made no impact on his followers' (Komoszewski *et al.* 2006, p. 49). Once the 'authentic' fragments of Jesus' teaching are identified according to this criterion, scholars are left with the task of weaving them into a coherent hypothesis. The result of this process looks something like this: though Jesus grew up in a Jewish milieu and imbibed the eschatological preaching of John the Baptist, when he embarked on his ministry he recast himself as a teacher of esoteric wisdom; following his death his disciples recollected, memorised and treasured his esoteric sayings; however, within a generation they fabricated an alternative body of eschatological teaching and superimposed it seamlessly upon the 'authentic' esoteric sayings; and, as it transpired, were prepared to suffer martyrdom rather than recant their fabricated claims. This hypothesis is counter-intuitive, and fails to explain the motivations of both Jesus and his disciples. A far simpler and more powerful hypothesis is the Jesus simply extended and developed John's eschatological message, and that his disciples sought to transmit his teaching as faithfully as possible.

The root problem lies in the New Quest's replication of the methodology of positivistic historiography: rather than seeking to construct a holistic retroductive explanatory hypothesis on the basis of the available evidence, it sets out to expunge supposedly inauthentic historical data from the Gospels in the light of pre-established criteria and then place the remaining data in some form of coherent order.

The New Quest: idealism

John Dominic Crossan's *The Historical Jesus: The Life of a Mediterranean Peasant* constitutes 'the highpoint of achievement in the new wave of the New Quest' (Wright 1996, p. 44; see also Crossan 1991). Wright argues that, for all its brilliance and originality, Crossan's historical reconstruction is highly speculative and intellectually unsustainable. Jesus is presented as a rural Galilean Cynic, a hybrid of the urban Hellenistic Cynic philosophers itinerantly wandering the Hellenistic basin at that time. The Cynics sought happiness by living simple, disciplined, self-sufficient lives in harmony with nature. This required the abandonment of any aspiration for social status, possessions, wealth and power. They viewed suffering as the product of misplaced desires generated by the hegemony of prevailing social norms, customs and institutions. The Roman Empire was a brokered realm, dominated by master/slave and patron/client relationships, and, whether through appeasement or resistance, the Jewish nation was embroiled in its power structures. According to Crossan, Jesus sought to inaugurate a 'brokerless kingdom', a community devoid of all authority and hegemony. He encouraged his followers, especially the dispossessed and marginalised peasantry, to turn aside from social norms and expectations, and put their trust exclusively in God, thereby circumnavigating the master/slave and patron/client relationships that dominated their lives. As a miracle worker he sought to free his followers from the powers of evil, and through his table fellowship with them sought to subvert the established socio-religious norms. Crossan's Jesus was an itinerant preacher, constantly on the move in order to avoid being set up as patron of this radical community, whose peasant members enjoyed unmediated relationships with one another and with God, grounded in 'a shared egalitarianism of spiritual and material resources' (Wright 1996, p. 58). For all the vivid luminosity of Crossan's reconstruction, closer inspection suggests it constitutes yet another example of the idealist emplotment of historical data within a pre-established vision.

To claim Jesus as a hybrid Cynic philosopher Crossan needs to play down both the Jewish background to the Gospels (especially their eschatological features), and their affirmation of the significance of specific historical events (especially Jesus' arrest and execution). Since it is only Jesus' practical wisdom teaching that has any lasting eternal significance, Jewish hopes for the establishment of the Kingdom of God through divine intervention, together with Christian beliefs about the significance of Jesus' death and resurrection, cannot have a secure basis in history. In order to sustain this thesis, Crossan produces a highly speculative treatment of the historical sources. He begins by reconstructing a set of hypothetical documents pre-dating the Gospels that highlight Jesus' wisdom teaching and miracles: a hypothetical collection of sayings drawn on by Matthew and Luke ('Q'), a hypothetical collection of miracles used by Mark and John, a hypothetical early version of the Gospel of Thomas, etc. He then places these hypothetical documents, along with extant historical documents, into a hypothetical chronological order in which documents containing sayings and miracle stories are deemed to precede documents describing historical events and reflecting Jewish influence. This documentary reconstruction and chronological reordering provides the basis for the identification of four hypothetical strands of primitive Christianity.

(1) Thomas Christianity. In continuity with the teachings of the historical Cynic Jesus and entirely disinterested in his life, this strand may be traced from a hypothetical early version of the non-canonical Gospel of Thomas, through selected aspects of the extant canonical Gospels, to the later Gnostic sects. The thesis requires the speculative affirmation of a direct link between the ideology of Egyptian Gnosticism in the mid-second century and the historical Jesus, a feat that can only be achieved by stripping Jesus of his Jewish identity and positing, despite the lack of any evidential support, the presence of itinerant Cynic teachers in rural Galilee.

(2) Pauline Christianity. In line with the Original Quest's tendency to valorise Jesus ('meek and mild') and demonise Paul ('arrogant and fanatic'), this strand was deemed responsible for the corruption of primal proto-Gnostic Christianity as a result of Paul's concern with the theological significance of Jesus' life in the context of Jewish eschatology. However, the positing of a Pauline Christianity dislocated from Jesus' own disciples is extremely difficult to sustain: the available data attest to Paul's close, albeit not consistently harmonious, relationship with Peter, James and Jesus' other immediate companions.

(3) Q Christianity. The product of a hypothetical community whose existence and theology is inferred from a hypothetical reconstruction of an early version of a hypothetical document, supposedly drawn on by the authors of Luke's and Matthew's Gospels, and deemed to contain only sayings of Jesus.

(4) Exegetical Christianity. Its members were skilled theological exegetes rather than uneducated peasants, who set about creating a historical framework within which to insert the extant sayings of Jesus by drawing on Old Testament texts. The passion narrative, rather than being broadly historical, is seen as the fictional creation of the author of a hypothetical 'Cross Gospel', which Crossan attempts to reconstruct from the mid-second-century apocryphal Gospel of Peter.

Crossan's speculative reconstructions and dating lack both inclusivity and simplicity: he excludes vast swathes of data and generates projected alternatives on the basis of a complex and convoluted hypothesis. Crossan holds that despite being born and brought up as a Palestinian Jew, Jesus' actions and teaching bore little relationship to his Jewish heritage, and that the Jewish cultural assumptions pervading the Gospels were inserted by the same group of Christians who constructed the accounts of Jesus' trial and crucifixion by cherry-picking selected verses from Jewish Scripture. Such a drastic and counter-intuitive thesis clearly requires solid evidential support if it is to prove sustainable; however, it is entirely dependent upon highly tendentious speculation regarding hypothetical sources and their dating. The traditional thesis that Jesus' actions and teaching were grounded in the worldview of first-century Palestinian Judaism within which he was born and grew up, that the historical traditions reflects that fact, and that the accounts of Jesus' arrest, trial and execution have their origins in actual events is simple, inclusive of the extant evidence, and does not require the construction of hypothetical documentary evidence.

According to Wright, the attempted 'reconstruction' of hypothetical sources depends 'wholly on Crossan's prior convictions both about Jesus himself and about the nature of early Christianity', and though he offers his chronological strata 'as starting points, they are in fact conclusions drawn from his basic thesis' (pp. 49ff.). Crossan's book was marketed as the 'first comprehensive determination of who Jesus was, what he did, what he said': here Wright discerns an inner tension between 'the stated claim to avoid spurious objectivism' and the 'implicit claim that here at last is the solid historical ground that generations of Jesus-researchers have been waiting for' (p. 55). Just as the Jesus reconstructed by the Original Quest possessed an uncanny resemblance to the ideal Liberal Protestant, so the Jesus reconstructed by Crossan – an itinerant deconstructor of both the received Jewish meta-narrative and the hegemony of the Roman Empire, and advocate of an ironic and esoteric wisdom whose true spiritual home stands on the counter-cultural margins of society – possesses an uncanny resemblance to the ideal participant in the postmodern New Age spiritual milieu: Crossan's Jesus is a figure with which both Derrida, Foucault and Baudrillard and adherents of New Age esotericism could easily empathise. Crossan appears to have emplotted the historical data within a narrative conducive to the ideology of postmodern relativism. His hermeneutic of suspicion supposes that the data as it stands is inherently untrustworthy, and disguises an alternative, and very different, story that the historian must strive to expose; however, as Wright point out, as soon as 'you doubt everything in the story, and postulate a chain of events by which someone might have taken it upon themselves to invent such a narrative from scratch, all things are possible' (p. 61). On this reading, Crossan utilises a positivistic historiographic methodology as a means to emplot his own postmodern ideological convictions onto the Gospel narratives.

> Crossan sees the need for appropriate reconstruction, but has not yet articulated a way by which this can actually be achieved. Unless we can do this – and I submit that a serious 'critical realism' goes at least some way towards such a goal – then history will remain a matter of brokerage, of the historian as patron and the reader as client. 'Critical realism', in other words, is an attempt to provide, in the sphere of historical method, what Crossan thinks Jesus was offering in the sphere of first-century peasant life: a brokerless kingdom.
>
> (Wright 1996, p. 55)

The Third Quest: critical realism

Wright draws attention to the existence of a 'Third Quest' for the historical Jesus, marked by a shared commitment, whether explicit or implied, to critically realistic historiography. He identifies twenty key writers, active since the early 1960s and without any common motivation or theological agenda, as representative of 'a new wave of historical seriousness about Jesus' (p. 89; cf. p. 84). The 'pursuit of truth – historical truth – is what the Third Quest is all about' (p. 87).

Their approach is marked by close attention to, and respect for, historical sources. Here a hermeneutic of trust runs alongside a hermeneutic of suspicion:

if trust without suspicion breeds naivety, then suspicion without trust breeds ideology. The assumption that historiography is a modern invention, and that ancient writers lacked historical insight and awareness, constitutes a modern myth, one that seeks to legitimate the cultural imperialism of the secular Enlightenment yet retains no purchase on the actual practice of historical writing in antiquity: 'the contemporary historians of the ancient world knew what history is as well as we do, and often a lot better' (Wright 1992, p. 84). In the fifth century BCE Herodotus and Thucydides established a genre of historical writing that, though in many respects different from modern historiography, nevertheless demonstrated significant levels of historical sensitivity and critical awareness (Fornara 1983). This mindset is reflected in early Christian writers, who possessed 'a historical awareness ... from the start, all the more so since the church was founded on the historical person of Jesus ... not on myth' (Drobner 2007, p. 225). The task of the ancient historian was to *explain* past events rather than merely describe them: fact and values were indistinguishable, and the notion of reducing history to a mere chronology of facts was unimaginable. Graham Stanton has argued persuasively that, though the Gospels are not biographies in the modern sense, comparison with similar Hellenistic and late Jewish literature suggests that the Gospel authors, in proclaiming the Jesus of history as the Christ of faith, were committed to representing and explaining the past as truthfully as possible within the historical conventions of their day (Stanton 1974). Ancient historiography was closely aligned to Aristotle's insistence that to understand an object is 'to know in a deep sense what it is and how it has come to be' (Lear 1988, p. 6). If, in his *Chronology*, the fourth-century historian Eusebius sought to locate secular history within a Christian timeline, this was merely a prolegomena to his *Ecclesiastical History*, which sought to provide a descriptive *explanation* of the emergence of the early Church. Whether through reference to the outworking of human greed and jealousy (Herodotus), the strengths and weaknesses of democratic polity (Thucydides) or divine providence (Eusebius), ancient historians were united by a desire to explain the past rather than merely describe it, and in so doing were actually closer to critical realist historiography than positivist and idealist historians (Thucydides 1972, pp. 35ff.; cf. Cochrane 1929). Further, methodological reflection was part and parcel of the ancient historian's craft: thus, for example, both Josephus, in the Introduction to his *Contra Apionem*, and the author of Luke's Gospel sought to explain and justify their working practices (Josephus 1987, p. 773; Luke 1:1ff.).

Wright suggests that the ancients were not deceived about the nature of history, since they possessed both historical sensibility and critical acumen. Rather it is we ourselves, as the heirs of 'the Enlightenment's rejection of reliance on auctores, "authorities" in a multiple sense', who 'have come to imagine ourselves to be the first to see the difference between subjects and objects, and so have both misjudged our fore-bearers and deceived ourselves' (Wright 1992, p. 85). Critically realist and ancient historians stand together in opposition to the naïve realism of positivists and emplotment strategies of idealists. If Wright's conclusions appear at times somewhat schematic and simplistic – relatively few major historians beyond the boundaries of biblical scholarship appear to have been entirely deceived by either positivism or idealism, and there can be little doubt as to the critical rigour of most contemporary historians – it is nevertheless abundantly clear that to dismiss ancient historiography, and therefore the historiography of the early Church, out of hand is a profoundly misguided and ideologically driven option. Like all of us, the Gospel writers were children of their age, dependent upon the prevailing expectations and standards of historical writing. If these standards fall short of our expectations, the difference is quantitative rather than qualitative. The Gospel writers sought to write theological history, not construct religious myth, and as such were cousins of Herodotus and Thucydides rather than Homer and Ovid. Insofar as they sought to ground their theology in the historical events of the life of Jesus, the Gospel writers stand poles apart from the authors of the later Gnostic Gospels who, in seeking to protect the heavenly figure of Jesus from contamination by the ugly contours of history, neglected the historical data concerning the person of Jesus and instead shrouded him in an ahistorical redeemer myth devoid of any historical grounding.

In dealing with ancient authors, the critically realistic historian refuses both the positivist task of working from atomistic parts in an attempt to distinguish between authentic and inauthentic testimony, and the idealist strategy of emplotting supposedly 'authentic' testimony within a pre-established ideological frame. Instead she works from wholes to wholes, seeking to establish large-scale retroductive narratives rooted in abductive insight and subject to iterative testing. The generation and testing of retroductive hypotheses takes precedence over a criterion-based pursuit of authenticity that opens the door to idealist emplotment. The holistic nature of historical explanation requires that Jesus is understood in the context of the prevailing worldviews encountered in first-century Palestine, above all that of Jesus' own Jewish heritage. Reference to Judaism does not merely provide background contextualisation: the clue to making sense of Jesus and explaining the emergence of Christianity lies in the dialectical task of identifying and explaining congruence, absence and originality at the interface of the two. To what extent were Jesus' actions and teachings congruent with Judaism? What aspects of Judaism are absent in Jesus' actions and teachings? What aspects of Jesus' actions and teachings were original? In order to achieving an appropriate explanatory depth, the historian must learn to formulate insightful questions capable of penetrating more deeply into the reality of the past than the positivist's 'Did this isolated event actually happen?' and the idealist's 'What, if anything, of lasting value remains once the historian has identified and isolated authentic events?' Wright suggests five integrated and overlapping questions, designed to enable the critical realist historian to establish a holistic retroductive explanation of the life of Jesus informed by a hermeneutical circle in which the parts illuminate the whole and the whole illuminates the parts: What did Jesus set out to achieve? Why did Jesus die? Who did Jesus think he was? How and why did the early Church begin? How does the Jesus we discover by doing 'history' relate to the world and our concerns?

15 Jesus Christ

A critical realist reading

The quality of historical writing is bound up with the quality of the questions historians ask of historical data. The critical realist historian's primary concern is to produce the best possible retroductive explanation of past events. This requires her to attend closely to the available data and ask appropriate questions of it, thereby prioritising exegeses and diminishing the dangers of forms of eisegeses that force the data to conform to a set of preconceived expectations. Both positivist and idealist historiography tend to fall into the trap of eisegesis. Positivists tend to foreground the question 'Could this event have happened?' Idealists tend to foreground the question 'Do the historical data fit with my own particular value system?' Both ultimately ask if the data conform to the prior commitments of the historian – whether to a naturalistic worldview in which miracles are deemed impossible or to a set of preconceived idealistic values – and in doing so tend to block out what the data are actually saying. Ultimately, both sets of questions fail to respect the integrity of two interrelated hermeneutical circles. (1) The hermeneutical circle linking the whole and the parts of the historical data. By seeking to interpret the parts in terms of the whole and the whole in terms of the parts, in an ongoing hermeneutical dialectic, the critical realist historian attempts to make retroductive sense of the historical data as it stands. A retroductive explanation that takes account of all of the available data is likely to be simpler and more coherent than one which seeks, from the start and on *a priori* grounds, to expunge specific testimonial evidence. The strategy of rejecting parts of the data as inauthentic is not the first port of call for the historian, but rather a last resort, turned to only when it proves impossible to make sense of the data as they stand. (2) The hermeneutical circle linking the horizons of meaning of the historical data and the historian. By recognising and acknowledging her own prior commitments, the critical realist historian opens up the possibility of allowing the horizons of meaning integral to the data to assert themselves against her presuppositions and prejudices. An uncritical emplotment of the data within the historian's own pre-established meta-narrative or values framework can only undermine a deep understanding of past events. The twin strategy, of first seeking to expunge inauthentic data and then seeking to locate the remaining data within a preconceived idealistic scheme, flows directly from asking questions of the text that are inadequate to the task of retroductive historical explanation. Wright's critically realistic exegesis of the historical data surrounding the life of Jesus proceeds by asking a very different set of questions, designed to allow a deep retroductive explanation of the events surrounding the life of Jesus to emerge: What did Jesus set out to achieve? Why did Jesus die? Who did Jesus think he was? How and why did the early Church begin? These questions force the historian to penetrate beneath the surface appearance of the data and attend to the hermeneutic circles between the parts and whole of the data and the horizons of meaning of text and interpreter. The primary task of the critical realist historian to seek to explain the whole of the data as it stands, both by taking into account the relationship between parts and wholes, and by acknowledging her own prior commitments in order to maximise the possibility of the data speaking for itself on its own terms. Before considering Wright's answers to his four questions, we will attend to a further question thrown up by previous discussions in the present book: How would Jesus' contemporaries have understood his miracles?

Miracles revisited

Given the frequency of first-hand reports of miracles in the Hellenistic era, it would have been highly unusual if eyewitnesses had *not* attributed some form of miraculous power to Jesus. Despite their ideological differences, both Wright and Crossan embrace the emerging scholarly consensus that Jesus' contemporaries, friend and foe alike, believed he performed 'mighty works' broadly similar to those attributed by eyewitnesses to other contemporary historical figures, ranging from Roman emperors through to itinerant preachers (Wright 1996, p. 194). Though this does not rule out later embellishment or imply a supranatural explanation, it does enable us to affirm the basic historicity of many of the miracle stories recounted in the Gospels.

Any retroductive explanation of miracles must appeal to underlying causal mechanisms rather than regular patterns of events. To define a miracle as a 'violation' of the laws of nature is to assume that these laws are basic to the ontological structure of reality. However, for Christians, it is the will of God – understood as unconditional love rather than raw power – that is ontologically basic: thus the claim that leprosy violates the will of God is more fundamental than the claim that miracles violate the laws of nature. On this reading a miracle is the product of God's providential power, not violating the supposedly sacrosanct laws of nature, but rather enabling creation to be more truly itself (p. 188). Christianity affirms a necessary Creator and a conditional creation. In accounting for the ordered-yet-contingent nature of creation, the early Christian theologians rejected both a 'theory of randomness that would have changed the notion of cosmos back into chaos' and 'a theory of cosmic necessity as an iron law over which the all-sovereign Creator was powerless' (Pelikan 1993, pp. 256ff.). Though God retains ultimate sovereign control over the natural order and the laws of nature, the divine will is neither deterministic nor arbitrary. Hence there was no place in the Christian scheme for Aristotelian necessitarianism. God's exercise of sovereignty over nature is driven by pure love, understood as the bond of reciprocal giving and receiving between the persons of the Trinity. Since love can never be coercive, God creates an ordered world in which human beings are free to love or not to love; since love can never be disinterested, God creates an ordered world with which he can proactively engage in the economy of salvation that culminates in the incarnation. Within this economy Jesus' miracles constitute the first fruits of the ontological actualisation of the reconciliation of Creator and creation rather than mere violations of the laws of nature.

Augustine's initial reserve towards miracles was driven by a pastoral concern to avoid nurturing forms of faith marked by immaturity, gullibility, incredulity and fear (Lancel 2002, p. 320). His later shift to a more positive attitude was the result of sustained theological reflection (pp. 467ff.). He argued that creation itself is a miracle that should never cease to amaze and astonish: miraculous events, which perfect rather than negate nature, simply serve to highlight this fact. In the act of creation God

> established 'causal reasons' – *causales rationes* – that work according to a double potential: some becoming and developing gradually – this is the natural order, the only one that men know – others eluding time and depending on the sovereign mastery which the Creator continues to exert over natural causality.
>
> (Lancel 2002, p. 468)

God governs his creation both indirectly through the laws of nature and directly through specific personal acts within the natural order.

> Thus God has, hidden within himself, the causes of certain elements that he has not placed in created things, and he actualizes these causes not through that providence by which he has made natures exist, but through the action by virtue of which he governs, as he wishes, those natures which he created, according to his will.
>
> (Lancel 2002, p. 468)

As Swinburne points out, it 'is immensely improbably that such an event [a miracle] would ever have happened if laws of nature were the fundamental determinants of what happens' (Swinburne 2010, p. 86). However, if God is more fundamental than the laws of nature, then 'all events occur only because God allows them to occur … all laws of nature operate only as long as God determines that they shall' (p. 85). In which case, it is perfectly reasonable to suppose that violations and quasi-violations of the laws of nature can be brought about by God. If we have good reason to believe in God and to trust the reliability of witnesses, and provided we are able to explain why God might have brought about a miraculous event, then we have good reason to believe in miracles.

In the classical Christian tradition miracles do not normally bear the burden of providing an epistemic warrant for Christian belief. This directly contradicts the Enlightenment critique of miracles, which mistakenly assumed that their primary function was to *demonstrate* the truth of Christianity. Thus according to Reimarus it 'is always a sign that a doctrine or history possesses no depth of authenticity when one is obliged to resort to miracles *in order to prove its truth*' (Voysey 1879, pp. 74ff.; emphasis added). John Mackie repeats this misapprehension in suggesting that the argument from the recognition of the actuality of miracles to the affirmation of the existence of God simply does not work, since Hume's critique makes it 'pretty well impossible that reported miracles should provide a worthwhile argument for theism' (Mackie 1982, p. 27). If the notion of miracles as epistemic warrants is a product of the Enlightenment, 'Christian apologetics has moved on ... "miracles" are not [now] advanced as a "proof" of anything much' (Wright 1996, p. 188). There is nothing particularly new in this; rather it marks the recovery of an earlier Christian understanding of miracles. If we attend to the historical evidence, we discover that:

> the strongest incarnational claims in the New Testament (e.g. those in Paul) have nothing to do with Jesus' mighty works, and the accounts of mighty works in the gospels are not usually offered as 'proof' of Jesus' 'divinity'.
>
> (Wright 1996, p. 186)

Though miracles were certainly appealed to as evidence of the truths of Christianity prior to the Enlightenment, such appeals tended to be used for popular apologetic purposes and were never central to Christian doctrine. The Gospel tradition itself adopts a reticent attitude towards miracles: Jesus refuses requests to perform miracles; treats such requests as temptations from a godless generation; warns that false prophets will work miracles; and questions those whose faith is dependent upon miraculous signs (Mark 8:11ff.; Matthew 12:38f., 24:24; John 4:48). As the third-century theologian Origen pointed out, belief in miracles was widespread in the Hellenistic world, embraced alike by 'genuine philosophers' and 'those who might be suspected of inventing legends' (Origen 1980, p. 308). What mattered to Jesus' contemporaries was not whether he performed miracles, since in the Hellenistic milieu their absence would have been more noteworthy than their presence, but the significance attributed to them. Thus Jesus' opponents attributed his ability to exorcise demons to the devil rather than to God (Mark 3:20ff.). Similarly, the critique of Christianity offered by the third-century philosopher Celsus views Jesus' miracles as evidence of black magic and sorcery rather than manifestations of divine power (Origen 1980, p. 105). In such an environment the claim that Jesus' miracles demonstrated the truth of Christianity would have generated more questions than answers.

Contemporary interpretations of miracles as violations of natural law tend to reject transcendence, embrace the hegemony of natural science, and affirm a naturalistic worldview. Ancient interpretations of miracles tend to accept transcendence as a given, and ask how a miraculous event relates to the supra-natural realm: 'Is this awe-full event the work of a good or an evil spirit?' The key issue for the ancients was not whether the laws of nature could be violated, but what a miraculous event might tell us about the supra-natural transcendent order. In other words, the issues raised by the ancients were ontological rather than epistemic. From their perspective, the proper question to ask was not, 'Did Jesus actually perform miracles?' but rather, 'What is their theological significance?' Thus John's Gospel identifies miracles as 'signs' through which the theological import of the person of Jesus is revealed (John 2:11). For Origen, what was truly miraculous about Jesus and his followers was the reality of transformed lives: 'the eyes of people blind in soul are always being opened, and the ears of those who are deaf to any talk of virtue eagerly hear about God and the blessed life with Him' (Origen 1980, p. 103). Similarly, for Augustine true resurrection lies not in the manipulation of dead matter but in spiritual regeneration: 'corpses were no longer restored to life, but one could see souls live again which had formerly lain in a living corpse' (Lancel 2002, p. 467).

The Hellenistic understanding of miracles was expressed in two overlapping word groups. (1) 'Wonder' (Greek Θαῦμα, thauma) and 'amazement' (Greek Θαυμάζω, thaumazō). Jesus' mighty works were greeted with astonishment: 'and the crowds were amazed and said, "Never has anything like this been seen in Israel"' (Matthew 9:33f.; cf. Mark 1:22). Such amazement was not a response to a violation of natural law, but to the manifestation of the activity of a transcendent agent. The crucial issue was whether this agent was divine or diabolical: did Jesus' power to cast out demons come from God, or from Satan (Mark 3:22ff.)? Ancient notions of divine agency did not carry with them the modern Humean-derived connotation of an arbitrary or irrational intervention from another realm.

> They indicate, rather, that something has happened, *within* what we would call the 'natural' world, which is not would have been anticipated, and which seems to provide evidence for the active presence of an authority, a power, at work, not invading the created order as an alien force, but rather enabling it to be more truly itself.
>
> (Wright 1996, p. 188)

(2) 'Signs' (sēmeion, σημεῖον) and 'portents' (teras, τέρας). In Hebrew Scripture the prophets frequently enacted symbolic actions intended as signs of God's response to Israel's actions (Lindblom 1962, pp. 165ff.). Apocalyptic literature abounds with references to 'signs' of the anticipated eschatological fulfilment of God's will for his creation (Russell 1964). The Gospel writers inhabit this tradition: Jesus' mighty works are signs of the in-breaking of the Kingdom of God (Luke 7:18ff.).

Luke recounts how Jesus, at the start of his ministry, entered the synagogue in Nazareth and read the following passage from the prophet Isaiah:

> The Spirit of the Lord is on me,
> because he has anointed me
> to bring good news to the poor.
> He has sent me to proclaim release to the captives
> and recovery of sight to the blind,
> to let the oppressed go free,
> to proclaim the year of the Lord's favour.
>
> (Luke 4:16ff.; Isaiah 61:1f.)

The verses constitute an anticipation of eschatological salvation, to which Jesus adds his own interpretation: 'Today this scripture has been fulfilled in your hearing' (Luke 4:21). Jesus' mighty works both *signify* the in-breaking of the Kingdom of God, and *constitute* the first fruits of the healing and restoration of God's creation.

Thus the Gospel writers operate within a hermeneutical horizon vastly different from that of Hume and his followers:

> When the biblical miracle stories excite serious and relevant wonderment, they intend to do this as *signals* of something fundamentally new, not as a violation of the natural order which is generally known and acknowledged. ... What took place were promises and intimations, anticipations of a redeemed nature, of a state of freedom, of a kind of life in which there will be no more sorrow, tears and crying, and where death as the last enemy will be no more.
>
> (Barth 1963, pp. 68ff.)

For the first Christians, the root meaning of Jesus' miracles was directly related to the restoration of a person's right relationship with self, society, nature and God. According to the Book of Leviticus in the Hebrew Bible, the disfiguration caused by leprosy served to 'destroy the wholeness that ought to characterise the creation' (Wenham 1979, p. 192; cf. Leviticus 13:1ff.). Consequently lepers faced both social segregation and exclusion from a spiritual relationship with God (13:46). In Jesus' own day, as the Dead Sea Scrolls make clear, the Qumran Community excluded from membership anyone 'paralysed in his feet or hands, or lame, or blind, or deaf, or dumb, or smitten in his flesh with a visible blemish' (Vermes 1995, p. 121). In sharp contrast, after healing a leper Jesus instructs him to show himself to the priest and offer the sacrifices laid down in the Torah in the event of such cleansing (Mark 1:44). The significance of Jesus' command was not epistemic: he did not order the leper to show himself to the priest in order to testify to Jesus' miraculous powers; on the contrary, he commanded the leper to 'say nothing to anyone' (Mark 1:44). The significance was rather ontological: in offering sacrifices from which he was previously excluded, the leper would enter into a restored relationship with God and with the Jewish community. Such healing and reconciliation was recognised in first-century Judaism as a mark of the Messianic Age, so that when John the Baptist sends messengers to enquire if Jesus is the expected Messiah, Jesus' indirect reply is unambiguously affirmative: 'Go and tell John what you have seen and heard: the blind receive their sight, the lame walk, the lepers are cleansed ... and the poor have good news brought to them' (Luke 7:22). Thus the healing miracles were 'seen clearly as bestowing the gift of *shalom*, wholeness, to those who lacked it, bringing not only physical health but renewed membership in the people of YHWH' (Wright 1996, p. 192).

The establishment of the Kingdom brings about the defeat of the forces of evil and chaos. Jesus' exorcisms take place on a cosmic stage: the battle is not merely against the wrongs of Roman occupation and perceived failings of Judaism's religious elite, but against the primal forces of evil that corrupt and disfigure the entire creation. When the disciples report their success in exorcising evil spirits, Jesus exclaims: 'I watched Satan fall from heaven like a flash of lightning' and – in a direct reference to the role of the serpent in the ancient myth of the fall of Adam and Eve – continues, 'I have given you authority to tread on snakes and scorpions, and over all the power of the enemy' (Luke 10:18f.; cf. Genesis 3:1ff.). This cosmic perspective is embraced by Saint Paul, who expresses hope 'that the creation itself will be set free from its bondage to decay and will obtain the freedom of the glory of the children of God' (Romans 8:21). When Jesus is reported to have walked on water, liberal theologians, guided by positivistic and naturalistic questions about 'what actually happened', speculated that he may actually have been walking on a sandbank. This misses entirely the theological point the Gospel writers were striving to make: in the Hebrew Bible the word for 'sea' is derived from the name of the destructive god in the Babylonian creation myth, and carried connotations of 'a mysterious and threatening force opposed to God', so that in walking on water Christ overcomes the primal forces of chaos and evil (Borg 1993, p. 68).

How then would Jesus' contemporaries have understood his miracles? Not as violations of the laws of nature, but as evidence of either divine of demonic agency. For the first Christians they constituted the first fruits of the long-awaited Messianic Age, in which God would restore the proper relationship between the Creator and his creation. Any retroductive attempt to understand and explain the life of Jesus in its historical context needs to address the dispute between his disciples and opponents regarding the nature of his miracles, and avoid the distractions of an alternative post-Enlightenment agenda. This is not to nullify the question as to 'what actually happened', but it is to suggest that any answer we might give will require access to philosophical and theological thinking beyond the remit of the historian, as well as access to the best available historical reconstructions of how Jesus' contemporaries, imbibed as they were by the culture of their time, made sense of them.

The Kingdom of God

What did Jesus set out to achieve? There is a scholarly consensus that the establishment of the Kingdom of God stands at the heart of Jesus' teachings and actions.

> Now after John was arrested, Jesus came to Galilee, proclaiming the good news of God, and saying, 'The time is fulfilled, and the kingdom of God has come near; repent, and believe in the good news'.
>
> (Mark 1:14f.)

To make sense of this verse, we need to understand what a first-century Jew would have made of it. (1) Jesus' proclamation, like all Jewish theology at that time, was essentially political. The notion of religion as a private spiritual activity is the product of the impact of Platonic dualism, Protestant pietism and liberal individualism on contemporary Christianity. The call to repent and believe referred not to a change in moral behaviour or acceptance of a new set of religious doctrines, but to a change in religio-political allegiance. (2) Jesus' proclamation was bound up with a God simultaneously transcendent of, yet immanently active within, the created order. This distinctively Jewish understanding of God had little to do with the utterly transcendent impersonal God of much Greek philosophy, later to re-emerge as the medieval God of classical theism and the deistic God of Enlightenment philosophers, and nothing whatsoever to do with the Platonic notion of the immanent presence of a spark of divinity in each human soul, as embraced by various syncretic Hellenistic cults and enjoying close parallels with contemporary forms of New Age spirituality. (3) Jesus' proclamation was intimately related to the Jewish understanding of divine election. From the beginning the Creator God had elected the people of Israel and formed a covenant with them, with the intention, not to save the Jews for eternal bliss and damn the Gentiles to eternal torment, but that the nation would be the instrument through which justice and mercy would flow out over all nations and transform the entire cosmos by bringing creation once more under the rule of God. (4) Jesus' proclamation was essentially eschatological, oriented towards the future climax of the history of Israel and culmination of God's plan for creation.

To proclaim the in-breaking of the Kingdom of God in a first-century Jewish milieu was to announce that history was reaching its climax as God worked through Israel to transform the current socio-political and cosmic order and establish once and for all the rule of divine justice and mercy. Israel was an occupied nation, and, for all the civilising goods the Romans might have brought with them (aqueducts and the like!), the Jews lived under a brutal yoke of martial law, iniquitous taxation and pagan gods. Just as, in 597 BCE, the Babylonians had invaded Israel, destroyed the Temple and driven the leaders of the people into geographic exile, so now the Jews lived in spiritual exile, subject to the rule of an occupying power and far removed from the rule of God. Hope for the future emancipation of Israel from foreign rule looked back to the archetypal account of the nation's emancipation from slavery in Egypt and journey through the wilderness to the Promised Land. In this situation there were a variety of different ways in which Jews sought to further the cause of God's Kingdom. (1) *Resistance*: the Pharisees sought to affirm the purity of Jewish identity in the face of pagan contamination as part of the political struggle to liberate the nation from

Roman oppression; the Zealots took this struggle a stage further, embarking upon a programme of armed resistance and insurrection that was to lead to the siege of Jerusalem and destruction of the Temple in 70 CE. (2) *Appeasement*: both the Chief Priests in charge of the Temple and Herod Antipas, ruler of Galilee and Perea, sought to work alongside the occupying forces, at best in order to protect the nation from further harm, and at worst – and this appears to have been the popular sentiment – for purposes of self-aggrandisement. (3) *Quietism*: this was the response of members of the Qumran community, who retreated from society into a monastic existence, from where they waited passively for God to vindicate their spiritual discipline and obedience by instigating an apocalyptic war that would lead to the destruction of God's enemies and establishment of his Kingdom.

Jesus' vision of the Kingdom of God was both rooted in the Judaism of his time and yet radically different: 'He was neither a quietist, nor a compromiser, nor a zealot', but instead 'went back to Israel's scriptures and found there another kingdom-model, equally Jewish if not more so' (Wright 2000, pp. 20ff.). Against the insurrectionists, his insistence upon non-violence reflected Isaiah's expectation that, at the coming of the Kingdom, swords would be beaten into ploughshares and spears into pruning hooks, and nations would no longer study the craft of war: 'Do not resist the evildoer. But if anyone strikes you on the right cheek, turn the other also' (Matthew 5:39; cf. Isaiah 2:1ff.). Against the compromisers, he advocated an ethic of radical obedience to God, insisting that it is impossible to serve both God and any human-derived economic, religious, social or political power structures (Matthew 6:24). This radical ethic cut to the very marrow of human motivation: 'You have heard it was said, "You shall not commit adultery". But I say to you that everyone who looks on a woman with lust [i.e. illicitly, in contemplating an adulterous relationship] has already committed adultery with her in his heart' (Matthew 5:27f.). It was also profoundly challenging: a call to abandon home, family, livelihood and security and take up the cross of discipleship (Matthew 16:24). Against the quietists, he rejects any rigorous self-denying spiritual discipline designed to invoke God's actions by way of reward. The in-breaking of the Kingdom is entirely the initiative of God, and a cause not for fasting but for celebration: 'the Son of Man came eating and drinking, and they say "Look, a glutton and a drunkard, a friend of tax collectors and sinners"' (Matthew 11:19).

Jesus calls his followers to trust in God and abandon themselves to divine providence: 'can any of you by worrying add a single hour to your span of life?' (Matthew 6:27). The realisation of the Kingdom lies in the power of God, not in human effort. It is the holiness of God that is transforming the world, and human efforts to protect the holiness of God, by living pure and righteous lives and affirming the purity of Jewish identity in the face of pagan contamination, are futile. Jesus rejected the received understanding that if anything impure comes into contact with anything holy it necessarily corrupts it. The holy God does not need protecting, since his sovereign power cleanses the impure: when Jesus touches the leper he does not become unclean, but rather – and contrary to

Jewish purity laws – the leper is made clean. Jesus came to call the impure, the outcasts and sinners, the lost sheep of the people of Israel, and through table-fellowship with them sought to transformed their lives: 'Those who are well have no need of a physician, but those who are sick; I have come to call not the righteous but sinners' (Mark 2:17). The Pharisees' quest for purity in order to accelerate the coming of the Kingdom, and concomitant criticism that Jesus' followers consisted primarily of social outcasts and the ritually impure, drew an angry response: in seeking to strengthen the boundaries between purity and impurity, righteousness and unrighteousness, they were blind to what is happening around them – the outpouring of the loving mercy of God on the ungodly that *constituted* the in-breaking of God's Kingdom. There is no separating Jesus' actions from his words: his miracles did not merely symbolise the coming Kingdom, but rather served to establish it; his parables constituted calls to action, invitations to embrace the proclamation of the in-breaking Kingdom and make it one's own.

The crucified Messiah

Why did Jesus die? Did he expect or intend to die as part of his vocation? Wright seeks to answer this question by focusing on three symbolic actions performed by Jesus in the last week of his life.

(1) Jesus' Triumphal Entry into Jerusalem. When Jesus stage-managed his dramatic entry into Jerusalem, in an apparently premeditated fulfilment of the prophecy of Zechariah, he was intending 'to symbolise and embody the long-awaited return of YHWH to Zion' (Wright 2000, p. 86; cf. Mark 11:1ff.; Zechariah 9:9f.). The symbolism was not lost on the onlookers, who chanted in response: 'Hosanna! Blessed is the one who comes in the name of the Lord. Blessed is the coming Kingdom of our ancestor David' (Mark 11:9f.). During his ministry, Jesus told a number of parables on the theme of a king returning to his kingdom to ascertain how his subjects had fulfilled their responsibilities during his absence. Traditionally, Christians have interpreted them as predictions of the second coming of Christ; however, Wright argues that they were concerned with the expected return of God to his Temple to establish his Kingdom. Read in the context of Jesus' proclamation of the coming of the Kingdom, the parables were referring to contemporary events rather than future expectations. In enacting the arrival of the Messiah in Jerusalem, he was effectively declaring the expected return of God to the Temple.

(2) Jesus' 'Cleansing' of the Temple. Once in Jerusalem Jesus entered the Temple, drove out the worshippers and officials, and for a time took control of the building (Mark 11:15ff.). The Temple was the symbolic heart of Judaism, the former and future dwelling place of God on earth. Given that Jesus had just symbolically enacted the return of God to Jerusalem, we would expect him to go on to affirm the received understanding of the place of the Temple in Israel's religio-political life. Why then did he engage in a direct attack on that institution instead? At the time the Temple was under the control of the High Priests and

political elite, who pursued a policy of appeasement towards the Roman occupiers. As a result, their role was subject to a current of internal criticism within Judaism: there was a widespread sentiment that the institution needed reforming owing to the corrupt practices of those in charge. According to one interpretation, in overturning the tables of the money changers Jesus was attacking the political and financial wrongdoing of those who followed the political strategy of appeasement, and thereby seeking to restore the institution to its traditional role. This would place Jesus on the side of the Pharisees and those who sought to bring about the Kingdom by affirming the purity of Jewish identity in the face of pagan contamination. However, his dual claim, that the Temple 'shall be called a house of prayer for all the nations' but that 'you have made it a den of robbers', suggests otherwise (Mark 11:17). The term 'robber' (lestai, λησταί) translates better as brigand, insurrectionist or terrorist, rather than common thief, and carries with it a connotation of violence. When linked to the issue of the access of Gentiles to Temple worship, this suggests that Jesus' attack was focused not upon the appeasers but upon those concerned with resisting the occupying Roman forces. As we have already seen, such resistance stressed ritual purity as a means to maintain national identity. On this reading, Jesus targeted the money changers, not because of financial corruption, but because their primary role was to ensure that coins marked with the face of the Roman Emperor were not brought into the inner courts of the Temple and allowed to contaminate the holy site. This being the case, Jesus' assault on the Temple was directed against those who stressed ritual purity and national identity. As such, it constituted a rejection of a strategy that placed the nation rather than God at the centre of the religio-political agenda, and ignored Israel's covenantal responsibility to call *all nations* to a state of righteousness under God. Jesus' '*kingdom-agenda for Israel demanded that Israel leave off her frantic and paranoid self-defence, reinforced as it now was by the ancestral codes, and embrace instead the vocation to be the light of the world, the salt of the earth*' (Wright 2000, p. 38; emphasis in original). Such a reading is consistent with Jesus' teaching and actions, which affirmed God's preferential inclusion of outcasts, sinners and the ritually unclean within the dawning Kingdom, and warned against the hypocrisy of the self-righteous. Jesus' attack on the Temple thus constituted more than an attempt at the reform of corrupt practices: with the eschatological dawning of the Kingdom, with the return of God to Jerusalem, the entire institution had become redundant. Having failed in its task to bring light to the Gentiles, and become instead a symbol of resistance to them, the entire Temple institution was now standing under the judgement of God, facing immanent destruction at the hands of the Romans, and in process of being replaced by a new Messianic community, one committed, not to the self-righteous protection of national identity, but to the actualisation of the rule of God's inclusive peaceable Kingdom across the whole of creation.

(3) The Last Supper. Jesus' final meal with his disciples was a celebration of Passover, which traditionally looked back to the exodus of Israel from Egypt and forward to the end of the nation's spiritual exile (Mark 14:12ff.). Wright

suggests that Jesus used the occasion to put a seal on a new exodus community: alongside the negative symbol of the destruction of the Temple stands the positive symbol of the constitution of the new Messianic Kingdom. Jesus was establishing a new Israel, a new inclusive community, in which the twelve disciples represented the twelve tribes of Israel and 'Israel's God was present and active in the same way as he normally was in the Temple' (p. 49). His insistence upon non-violence inevitably left members of this new community defenceless and vulnerable, and he warned his disciples that they must be prepared to take up the Cross and tread the path of suffering in defence of the dawning kingdom. As contemporary movements of non-violent resistance recognise, violence cannot ultimately be defeated by violence: non-violent resistance must first draw the sting of brutality, submit to it and allow it to do its worst if the futility of violence is to be exposed and the way of peace is ultimately to be victorious. It is reasonable to suppose that Jesus anticipated the possibility of his arrest and execution as a direct result of his attack on the Temple: a fate not unknown for dissidents in totalitarian regimes, especially those who stage coups at the symbolic heart of the systems they seek to overthrow. Jesus distributes bread and wine, identifying it with his own body and blood: 'this is my blood of the new covenant, which is poured out for many for the forgiveness of sins' (Matthew 26:28). 'Jesus seems to have regarded his own impending death as being part, indeed the climax, of the vocation in which his work and Israel's fate were bound up together' (Wright 2000, p. 62). There was a widespread belief among first-century Jews that the sufferings of Israel under foreign tutelage were redemptive:

> a belief, hammered out not in abstract debate but in and through poverty, torture, exile and martyrdom, that Israel's sufferings might be, not merely a state from which she would, in YHWH's good time, be redeemed, but paradoxically, under certain circumstances and in certain senses, be part of the means by which redemption would be effected.
>
> (Wright 2000, p. 64; emphasis in original)

Jesus sought to identify himself with Israel and to fulfil, through his own death, her vocation as the covenant people who, in suffering at the hands of the men of violence, would establish the righteousness rule of God over creation.

Why then did Jesus die? During the interrogation following his arrest, he is asked directly by the High Priest Caiaphas: 'Are you the Messiah, the Son of the Blessed One?' Jesus replies: 'I am; and "you will see the Son of Man seated at the right hand of the Power", and "coming with the clouds of heaven"' (Mark 14:61f.). Jesus here conflates two texts from Hebrew Scripture that refer directly to the vindication and enthronement in glory of God's agent of salvation (Psalm 110:1f.; Daniel 7:13f.). Caiaphas responds by accusing Jesus of blasphemy. Claiming to be the Messiah or threatening the Temple may not in themselves have been seen as blasphemous; however their combination, when accompanied by the claim that God would vindicate Jesus' actions and enthrone him in glory in heaven, certainly would have been. Jesus died because he enacted God's return to Jerusalem, attacked and discarded the institution of the Temple, sought to establish a new messianic community centred on his person, and claimed that he would be vindicated by God regardless of the actions of the men of violence. He died because in the eyes of the ruling elite he was a blasphemer, claiming in both word and deed to possess the authority to represent the will of God on earth. By representing Jesus to the Roman authorities as the leader of an insurrectionist movement rather than a religious blasphemer, it was a simple step to transmute 'that theological verdict into a political one that Pilate would have taken notice of' (Wright 2000, p. 89).

Did Jesus expect or intend to die as part of his vocation? The gospel writers make it clear that Jesus had no wish to die a martyr's death, and that he pleaded with his heavenly Father to 'remove this cup from me' (Mark 14:36). Nevertheless, he could not possibly have been ignorant of the fact that his symbolic entry into Jerusalem and intervention in the Temple would inevitably lead him into direct conflict with the ruling religious elite and their Roman partners; and – as his references to his impending betrayal and death during the Last Supper, and prayer to his Father in the garden at Gethsemane made plain – he clearly anticipated the likelihood of his arrest and subsequent execution.

Jesus and God

Who did Jesus think he was? The theological question 'Was Jesus God?' raises ontological, metaphysical and theological questions that cannot be resolved by the historian. Nevertheless, the question 'Did Jesus believe he was God?' clearly falls within the remit of historical investigation.

Though the claim of the first Christians that 'Jesus is God' was unthinkable for first-century Jews and anathema to the entire Jewish tradition, belief in the possibility of the intimate presence of the Creator within creation was normative. God could be encountered directly in the study of the Torah, his glory had once filled the Temple, his spirit had once rested on Moses and the prophets, through his covenant he participated actively in the history of Israel and the nations of the world, and creation itself reflected his wisdom. The Babylonian exile was experienced as the abandonment of Israel by God, the self-imposed fruits of their failure to fulfil their part in the covenantal agreement. Ezekiel and Deutero-Isaiah looked forward to the homecoming of Israel from exile, the return of God to the Temple, the renewal of the Covenant, the transformation of creation. However, the exodus from Babylon, the repossession of the land, and the rebuilding of the Temple had not produced the expected fruits: the Kingdom had not yet been established, God remained absent from the Temple, and Israel once again found herself under foreign domination. Living in spiritual exile, the nation looked to the future for the establishment of the Kingdom. One nebulous cluster of hopes for the future came to solidify around the notion of the Son of Man, an angelic or human figure who would act as God's agent in the restoration of the Kingdom and be exalted and enthroned in glory at God's right hand

(Daniel 7:13ff.; Borsch 1967). Thus the notion of God's immanent presence in the world, and of his intimate relationship with those called to specific tasks in the divine economy of salvation, was a central feature of first-century Judaism: if incarnation was not an option, it was certainly far closer to Jewish theology than pantheism and classical theism.

The most distinctive feature of early Christianity was the worship of Jesus and affirmation of his divinity:

> He is the image of the invisible God, the firstborn of all creation; for in him all things in heaven and on earth were created.... He himself is before all things, and in him all things hold together.... For in him all the fullness of God was pleased to dwell, and through him God was pleased to reconcile to himself all things, whether on earth or in heaven, by making peace through the blood of his cross.
>
> (Colossians 1:15ff.)

In his first extant letter to the Christian community in Corinth, Paul, having explicitly rejected polytheism, offers a drastically recast version of the Jewish Shema ('Hear, O Israel, the Lord our God, the Lord is One'): 'there is one God, the Father, from whom are all things and for whom we exist, and one Lord, Jesus Christ, through whom are all things and through whom we exist' (1 Corinthians 8:6).

> This stunning adaptation of the Jewish prayer ... emphasizing creation and redemption as equally originating in the Father and equally implemented through Jesus, encapsulates, at the earliest stage of Christianity for which we have hard evidence, everything that later generations and centuries would struggle to say about Jesus and God. From here on, we must say that if trinitarian theology had not existed it would be necessary to invent it.
>
> (Wright 2000, p. 79)

In worshipping Jesus as God the first Christians violated the boundaries of Jewish theology and necessitated the construction of an alternative worldview and new incarnational theology. The critical realist historian needs to look for a historical explanation of this unexpected and entirely unpredictable turn of events. The suggestion that belief in Jesus' divinity was stimulated by contact with pagan polytheism makes little sense: Judaism, though cognisant of polytheism, had never been tempted to assert the divinity of her prophets, and there is no evidence that Christianity responded any differently. In the absence of any compounding reason to think otherwise, it is the first Christians' own explanation, taken as face value, which appears to have the edge over other speculative historical reconstructions: they present their proclamation of Jesus' divinity as a direct outcome of their experience of his life and death and of their belief in his subsequent resurrection. What was it *specifically* about Jesus that led the first Christians to proclaim his divinity? And how did Jesus himself understand his relationship with God? In seeking to answer these questions it is important to recognise that neither the identification of Jesus as 'Messiah' and 'Son of God', nor the belief in his resurrection from the dead, taken by themselves, imply his divinity: David was recognised as both God's anointed king and as a son of God, and Lazarus rose from the dead – yet neither was considered to be divine. The current practice of using 'Son of God' to refer to Jesus' divinity does not have any historical precedent.

As we have already noted, the three symbolic acts performed by Jesus in the last week of his life – the Triumphal Procession into Jerusalem, the Attack on the Temple and the Last Supper with the disciples – enacted the return of God to Jerusalem, God's rejection of the institution of the Temple, and God's establishment through the person of Jesus of a new messianic community.

> Jesus was claiming that he, rather than the Temple, was the place where, and the means by which, the living God was present with Israel ... in his entire public career, [he] was acting as if he were bringing about the new Exodus. God's people were in slavery; he had heard their cry, and was coming to rescue them.... He would bring about the final redemption of God's people, and thereby set in motion the fulfilment of Israel's destiny to be the light of the whole world.
>
> (Wright 2000, pp. 82, 85f.)

A recurrent feature of the Gospel narratives is the recognition, by friend and foe alike, of Jesus' claim to speak and act with divine authority. Thus, with reference to the Law of Moses, he announces: 'You have heard that it was said to those of ancient times ... *But I say to you....*' (Matthew 5:21, 27, 33, 38, 43; emphasis added). Similarly, when a paralysed man is brought to him, Jesus declares, 'Son, your sins are forgiven'; the scribes respond, 'Why does this fellow speak in this way? It is blasphemy! Who can forgive sins but God alone?'; Jesus then heals the paralytic, 'so that you know that the Son of Man has authority on earth to forgive sins' (Mark 2:1ff.). Jesus spoke and acted in the belief that God was working through him to establish the Kingdom and return creation to a proper relationship with its Creator, that he had the divine authority to do so, and that, following his probable death, he would be vindicated by God and enthroned in glory at God's right hand: as people related to him, so they related to God.

> In Jesus himself ... we see the biblical portrait of YHWH come to life ... the creator God, giving new life ... the faithful God, dwelling in the midst of his people ... the stern and tender God, relentlessly opposed to all that destroys and distorts the good creation ... but recklessly loving all those in need and distress.
>
> (Wright 2000, p. 90)

What, then, of Jesus' own sense of self? There is no suggestion in the historical sources that Jesus was omniscient, supra-naturally aware that he was the second person of the Trinity who had left the eternal heavenly realm to enter into a temporal earthly existence. To propose otherwise is to import a later, distorted and thoroughly Docetic understanding of incarnation into the historical data. There are, of course, important theological and philosophical issues raised by the orthodox claim that Jesus was simultaneously both fully man and fully God; however, these are not the primary concern of the historian, who must remain faithful to the historical evidence. The Gospels portray a human Jesus who wrestles with his vocation and grows in understanding, not as some otherworldly supra-natural being. Jesus clearly does not know he is divine, either in the positivistic sense in which we might know an object under direct observation, or in the idealist sense of immediately apprehending a self-evident truth. Rather, the Gospel evidence suggests that Jesus' knowledge was tacit: he knew he was God in the same way in which we might know that we are loved by those closest to us, or recognise the beauty of a sunset, or discern the inherent goodness of a moral action. Most of our knowledge is tacit in this sense: we know deeply and profoundly, yet without being able to fully explain, account for or verify our knowledge. Wright suggests that Jesus' self-understanding was akin to vocational knowledge: just as an artist is *vocationally driven* to paint, a dancer is *vocationally driven* to dance, or a priest is *vocationally driven* to ordination by impulses that they might not fully understand, so Jesus

> as part of his human vocation, grasped in faith, sustained in prayer, tested in confrontation, agonized over in further prayer and doubt, and implemented in action ... believed he had to do and be, for Israel and the world, that which according to scripture only YHWH himself could do and be.
>
> (Wright 2000, p. 91)

The resurrection

How and why did the Christian Church begin? What generated its innovative worldview? Since the crucifixion revealed Jesus to be a false Messiah who had clearly failed to gain the expected victory over the Roman oppressors, we would expect Jesus' disciples either to abandon or modify their belief, or else seek out a new messianic leader – James the brother of Jesus, active in the leadership of the first Christian community in Jerusalem, being the obvious candidate to succeed him in a hereditary dynasty. Yet they continued to proclaim Jesus as the true Messiah who had successfully inaugurated the Kingdom of God, despite the fact that Israel remained in spiritual exile, the Temple had not been restored, and the Romans continued to occupy the land. What drove them to radically transform the Jewish worldview in such a counter-intuitive manner? According to one popular view, the true founder of Christianity was not Jesus but Paul, who, in a fundamental misunderstanding of Jesus intentions, replaced Jesus' message of the in-breaking Kingdom with a religion designed to venerate the messenger himself (Wenham 1995). But this position lacks evidential support and simply relocates – rather than answers – our original questions: Why did *Paul* found

Christianity? Where did *Paul's* innovative worldview come from? The fundamental problem remains: of explaining why the early Church – whether at the behest of Jesus' disciples or of Paul – insisted on proclaiming Jesus as the Messiah when, by all received Jewish criteria, this was palpably not the case. The extant literature of early Christianity gives a clear and consistent answer: 'there is no form of early Christianity known to us ... that does not affirm at its heart that after Jesus' shameful death God raised him to life again', thereby vindicating him and confirming his Messiahship (Wright 2000, p. 94).

We need to understand the Christian belief in Jesus' resurrection in the light of the culture within which it originated (Wright 2003). Though Greek, Roman and Hellenistic culture allowed for the possibility of immortality, bodily resurrection was deemed impossible. Within Judaism there was a wide spectrum of beliefs regarding the afterlife: eternal bliss in a spiritually disembodied state; eternal bliss in a physically embodied state; an interim period of disembodiment prior to re-embodiment once the Kingdom of God was established; and the denial of any form of life after death. In first-century Judaism, 'resurrection' (Greek ανάστασις, anástasis) referred exclusively to the re-embodied state of the righteous following the final establishment of God's Kingdom. Ezekiel's vision, written while in exile in Babylon, of a valley of dry bones and their transformation into living bodies symbolised the return from exile, the establishment of the Kingdom, and the transformation of fallen creation (Ezekiel 37:1ff.). Thus, when the first Christians announced the bodily resurrection of Jesus, they were effectively claiming the final victory of God over the forces of evil. This would have been puzzling for contemporary Jews on two counts: (1) because there was no expectation of the resurrection of an individual prior to the general resurrection of the dead following the establishment of the Kingdom; (2) because there were no obvious signs that the Kingdom of God had actually arrived.

Wright suggests the most powerful explanation of the emergence of Christianity currently available to the historian is that proffered by the first Christians themselves: namely that their radical rewriting of the Jewish understanding of Messiahship and the Kingdom of God was a direct result of (their belief in) the resurrection of Jesus. The academic historian is at liberty to dismiss a raft of unfounded and entirely speculative conspiracy theories: for example, that the disciples fraudulently constructed the Christian faith fully aware that Jesus had survived his crucifixion and was living secretly in domestic bliss with Mary Magdalene. Having done so, only one serious alternative to the hypothesis that Christianity emerged as a direct result of the first Christians' belief in the resurrection holds sway in the academic world: namely that the first disciples became convinced of the continuing presence of the spirit of Jesus in their lives (in much the same way as, say, the spirit of Che Guevara lives on in some revolutionary movements), and that the later Gospel writers transformed this spiritual awareness into the physical 'event' of Christ's resurrection. This alternative hypothesis is naturally attractive to post-Enlightenment minds: it can be made to fit with contemporary psychology, does not require belief in miracles, allows the first Christians to be well-intentioned individuals rather than deliberate fraudsters, and enables liberal Christians to discard a raft of unwanted supra-natural, metaphysical and dogmatic baggage. Wright's objection is that the hypothesis is seriously lacking in both evidential support and explanatory power.

There is no evidence in the extant historical documents of (1) an early stratum of belief in the continuing presence of the spirit of Jesus among the first Christians, and (2) a later stratum of belief in the empty tomb and Jesus' embodied post-resurrection appearances. On the contrary, Paul – whose letters constitute the earliest available evidence of the emergent Christian community – makes it abundantly clear that his faith is grounded in the resurrection rather than a sense of Christ's continuing spiritual presence in the community; indeed, he appeals to a formal list of witnesses to the resurrection, many of whom were still alive when he wrote, and known to him personally. It makes little sense to suggest that Paul's insistence upon a bodily resurrection was established *before* the accounts of the empty tomb and embodied resurrection appearances had been generated by the later Gospel writers; nor that the latter 'invented the empty tomb and the "meetings" or "sightings" of the risen Jesus in order to explain a faith they already had' (Wright 2003, p. 707). There was little to be gained theologically for a community constituted by a belief in the continuing presence of the spirit of Jesus to shift its position fundamentally (1) by insisting on the central significance of the resurrection, and (2) subsequently constructing the resurrection accounts in support of that shift. Though the spirit of the Maccabaean martyrs continued to inspire Jewish insurrectionists during Jesus' lifetime, there was no need to posit their bodily resurrection in order to affirm 'that their cause was indeed righteous and that they were alive in a place of honour in the presence of God', and no attempt was made to do so (Wright 2000, p. 102). Further, the hypothesis of a transition from spiritual awareness to resurrection stories does nothing to explain why the first Christians chose to take such a drastic and counter-intuitive course of action, and so fails to account for the origins of the distinctive Christian worldview. Since there is no reason to suppose that the early Church could not have continued to sustain itself on the basis of an original belief in the continuing spiritual presence of Jesus within the community, the historian is bound to explain why some Christians began to proclaim Christ's resurrection and why the later Gospel writers followed them by constructing fictional resurrection narratives. There is a consensus that the resurrection accounts constitute the central 'puzzle' of the New Testament. The theological reflections of the New Testament writers make eminent sense as attempts to resolve this puzzle, but little sense as attempts to resolve problems in Pauline theology (i.e. his unprecedented and theologically troublesome belief in the resurrection as a physical reality rather than a spiritual ideal), or as part of an apologetic drive to make Christianity more attractive to outsiders (there is considerable evidence that belief in the resurrection was a source of both popular derision and philosophical scepticism).

If the accounts of the resurrection and empty tomb are later creations of the Church, then they contain elements that are extremely difficult to explain (Wright 2003, pp. 599ff.). (1) Unlike the rest of the Gospel narratives, they do not cross-reference Old Testament texts, despite the widespread claim that Christ rose from the dead 'in accordance with the scriptures': if they were constructed by the Christian community we should expect to see an increased level of cross-referencing rather than its virtual absence. (2) There is no expression in the narratives of hope in an afterlife: in a cultural context in which the Sadducees and others denied the reality of life after death, it is difficult to understand why a narrative constructed for apologetic purposes missed the opportunity to relate the resurrection of Jesus to the future resurrection of the narratives' readers. (3) The descriptions of the risen Jesus are remarkably 'matter of fact': the Jewish apocalyptic tradition provided a wealth of imagery and symbolism that would almost certainly have been drawn upon if the accounts were literary creations rather than attempts at historical reportage. (4) The fact that the formal list of witnesses to the resurrection reported by Paul includes only men is understandable, given the lack of credibility, and in some cases legal status, attributed to the testimony of women in the first-century milieu: this being the case, the priority given to the testimony of women in the resurrection stories is inexplicable if the authors were seeking to provide apologetic support for Christianity rather than recount historical testimony. (5) There are a number of inconsistencies in the different resurrection accounts which we would expect to have been ironed out if they were literary creations designed for apologetic purposes; the fact that the accounts independently 'tell a tale which, despite the multiple surface inconsistencies, succeeds in hanging together', points towards a concern on the part of the Gospel writers to be faithful to their historical sources (p. 614).

Wright argues that it makes no sense to 'run the movie in reverse': the resurrection stories do not read as explanations of a theological conundrum created by Paul, or as exercises in Christian apologetics; rather, they 'have the puzzled air of someone saying, "I didn't understand it at the time, and I'm not sure I do now, but this is more or less how it was"' (p. 611). Wright affirms the 'very strong historical probability … that, when Matthew, Luke and John describe the risen Jesus, they are writing down very early oral tradition, representing three different ways in which the original astonished participants told the stories' (p. 611).

How and why did the Christian Church begin? What generated its innovative worldview? Wright suggests that the best retroductive hypothesis currently available to us is that Christianity began because the first Christians believed that they had encountered the living Jesus resurrected from the dead. The innovative worldview of Christian emerged as the first Christians responded to their abductive belief in the resurrection by a process of retroductive-iterative historical and theological reflection on the person, actions and teachings of Jesus in the light of Jewish Scripture and their post-Easter experiences. Regardless of the veracity of Christian belief in the resurrection, the hypothesis that Christianity emerged as a result of this belief is consistent with the available historical data – including both the life and teaching of Jesus himself, and the various lines of theological reflection present in the New Testament documents – and possesses an explanatory power that, in Wright's judgement, effectively trumps alternative explanations. If the hypothesis is faulty, then an alternative retroductive hypothesis must be put in its place. The only viable alternative is the suggestion that the first Christians believed in the ongoing presence of the spirit of Jesus within the community; however, this does nothing to explain why the first Christians worshipped him as God, or why they subsequently generated an alternative belief in Jesus' embodied resurrection in a manner entirely at odds with received Jewish understanding. As such, this alternative hypothesis is unable to explain the origins of the distinctiveness of early Christianity vis-à-vis Judaism, or why the first Christians were unable to remain within the Jewish community – an option that belief in the spiritual presence of Jesus would have allowed. Until such time as a more powerful and comprehensive hypothesis emerges, the belief of the first Christians in the embodied resurrection of Jesus is both sufficient and necessary to explain the emergence of the Church and the origins of the distinctive Christian worldview. Faithful to his critically realistic understanding of historiography, Wright insists that the resurrection stories 'provide evidence not directly for what happened but for what several different people thought had happened' (p. 614).

Wright recognises that, in moving from the earliest Christians' *belief* in Christ's resurrection to the ontological question of the truth and veracity of that belief, historians find themselves 'at the borders of language, of philosophy, of history and of theology' (Wright 2000, p. 112). But, then, so do defenders of naturalism and the philosophy of meta-Reality: the exercise of judgemental rationality with regard to the ultimate order-of-things and our place within it is notoriously difficult for all parties involved. In exposing the poverty of idealism, empiricism and pragmatism, critical realism has shown the intellectual necessity of asking ultimate questions of meaning and truth. It has also demonstrated the importance of doing so attentively, intelligently, reasonably and responsibly. The ambiguity surrounding the ontological truth claims of Christianity, naturalism, meta-Reality and a host of other worldviews demands the careful and reflective exercise of judgemental rationality. Without it, we are condemned to live in a spiritually, religiously and theologically illiterate society – one in which the economy of power and persuasion takes precedence over the economy of the pursuit of truth and truthful living.

Bibliography

Anatolios, Khaled (1998) *Athanasius: The Coherence of His Thought*, London: Routledge.

Anglicans Online (2011) *Athanasian Creed*, http://anglicansonline.org/basics/athanasian.html, accessed 27 April 2011.

Anselm (1962) *Basic Writings*, La Salle, IL: Open Court.

Aquinas, Thomas (1920) *Summa Theologica*, 22 vols, London: Burns Oates & Washbourne.

Archer, Margaret S., Bhaskar, Roy and Collier, Andrew *et al.* (1998) *Critical Realism: Essential Readings*, London: Routledge.

Archer, Margaret S., Collier, Andrew and Porpora, Douglas V. (2004) *Transcendence: Critical Realism and God*, London: Routledge.

Aristotle (1984) *The Complete Works of Aristotle*, vol. 2, Princeton, NJ: Princeton University Press.

Astley, Jeff, Francis, Leslie J., Sullivan, John and Walker, Andrew (eds) (2004) *The Idea of a Christian University: Essays on Theology and Higher Education*, Milton Keynes, Buckinghamshire: Paternoster Press.

Athanasius (1996) *On the Incarnation*, Crestwood, NY: St Vladimir's Seminar Press.

Auerbach, Erich (2003) *Mimesis: The Representation of Reality in Western Literature*, Princeton, NJ: Princeton University Press.

Augustine (1961) *Confessions*, Harmondsworth, Middlesex: Penguin Classics.

Aulén, Gustaf (2010) *Christus Victor: An Historical Study of the Three Main Types of the Idea of the Atonement*, London: SPCK.

Babcock, William S. (1991) 'A Changing of the Christian God: The Doctrine of the Trinity in the Seventeenth Century', *Interpretation*, 45, pp. 133–146.

Balthasar, Hans Urs Von (1989) *The Glory of the Lord: A Theological Aesthetics, Volume IV: The Realm of Metaphysics in Antiquity*, Edinburgh: T. & T. Clark.

Barbour, Ian G. (1966) *Issues in Science and Religion*, Englewood Cliffs, NJ: Prentice-Hall.

—— (1968) 'Science and Religion Today', in Barbour, Ian G. (ed.) *Science and Religion: New Perspectives on the Dialogue*, London: SCM, pp. 3–29.

—— (1974) *Myths, Models and Paradigms: A Comparative Study in Science and Religion*, New York: Harper & Row.

—— (1990) *Religion in an Age of Science*, London: SCM.

—— (1998) *Religion and Science: Historical and Contemporary Issues*, London: SCM.

—— (2000) *When Science Meets Religion: Enemies, Strangers or Partners?*, London: HarperOne.

—— (2002) *Nature, Human Nature and God*, London: SPCK.
Barr, James (1981) *Fundamentalism*, London: SCM.
Barth, Karl (1936–1952) *Church Dogmatics*, 14 vols, Edinburgh: T. & T. Clark.
—— (1963) *Evangelical Theology: An Introduction*, London: Weidenfeld & Nicolson.
—— (2001) *Protestant Theology in the Nineteenth Century: Its Background and History*, London: SCM.
Bauer, Walter (1971) *Orthodoxy and Heresy in Earliest Christianity*, Philadelphia, PA: Fortress Press.
Behr, John (2004) *The Nicene Faith*, 2 vols, Crestwood, NY: St Vladimer's Seminary Press.
Bentley, Michael (1999) *Modern Historiography: An Introduction*, London: Routledge.
Bernstein, Richard J. (1983) *Beyond Objectivism and Relativism: Science, Hermeneutics and Praxis*, Oxford: Blackwell.
Bhaskar, Roy (1986) *Scientific Realism and Human Emancipation*, London: Verso.
—— (1989) *Reclaiming Reality*, London: Verso.
—— (1991) *Philosophy and the Ideal of Freedom*, Oxford: Blackwell.
—— (1993) *Dialectic: The Pulse of Freedom*, London: Verso.
—— (1994) *Plato Etc.: The Problems of Philosophy and their Resolution*, London: Verso.
—— (1997) *A Realist Theory of Science*, London: Verso.
—— (1998) *The Possibility of Naturalism: A Philosophical Critique of the Contemporary Human Sciences*, London: Routledge.
—— (2000) *From East to West: Odyssey of a Soul*, London: Routledge.
—— (2002a) *From Science to Emancipation: Alienation and the Actuality of Enlightenment*, New Delhi: Sage.
—— (2002b) *Reflections on Meta-Realty: Transcendence, Emancipation and Everyday Life*, New Delhi: Sage.
—— (2002c) *Meta-Reality: Creativity, Love and Freedom (The Philosophy of Meta-Reality, Volume 1)*, New Delhi: Sage.
Bhaskar, Roy with Hartwig, Mervyn (2010) *The Formation of Critical Realism: A Personal Perspective*, London: Routledge.
Bicknell, E.J. (1955) *A Theological Introduction to the Thirty-Nine Articles of the Church of England*, London: Longmans, Green & Co.
Blond, Philip (1999) 'Perception: From Modern Painting to the Vision of Christ', in Milbank, John, Pickstock, Catherine and Ward, Graham (eds) *Radical Orthodoxy: A New Theology*, London: Routledge, pp. 220–241.
Borg, Marcus J. (1993) *Jesus, A New Vision: Spirit, Culture and the Life of Discipleship*, London: SPCK.
—— (1998) *Conflict, Holiness and Politics in the Teachings of Jesus*, Harrisburg, PA: Trinity Press International.
Borsch, Frederick H. (1967) *The Son of Man in Myth and History*, London: SCM.
Bowen, John (1981) *A History of Western Education, Volume 3: The Modern West*, London: Methuen.
Braaten, Carl E. (1995) 'The Gospel for a Neopagan Culture', in Braaten, Carl E. and Jenson, Robert W. (eds) *Either/Or: The Gospel or Neopaganism*, Grand Rapids, MI: William B. Eerdmans, pp. 7–21.
Bretherton, Luke (2010) *Hospitality as Holiness: Christian Witness Amid Moral Diversity*, Farnham, Surrey: Ashgate.
Brown, James (1953) *Subject and Object in Modern Theology*, London: SCM.

Brueggemann, Walter (1997) *Theology of the Old Testament: Testimony, Dispute, Advocacy*, Minneapolis MN: Fortress Press.
Buckley, Michael J. (1987) *At the Origins of Modern Atheism*, New Haven, CT: Yale University Press.
Bultmann, Rudolph (1958) *Jesus Christ and Mythology*, New York: Scribners.
—— (1966) *Faith and Understanding*, London: SCM.
—— (1972) *The History of the Synoptic Tradition*, Oxford: Blackwell.
Burns, Robert M. and Rayment-Pickard, Hugh (eds) (2000) *Philosophies of History: From Enlightenment to Postmodernity*, Oxford: Blackwell.
Busch, Eberhard (1976) *Karl Barth: His Life from Letters and Autobiographical Texts*, London: SCM.
—— (2004) *The Great Passion: An Introduction to Karl Barth's Theology*, Grand Rapids, MI: William B. Eerdmans.
Byrne, Peter (2003) *God and Realism*, Aldershot, Hampshire: Ashgate.
Cady, Linell E. and Brown, Delwin (2002) 'Introduction', in Cady, Linell E. and Brown, Delwin (eds) *Religious Studies, Theology, and the University: Conflicting Maps, Changing Terrain*, New York: State University of New York Press, pp. 1–12.
Calvin, John (1960) *Institutes of the Christian Religion*, 2 vols, Philadelphia, PA: Westminster Press.
Cassirer, Ernst (1951) *The Philosophy of the Enlightenment*, Princeton, NJ: Princeton University Press.
Clayton, Philip (ed.) (2007) *All That Is: A Naturalistic Faith for the Twenty First Century*, Minneapolis, MN: Fortress Press.
Cobb, John B. and Griffin, David R. (1976) *Process Theology: An Introductory Exposition*, Louisville, KY: Westminster John Knox.
Cochrane, Charles N. (1929) *Thucydides and the Science of History*, Oxford: Oxford University Press.
—— (1944) *Christianity and Classical Culture*, Oxford: University of Oxford Press.
Collier, Andrew (1994) *Critical Realism: An Introduction to Roy Bhaskar's Philosophy*, London: Verso.
—— (1999) *Being and Worth*, London: Routledge.
—— (2001) *Christianity and Marxism: A Philosophical Contribution to their Reconciliation*, London: Routledge.
—— (2003) *On Christian Belief: A Defence of a Cognitive Conception of Religious Belief in a Christian Context*, London: Routledge.
Collingwood, Robin. G. (1946) *The Idea of History*, Oxford: Oxford University Press.
Collins, James (1983) *The Mind of Kierkegaard*, Princeton, NJ: Princeton University Press.
Creaven, Sean (2001) 'Materialism, Agnosticism, and God', *Journal for the Theory of Social Behaviour*, 31(4), pp. 419–448.
—— (2010) *Against the Spiritual Turn: Marxism, Realism and Critical Theory*, London: Routledge.
Crossan, J. Dominic (1991) *The Historical Jesus: The Life of a Mediterranean Peasant*, San Francisco, CA: HarperCollins.
Cumin, Paul (2010) 'The Taste of Cake: Relation and Otherness with Colin Gunton and the Second Strong Hand of God', in Harvey, Lincoln (ed.) *The Theology of Colin Gunton*, Edinburgh: T. & T. Clark, pp. 65–85.
Cunningham, David S. (1998) *These Three are One: The Practice of Trinitarian Theology*, Oxford: Blackwell.
Cupitt, Don (1980) *Taking Leave of God*, London: SCM.

Davies, Brian (1993) *The Thought of Thomas Aquinas*, Oxford: Clarendon Press.
Dawkins, Richard (2006) *The God Delusion*, Boston, MA: Houghton Mifflin.
Day, Mark (2008) *The Philosophy of History*, London: Continuum.
D'Costa, Gavin (2000) *The Meeting of Religions and the Trinity*, Edinburgh: T. & T. Clark.
—— (2005) *Theology in the Public Square: Church, Academy and Nation*, Oxford: Blackwell.
Dennett, Daniel C. (2006) *Breaking the Spell: Religion as a Natural Phenomenon*, New York: Viking.
Denton, Donald L. (2004) *Historiography and Hermeneutics in Jesus Studies*, London: T. & T. Clark International.
Descartes, René (1969) *Discourse on Method and The Meditations*, Harmondsworth, Middlesex: Penguin.
—— (1970) *Philosophical Writings*, London: Nelson.
Diekema, Anthony J. (2000) *Academic Freedom and Christian Scholarship*, Grand Rapids, MI: William B. Eerdmans.
Dodd, Charles H. (1978) *The Authority of the Bible*, London: Collins.
Drake, Durant (ed.) (1920) *Essays in Critical Realism: A Co-operative Study of the Problem of Knowledge*, London: Macmillan.
Drobner, Hubertus R. (2007) *The Fathers of the Church: A Comprehensive Introduction*, Peabody, MA: Hendrickson Publishers.
Dunn, James D.G. (1977) *Unity and Diversity in the New Testament: An Inquiry into the Character of Earliest Christianity*, London: SCM.
Dunne, Joseph (1997) *Back to the Rough Ground: Practical Judgement and the Lure of Technique*, Notre Dame, IN: University of Notre Dame Press.
Early Christian Texts (2011) *The Chalcedonian Definition*, www.earlychurchtexts.com, accessed 27 May 2011.
Everington, Judith (1996) 'A Question of Authenticity: The Relationship between Educators and Practitioners in the Representation of Religious Traditions', *British Journal of Religious Education*, 18(2), pp. 69–77.
Farley, Edward (1994) *Theologia: The Fragmentation and Unity of Theological Education*, Eugene, OR: Wipf & Stock.
Feuerbach, Ludwig (1957) *The Essence of Christianity*, London: Harper & Row.
Fichte, Johann G. (2009) *Addresses to the German Nation*, Cambridge: Cambridge University Press.
—— (2010) *Attempt at a Critique of All Revelation*, Cambridge: Cambridge University Press.
Fiddes, Paul S. (1992) *The Creative Suffering of God*, Oxford: Clarendon Press.
Fisher, Ben C. (1989) *The Idea of a Christian University in Today's World*, Macon, GA: Mercer University Press.
Fitzgerald, Timothy (2000) *The Ideology of Religious Studies*, Oxford; Oxford University Press.
Flood, Gavin (1996) *An Introduction to Hinduism*, Cambridge: Cambridge University Press.
—— (1999) *Beyond Phenomenology: Rethinking the Study of Religion*, London: Cassell.
Fornara, Charles W. (1983) *The Nature of History in Ancient Greece and Rome*, San Francisco, CA: University of California Press.
Foster, Michael B. (1973) 'The Christian Doctrine of Creation and the Rise of Modern Natural Science', in Russell, C. A. (ed.) *Science and Religious Belief: A Selection of Recent Historical Studies*, London: Hodder & Stoughton, pp. 294ff.

Foucault, Michel (1989) *The Archaeology of Knowledge*, London: Routledge.

Funk, Robert W. (1997) *The Acts of Jesus: The Search for the Authentic Deeds of Jesus*, San Francisco, CA: HarperCollins.

—— (1999) *The Gospel of Jesus: According to the Jesus Seminar*, New York: Polebridge Press.

Funk, Robert W. and Hoover, Roy W. (1993) *The Five Gospels: The Search for the Authentic Words of Jesus*, New York: Macmillan.

Gadamer, Hans-Georg (1979) *Truth and Method*, London: Sheed & Ward.

Gerson, Lloyd P. (1994) *God and Greek Philosophy: Studies in the Early History of Natural Theology*, London: Routledge.

Gillespie, Michael A. (1996) *Nihilism before Nietzsche*, Chicago, IL: Chicago University Press.

Gregory of Nazianzus (2002) *On God and Christ: The Five Theological Orations and Two Letters to Cledonius*, Crestwood, NY: St Vladimir's Seminary Press.

Grenz, Stanley J. (2004) *Rediscovering the Triune God: The Trinity in Contemporary Theology*, Minneapolis, MN: Fortress Press.

Griffin, David R. and Hough, Joseph C. (eds) (1991) *Theology and the University*, Albany, NY: State University of New York Press.

Groff, Ruth (2004) *Critical Realism, Post-positivism and the Possibility of Knowledge*, London: Routledge.

Gruchy, John W. De (2002) *Reconciliation: Restoring Justice*, London: SCM.

Gunton, Colin E. (1978) *Becoming and Being: The Doctrine of God in Charles Hartshorne and Karl Barth*, Oxford: Oxford University Press.

—— (1983) *Yesterday and Today: A Study in Continuities in Christology*, London: Darton, Longman & Todd.

—— (1985) *Enlightenment and Alienation: An Essay Towards a Trinitarian Theology*, Basingstoke: Marshall, Morgan & Scott.

—— (1988) *The Actuality of Atonement: A Study of Metaphor, Rationality and the Christian Tradition*, Edinburgh: T. & T. Clark.

—— (1990) 'Augustine, the Trinity, and the Theological Crisis of the West', *Scottish Journal of Theology*, 43(1), pp. 33–58.

—— (1991) *The Promise of Trinitarian Theology*, Edinburgh: T. & T. Clark.

—— (1993) *The One, the Three and the Many: God, Creation and the Culture of Modernity*, Edinburgh: T. & T. Clark.

—— (1995) *A Brief Theology of Revelation*, Edinburgh: T. & T. Clark.

—— (1998) *The Triune Creator: A Historical and Systematic Study*, Edinburgh: Edinburgh University Press.

—— (2002) *The Christian Faith: An Introduction to Christian Doctrine*, Oxford: Blackwell.

—— (2003) *Father, Son and Holy Spirit: Toward a Fully Trinitarian Theology*, Edinburgh: T. & T. Clark.

Guthrie, William K.C. (1968) *The Greeks and their Gods*, London: Methuen.

Hamidullah, Muhammad (1979) *Introduction to Islam*, London: MWH.

Hanby, Michael (2003) *Augustine and Modernity*, London: Routledge.

Harman, Gilbert (1965) 'The Inference to the Best Explanation', *Philosophical Review*, 74(1), pp. 88–95.

Harnack, Adolf Von (1904) *What is Christianity?* London: Williams & Norgate.

Hart, Trevor (1997) 'Redemption and Fall', in Gunton, Colin E. (ed.) *The Cambridge Companion to Christian Doctrine*, Cambridge: Cambridge University Press, pp. 189–206.

Hartshorne, Charles (1941) *Man's Vision of God and the Logic of Theism*, New York: Harper & Brothers.
—— (1948) *The Divine Relativity: A Social Conception of God*, New Haven, CT: Yale University Press.
—— (1953) *Reality as Social Process: Studies in Metaphysics and Religion*, Glencoe, IL: Free Press.
—— (1965) *Anselm's Discovery: A Re-examination of the Ontological Proof of God's Existence*, La Salle, IL: Open Court.
—— (1967) *A Natural Theology for Our Time*, La Salle, IL: Open Court.
Hartwig, Mervyn (2000) 'God and Bhaskar: Response to Gary MacLennan's Review of "From East to West" ', *Marxism Mailing List*, www.raggedclaws.com/critical realism/archive/gmac_abtf.html, accessed 21 February 2008.
—— (2001) 'New Left, New Age, New Paradigm? Roy Bhaskar's "From East to West" ', *Journal for the Theory of Social Behaviour*, 31(2), pp. 139–165.
—— (ed.) (2007) *Dictionary of Critical Realism*, London: Routledge.
Hauerwas, Stanley (2007) *The State of the University: Academic Knowledges and the Knowledge of God*, Oxford: Blackwell.
Heelas, Paul (1996) *The New Age Movement: The Celebration of the Self and the Sacralization of Modernity*, Oxford: Blackwell.
Hegel, Georg W.F. (1949) *The Phenomenology of Mind*, London: George Allen & Unwin.
—— (1956) *The Philosophy of History*, New York: Dover Publications.
—— (1977) *Hegel's Phenomenology of Spirit*, trans. A.V. Miller with analysis of text, Oxford: Oxford University Press.
Hemming, Laurence P. (ed.) (2000) *Radical Orthodoxy: A Catholic Enquiry*, Aldershot, Hampshire: Ashgate.
Herberg, Will (1976) *Faith as Enacted in History: Essays in Biblical Theology*, Philadelphia. PA: Westminster Press.
—— (1977) *Phenomenology of Spirit*, Oxford: Oxford University Press.
Hermes Trismegistus (1999) *The Way of Hermes: The Corpus Hermeticum and The Definitions of Hermes Trismegistus to Asclepius*, London: Gerald Duckworth.
Hick, John (1977) *God and the Universe of Faiths*, London: Collins.
—— (1989) *An Interpretation of Religion: Human Responses to the Transcendent*, London: Macmillan.
Hitchens, Christopher (2010) *Christopher Hitchens v Tony Blair: The Full Transcript*, www.newstatesman.com/blogs/the-staggers/2010/11/christopher-hitchens-tony-blair, accessed 21 March 2011.
Hodgson, Peter E. (2005) *Theology and Modern Physics*, Aldershot, Hampshire: Ashgate.
Holdrege, Barbara A. (2004) 'Dharma', in Mittal, Sushil and Thursby, Gene (eds) *The Hindu World*, London: Routledge, pp. 213–248.
Holmes, Stephen R. (2010) 'Towards the *Analogia Personae et Relationis*: Developments in Gunton's Trinitarian Thinking', in Harvey, Lincoln (ed.) *The Theology of Colin Gunton*, Edinburgh: T. & T. Clark, pp. 32–48.
Hooykaas, Reijer (1973) *Religion and the Rise of Modern Science*, Edinburgh: Scottish Academic Press.
Howard, Thomas A. (2006) *Protestant Theology and the Making of the Modern German University*, Oxford: Oxford University Press.
Hughes, Graham (1979) *Hebrews and Hermeneutics: The Epistle to the Hebrews as a New Testament Example of Biblical Interpretation*, Cambridge: Cambridge University Press.

Hume, David (1902) *Enquiries Concerning the Human Understanding and Concerning the Principles of Morals*, Oxford: Clarendon Press.

Irenaeus (2004) 'Against Heresies', in Roberts, Alexander and Donaldson, James (eds) *Ante-Nicene Fathers, Volume One: The Apostolic Fathers, Justin Martyr and Irenaeus*, Peabody, MA: Hendrickson, pp. 309–567.

Jackson, Robert (1997) *Religious Education: An Interpretative Approach*, London: Hodder & Stoughton.

—— (2004) *Rethinking Religious Education and Plurality: Issues in Diversity and Pedagogy*, London: RoutledgeFalmer.

—— (2008) 'Contextual Religious Education and the Interpretative Approach', *British Journal of Religious Education*, 30(1), pp. 13–24.

Jaeger, Werner (1945) *Paideia: The Ideals of Greek Culture*, 3 vols, Oxford: Oxford University Press.

—— (1961) *Early Christianity and Greek Paideia*, Cambridge, MA: Belknap/Harvard University Press.

Jaki, Stanley L. (1980) *The Road of Science and the Ways to God*, Edinburgh: Scottish Academic Press.

Jammer, Max (1999) *Einstein and Religion: Physics and Theology*, Princeton, NJ: Princeton University Press.

Jenson, Robert W. (1997) 'Karl Barth', in Ford, David F. (ed.) *The Modern Theologians: An Introduction to Christian Theology in the Twentieth Century*, 2nd edn, Oxford: Blackwell, pp. 21–36.

—— (2010) 'A Decision Tree of Colin Gunton's Thinking', in Harvey, Lincoln (ed.) *The Theology of Colin Gunton*, Edinburgh: T. & T. Clark, pp. 8–16.

Jones, Gareth (1991) *Bultmann: Towards a Critical Theology*, Cambridge: Polity Press.

Josephus (1987) *The Works of Josephus*, Peabody, MA: Hendrickson Publishers.

Kant, Immanuel (1992) *Conflict of the Faculties*, Lincoln, NE: University of Nebraska Press.

Kärkkäinen, Veli-Matti (2009) 'The Trajectories of the Contemporary "Trinitarian Renaissance" in Different Contexts', *Journal of Reformed Theology*, 3, pp. 7–21.

Käsemann, Ernst (1964) *Essays on New Testament Themes*, London: SCM.

Kekes, John (1999) *Against Liberalism*, New York: Cornell University Press.

Kelly, John N.D. (1950) *Early Christian Creeds*, London: Longmans, Green & Co.

—— (1977) *Early Christian Doctrines*, London: Adam & Charles Black.

Kelsey, David H. (1992) *To Understand God Truly: What's Theological about a Theological School*, Louisville, KY: Westminster/John Knox Press.

Kendrick, Graham (2011) *The Servant King*, www.grahamkendrick.co.uk/songs/lyrics/the_servant_king.php, accessed 6 June 2011.

Kenny, Anthony (2008) *The Five Ways: St Thomas Aquinas' Proofs of God's Existence*, London: Routledge.

Kierkegaard, Søren (1967) *Philosophical Fragment, or A Fragment of Philosophy, by Johannes Climacus*, Princeton, NJ: Princeton University Press.

—— (1968) *Concluding Unscientific Postscript to the Philosophical Fragments, by Johannes Climacus*, Princeton, NJ: Princeton University Press.

Komoszewski, J. Ed, Sawyer, M. James and Wallace, Daniel B. (2006) *Reinventing Jesus: How Contemporary Skeptics Miss the Real Jesus and Mislead Popular Culture*, Grand Rapids, MI: Kregel Publications.

Kristeva, Julia (1994) *Strangers to Ourselves*, New York: Columbia University Press.

Kuhn, Thomas S. (1970) *The Structure of Scientific Revolutions*, Chicago, IL: University of Chicago Press.

Kümmel, Werner G. (1973) *The New Testament: The History of the Investigation of its Problems*, London: SCM.

Küng, Hans (1971) *Infallible? An Enquiry*, London: Collins.

—— (1981) *Justification: The Doctrine of Karl Barth and a Catholic Reflection*, London: Burns & Oates.

—— (1991) *Global Responsibility: In Search of a New World Ethic*, New York: Crossroad/Herder & Herder.

—— (1995) 'A Global Ethic and Education', *British Journal of Religious Education*, 18(1), pp. 6–21.

Lackey, Jennifer and Sosa, Ernest (eds) (2006) *The Epistemology of Testimony*, Oxford: Clarendon Press.

Lancel, Serge (2002) *St Augustine*, London: SCM.

Lear, Jonathan (1988) *Aristotle: The Desire to Understand*, Cambridge: Cambridge University Press.

Lemon, Michael C. (2003) *Philosophy of History: A Guide for Students*, London: Routledge.

Lessing, Gotthold (1957) *Theological Writings*, Stanford, CA: Stanford University Press.

Levin, David M. (1988) *The Opening of Vision: Nihilism and the Postmodern Situation*, London: Routledge.

Lindbeck, George A. (1984) *The Nature of Doctrine: Religion and Theology in a Postliberal Age*, Philadelphia, PA: Westminster Press.

Lindblom, Johannes (1962) *Prophecy in Ancient Israel*, Oxford: Blackwell.

Lipton, Peter (2004) *Inference to the Best Explanation*, London: Routledge.

Locke, John (1975) *An Essay Concerning Human Understanding*, Oxford: Clarendon Press.

Lonergan, Bernard J.F. (1957) *Insight: A Study of Human Understanding*, London: Longmans.

—— (1973) *Method in Theology*, London: Darton, Longman & Todd.

Longenecker, Richard (1975) *Biblical Exegesis in the Apostolic Period*, Grand Rapids, MI: William B. Eerdmans.

Lopez, Jose and Potter, Garry (eds) (2001) *After Postmodernism: An Introduction to Critical Realism*, London: Athlone Press.

Lossky, Vladimir (1991) *The Mystical Theology of the Eastern Church*, Cambridge: James Clarke.

Louth, Andrew (1983) *Discerning the Mystery: An Essay on the Nature of Theology*, Oxford: Clarendon Press.

Lowrie, Walter (1974) *A Short Life of Kierkegaard*, Princeton, NJ: Princeton University Press.

Lubac, Henri de (1988) *Catholicism: Christ and the Common Destiny of Man*, San Francisco, CA: Ignatius Press.ööö

Lyotard, Jean-François (1984) *The Postmodern Condition: A Report on Knowledge*, Manchester: Manchester University Press.

MacIntyre, Alasdair (1985) *After Virtue: A Study in Moral Theory*, London: Duckworth.

MacKenzie, Iain M, (2002) *Irenaeus's Demonstration of the Apostolic Preaching: A Theological Commentary and Translation*, Aldershot: Ashgate.

Mackey, James P. (1983) *The Christian Experience of God as Trinity*, London: SCM.

Mackie, John L. (1982) *The Miracle of Theism: Arguments For and Against the Existence of God*, Oxford: Clarendon Press.

MacLennan, Gary (2000) '"From East to West": A Book Too Far?' *Marxism Mailing List*, www.raggedclaws.com/critical realism/archive/gmac_abtf.html, accessed 21 February 2008.

Malinvaud, Edmond and Archer, Margaret S. (2003) *Work and Human Fulfilment*, Ypsilanti, MI: Sapientia Press. Malinvaud, Edmond & Archer, Margaret S. (2003) *Work and Human Fulfilment*, Ypsilanti, MI: Sapientia Press.

Markham, Ian (1998) *Truth and the Reality of God*, Edinburgh: T. & T. Clark.

Marsden, George M. (1994) *The Soul of the American University: From Protestant Establishment to Established Nonbelief*, Oxford: Oxford University Press.

—— (1998) *The Outrageous Idea of Christian Scholarship*, Oxford: Oxford University Press.

Marshall, Bruce D. (2000) *Trinity and Truth*, Cambridge: Cambridge University Press.

Marshall, I. Howard (2004) *New Testament Theology: Many Witnesses, One Gospel*, Downers Grove, IL: IVP Academic.

Martin, Sean (2005) *The Gnostics: The First Christian Heretics*, London: Pocket Essentials.

McCormack, Bruce L. (1997) *Karl Barth's Critically Realistic Dialectical Theology: Its Genesis and Development 1909–1936*, Oxford: Clarendon Press.

McEvenue, Sean and Meyer, Ben F. (1992) *Lonergan's Hermeneutics: Its Development and Application*, Washington, DC: The Catholic University of America Press.

McGrath, Alister E. (1999) *T. F. Torrance: An Intellectual Biography*, Edinburgh: T. & T. Clark.

—— (2001) *A Scientific Theology, Volume One: Nature*, London: Continuum/T. & T. Clark.

—— (2002) *A Scientific Theology, Volume Two: Reality*, London: Continuum/T. & T. Clark.

—— (2003) *A Scientific Theology, Volume Three: Theory*, London: Continuum/T. & T. Clark.

—— (2004a) *The Science of God: An Introduction to Scientific Theology*, London: Continuum.

—— (2004b) *Dawkins' God: Genes, Memes and the Meaning of Life*, Oxford: Blackwell.

—— (2006) *The Order of Things: Explorations in Scientific Theology*, Oxford: Blackwell.

—— (2008) *The Open Secret: A New Vision for Natural Theology*, Oxford: Blackwell.

—— (2009a) *A Fine-Tuned Universe: The Quest for God in Science and Theology*, Louisville, KY: Westminster John Knox.

—— (2009b) *Heresy: A History of Defending the Truth*, London: SPCK.

—— (2012) 'Transcendence and God: Reflections on Critical Realism, the "New Atheism", and Christian Theology', in Hartwig, Mervyn and Morgan, Jamie (eds) *Critical Realism and Spirituality*, London: Routledge, pp. 188–203.

McGrath, Alister and Collicutt, Joanna (2007) *The Dawkins Delusion? Atheist Fundamentalism and the Denial of the Divine*, London: SPCK.

McLennan, Gregor (1981) *Marxism and the Methodologies of History*, London: Verso.

Meyer, Ben F. (1979) *The Aims of Jesus*, London: SCM.

—— (1989) *Critical Realism and the New Testament*, Eugene, OR: Pickwick Publications.

—— (1995) *Reality and Illusion in New Testament Scholarship: A Primer in Critical Realist Hermeneutics*, Collegeville, MN: Michael Glazier.

Miegge, Giovanni (1960) *Gospel and Myth in the Thought of Rudolf Bultmann*, London: Lutterworth Press.

Milbank, John (1990a) *Theology and Social Theory: Beyond Secular Reason*, Oxford: Blackwell.

—— (1990b) 'The End of Dialogue', in D'Costa, Gavin (ed.) *Christian Uniqueness Reconsidered: The Myth of a Pluralistic Theology of Religions*, Maryknoll, NY: Orbis Books, pp. 174–191.
—— (1997) *The World Made Strange: Theology, Language, Culture*, Oxford: Blackwell.
—— (2003) *Being Reconciled: Ontology and Pardon*, London: Routledge.
Milbank, John and Pickstock, Catherine (2001) *Truth in Aquinas*, London: Routledge.
Milbank, John, Pickstock, Catherine and Ward, Graham (eds) (1999) *Radical Orthodoxy: A New Theology*, London: Routledge.
Miner, Robert (2004) *Truth in the Making: Creative Knowledge in Theology and Philosophy*, London: Routledge.
Mitchell, Basil (1973) *The Justification of Religious Belief*, London: Macmillan.
Moltmann, Jürgen (1974) *The Crucified God: The Cross of Christ as the Foundation and Criticism of Christian Theology*, London: SCM.
Moore, Andrew (2003) *Realism and Christian Faith: God, Grammar, and Meaning*, Cambridge: Cambridge University Press.
Moore, Andrew and Scott, Michael (eds) (2007) *Religion and Realism: Philosophical and Theological Perspectives*, Aldershot, Hampshire: Ashgate.
Mößner, Nicola, Schmoranzer, Sebastian and Weidemann, Christian (2008) *Richard Swinburne: Christian Philosophy in a Modern World*, Frankfurt: Ontos Verlag.
Nagel, Thomas (1986) *The View from Nowhere*, Oxford: Oxford University Press.
Neill, Stephen (1966) *The Interpretation of the New Testament 1861–1961*, Oxford: Oxford University Press.
Neurath, Otto (1973) *Empiricism and Sociology*, Dordrecht: Reidel.
Newman, Carey C. (ed.) (1999) *Jesus and the Restoration of Israel: A Critical Assessment of N.T. Wright's 'Jesus and the Victory of God'*, Westmont, IL: InterVarsity Press.
Newman, John H. (1979) *An Essay in Aid of a Grammar of Assent*, Notre Dame, IN: University of Notre Dame Press.
—— (1982) *The Idea of a University*, Notre Dame, IN: University of Notre Dame Press.
Nietzsche, Friedrich (1993) *The Birth of Tragedy*, London: Penguin Books.
Norrie, Alan (2010) *Dialectic and Difference: Dialectical Critical Realism and the Grounds of Justice*, London: Routledge.
Ockham, William (1983) *Predestination, God's Foreknowledge and Future Contingencies*, Cambridge, MA: Hackett.
—— (1990) *Philosophical Writings*, Cambridge, MA: Hackett.
O'Collins, Gerald (1999) *The Tripersonal God: Understanding and Interpreting the Trinity*, London: Geoffrey Chapman.
O'Meara (1993) *Plotinus: An Introduction to the Enneads*, Oxford: Clarendon Press.
Origen (1980) *Contra Celsum*, Cambridge: Cambridge University Press.
Osborn, Eric (2001) *Irenaeus of Lyons*, Cambridge: Cambridge University Press.
—— (2008) *Clement of Alexandria*, Cambridge: Cambridge University Press.
Otto, Rudolf (1931) *The Idea of the Holy: An Inquiry into the Non-rational Factor in the Idea of the Divine and its Relation to the Rational*, London: Oxford University Press.
Pannenberg, Wolfhart (1976) *Theology and the Philosophy of Science*, London: Darton, Longman & Todd.
Parmenides (1984) *Fragments*, Toronto: University of Toronto Press.
Pascal, Blaise (1966) *Pensées*, Harmondsworth, Middlesex: Penguin Classics.
Patterson, Sue (1999) *Realist Christian Theology in a Postmodern Age*, Cambridge: Cambridge University Press.

Pattison, George (1999) *Kierkegaard: The Aesthetic and the Religious – From the Magic Theatre to the Crucifixion of the Imagination*, London: SCM Press.

Peacocke, Arthur R. (1971) *Science and the Christian Experiment*, London: Oxford University Press.

—— (1979) *Creation and the World of Science*, Oxford: Oxford University Press.

—— (1984) *Intimations of Reality: Critical Realism in Science and Religion*, Notre Dame, IN: University of Notre Dame Press.

—— (1986) *God and the New Biology*, London: Everyman.

—— (1990) *Theology for a Scientific Age*, Oxford: Blackwell.

—— (2001) *Paths from Science Towards God: The End of All Our Exploring*, Oxford: Oneworld Publications.

—— (2004) *Evolution: The Disguised Friend of Faith. Collected Essays*, West Conshohocken, PA: Templeton Foundation Press.

Pelikan, Jaroslav (1985) *The Christian Tradition: A History of the Development of Doctrine, Volume Four, Reformation of Church and Dogma (1300–1700)*, Chicago, IL: University of Chicago Press.

—— (1993) *Christianity and Classical Culture: The Metamorphosis of Natural Theology in the Christian Encounter with Hellenism*, New Haven, CT: Yale University Press.

Pickstock, Catherine (1998) *After Writing: On the Liturgical Consummation of Philosophy*, Oxford: Blackwell.

Placher, William C. (1996) *The Domestication of Transcendence: How Modern Thinking about God Went Wrong*, Louisville, KY: Westminster John Know Press.

Plantinga, Alvin (1968) *The Ontological Argument from St. Anselm to Contemporary Philosophers*, London: Macmillan.

Plato (1961) *Collected Dialogues*, Princeton, NJ: Princeton University Press.

Plotinus (1991) *The Enneads*, London: Penguin Classics.

Polanyi, Michael (1958) *Personal Knowledge: Towards a Post-critical Philosophy*, London: Routledge & Kegan Paul.

—— (1974) *Scientific Thought and Social Reality: Essays by Michael Polanyi*, Madison, CT: International Universities Press.

Polkinghorne, John C. (1986) *One World: The Interaction of Science and Theology*, London: SPCK.

—— (1988) *Science and Creation: The Search for Understanding*, London: SPCK.

—— (1994) *Science and Christian Belief*, London: SPCK.

—— (1996) Scientists as *Theologians: A Comparison of the Writings of Ian Barbour, Arthur Peacocke and John Polkinghorne*, London: SPCK.

—— (1998) *Belief in God in an Age of Science*, New Haven, CT: Yale University Press.

Pope Paul VI (1996) 'Pastoral Constitution on the Church in the Modern World, Gaudium et Spes, 7 December 1965', in Flannery, Austin (ed.) *Vatican Council II: Constitutions, Decrees and Declarations*, Northport, NY: Costello Publishing, pp. 163–282.

Popper, Karl R. (1957) *The Poverty of Historicism*, London: Routledge & Kegan Paul.

—— (1966) *The Logic of Scientific Discovery*, London: Hutchinson.

Porpora, Douglas V. (2001) *Landscapes of the Soul: The Loss of Moral Meaning in American Life*, New York: Oxford University Press.

—— (2004) 'A Propaedeutic to a Propaedeutic on Inter-religious Dialogue', in Archer, Margaret S., Collier, Andrew and Porpora, Douglas V. (eds) *Transcendence: Critical Realism and God*, London: Routledge, pp. 109–128.

—— (2005) 'The Spiritual Turn in Critical Realism', *New Formations*, (56), pp. 147–161.

Pritchard, James B. (ed.) (1958) *The Ancient Near East: An Anthology of Texts and Pictures*, vol. 1, Princeton, NJ: Princeton University Press.

Quine, Willard Van and Ullian, Joseph S. (1978) *The Web of Belief*, New York: McGraw-Hill.

Race, Alan (1994) *Christians and Religious Pluralism*, London: SCM.

Rahner, Karl (1970) *The Trinity*, London: Burns & Oates.

—— (1998) *Foundations of Christian Faith: An Introduction to the Idea of Christianity*, New York: Crossroad.

Ricoeur, Paul (1976) *Interpretative Theory: Discourse and the Surplus of Meaning*, Forth Worth, TX: Texas Christian University Press.

—— (1995) *Figuring the Sacred: Religion, Narrative and Imagination*, Minneapolis, MA: Fortress Press.

Robinson, James M. (ed.) (1996) *The Nag Hammadi Library in English*, Leiden: E. J. Brill.

Rorty, Richard (1980) *Philosophy and the Mirror of Nature*, Oxford: Blackwell.

—— (1989) *Contingency, Irony and Solidarity*, Cambridge: Cambridge University Press.

Ross, Alistair (2000) *Curriculum: Construction and Critique*, London: RoutledgeFalmer.

Rousseau, Jean-Jacques (1986) *Émile*, London: Dent.

Ruether, Rosemary R. (1987) 'Feminism and Jewish–Christian Dialogue: Particularism and Universalism in the Search for Religious Truth', in Knitter, Paul and Hick, John (eds) *The Myth of Christian Uniqueness: Toward a Pluralistic Theology of Religions*, Maryknoll, NY: Orbis, pp. 137–142.

Ruether, Rosemary R. and Grau, Marion (2006) *Interpreting the Postmodern: Responses to 'Radical Orthodoxy'*, London: T. & T. Clark.

Rumscheidt, Martin (ed.) (1989) *Adolf von Harnack: Liberal Theology at its Height*, London: Collins.

Russell, David S. (1964) *The Method and Message of Jewish Apocalyptic: 200 BC–AD 100*, London: SCM.

Russell, Robert J., (ed.) (2004) *Fifty Years in Science and Religion: Ian G. Barbour and His Legacy*, Aldershot, Hampshire: Ashgate.

Sanders, Edward P. (1985) *Jesus and Judaism*, London: SCM.

Sayer, Andrew (1992) *Method in Social Science: A Realist Approach*, London: Routledge.

Schleiermacher, Friedrich D.E. (1958) *On Religion: Speeches to its Cultured Despisers*, London: Harper & Row.

—— (1976) *The Christian Faith*, Edinburgh: T. & T. Clark.

—— (2011) *Brief Outline of Theology as a Field of Study*, Louisville, KY: Westminster/John Knox Press.

Schweitzer, Albert (1945) *The Quest of the Historical Jesus: A Critical Study of its Progress from Reimarus to Wrede*, London: Adam & Charles Black.

Schwöbel, Christoph (1992) *God, Action and Revelation*, Kampen: Kok Pharos.

—— (1995) 'Imago Libertatis: Human and Divine Freedom', in Gunton, C.E. (ed.) *God and Freedom: Essays on Historical and Systematic Theology*, Edinburgh: T. & T. Clark, pp. 57–81.

—— (2010) 'The Shape of Colin Gunton's Theology: On the Way towards a Fully Trinitarian Theology', in Harvey, Lincoln (ed.) *The Theology of Colin Gunton*, Edinburgh: T. & T. Clark, pp. 182–208.

Searle, John R. (1996) *The Construction of Social Reality*, Harmondsworth, Middlesex: Penguin Books.

Shakespeare, Steven (2007) *Radical Orthodoxy: A Critical Introduction*, London: SPCK.
Sharpe, Eric J. (1986) *Comparative Religion: A History*, London: Duckworth.
Shipway, Brad (2000) 'Critical Realism and Theological Critical Realism: Opportunities for Dialogue', *Alethia*, 3(2), pp. 29–33.
—— (2011) *A Critical Realist Perspective of Education*, London: Routledge.
Sia, Santiago (ed.) (1990) *Charles Hartshorne's Concept of God: Philosophical and Theological Responses*, Dordrecht: Kluwer.
Smart, Ninian (1979) *The Phenomenon of Christianity*, London: Collins.
Smith, James K.A. (2004) *Introducing Radical Orthodoxy: Mapping a Post-secular Theology*, Grand Rapids, MI: Baker Academic.
Soskice, Janet Martin (1987) *Metaphor and Religious Language*, Oxford: Clarendon Press.
Spade, Paul V. (ed.) (2000) *The Cambridge Companion to Ockham*, Cambridge: Cambridge University Press.
Sponheim, Paul (1968) *Kierkegaard on Christ and Christian Coherence*, London: SCM.
Stanton, Graham N. (1974) *Jesus of Nazareth in New Testament Preaching*, Cambridge: Cambridge University Press.
Stevenson, James (1960) *A New Eusebius: Documents Illustrative of the History of the Church to A.D. 337*, London: SPCK.
Stewart, Robert B. (2008) *The Quest of the Hermeneutical Jesus: The Impact of Hermeneutics on the Jesus Research of John Dominic Crossan and N.T. Wright*, Lanham, MD: University Press of America.
Strauss, David F. (1873) *The Old Faith and the New: A Confession*, London: Asher & Co.
—— (1973) *The Life of Jesus Critically Examined*, London: SCM.
—— (1977) *The Christ of Faith and the Jesus of History*, Philadelphia, PA: Fortress Press.
Swinburne, Richard (1993) *The Coherence of Theism*, Oxford: Clarendon Press.
—— (1994) *The Christian God*, Oxford: Clarendon Press.
—— (1998) *Providence and The Problem of Evil*www.oxfordscholarship.com/oso/public/content/philosophy/0198237987/toc.html?q=god?%7Cswinburnewww.oxfordscholarship.com/oso/public/content/philosophy/0198237987/toc.html?q=god?%7Cswinburnewww.oxfordswww.oup.co.uk/isbn/0–19–823798–7, Oxford: Clarendon Press.
—— (2001) *Epistemic Justification*, Oxford: Clarendon Press.
—— (2004) *The Existence of God* www.oxfordscholarship.com/oso/public/content/philosophy/0198239637/toc.html?q=existence%7Cgod, Oxford: Clarendon Press.
—— (2005) *Faith and Reason*, Oxford: Clarendon Press.
—— (2007) *Revelation*, Oxford: Clarendon Press.
—— (2010) *Was Jesus God?* Oxford: Oxford University Press.
Tanner, Kathryn (2001) *Jesus, Humanity and the Trinity: A Brief Systematic Theology*, Edinburgh: T. & T. Clark.
Taylor, Charles (1992) *Sources of the Self: The Making of the Modern Identity*, Cambridge: Cambridge University Press.
Taylor, Mark C. (1982) *De-constructing Theology*, New York: Crossroad.
—— (1987) *Erring: A Postmodern A/theology*, Chicago, IL: University of Chicago Press.
Terry, Justyn (2010) 'Colin Gunton's Doctrine of Atonement: Transcending Rationalism by Metaphor', in Harvey, Lincoln (ed.) *The Theology of Colin Gunton*, Edinburgh: T. & T. Clark, pp. 130–145.
Thatcher, Margaret (1993) *The Thatcher Years*, London: HarperCollins.

Thiessen, Elmer (2011) *The Ethics of Evangelism: A Philosophical Defence of Ethical Proselytizing and Persuasion*, Milton Keynes: Paternoster.

Thiselton, Anthony C. (1980) *The Two Horizons: New Testament Hermeneutics and Philosophical Description with Special Reference to Heidegger, Bultmann, Gadamer and Wittgenstein*, Exeter: Paternoster.

—— (1992) *New Horizons in Hermeneutics: The Theory and Practice of Transforming Biblical Reading*, London: HarperCollins.

—— (2007) *The Hermeneutics of Doctrine*, Grand Rapids, MI: Eerdmans,

Thucydides (1972) *History of the Peloponnesian War*, London: Penguin Classics.

Tillich, Paul (1978) *Systematic Theology, Part 3: Existence and the Christ*, London: SCM.

Torrance, Alan J. (1996) *Persons in Communion: Trinitarian Description and Human Participation*, Edinburgh: T. & T. Clark.

Torrance, James B. (1996) *Worship, Community, and the Triune God of Grace*, Carlisle: Paternoster.

Torrance, Thomas F. (1969) *Theological Science*, London: Oxford University Press.

—— (1976) *Space, Time and Resurrection*, Edinburgh: T. & T. Clark.

—— (1980a) *Christian Theology and Scientific Culture*, Belfast: Christian Journals.

—— (1980b) *The Ground and Grammar of Theology*, Belfast: Christian Journals.

—— (1984) *Transformation and Convergence in the Frame of Knowledge: Explorations in the Interrelations of Scientific and Theological Enterprise*, Belfast: Christian Journals.

—— (1988) *The Hermeneutics of John Calvin*, Edinburgh: Scottish Academic Press.

—— (1990) *Karl Barth: Biblical and Evangelical Theologian*, Edinburgh: T. & T. Clark.

—— (1993) *The Trinitarian Faith: The Evangelical Theology of the Ancient Catholic Church*, Edinburgh: T. & T. Clark.

—— (1995) *Divine Meaning: Studies in Patristic Hermeneutics*, Edinburgh: T. & T. Clark.

—— (1996) *The Christian Doctrine of God, One Being Three Persons*, Edinburgh: T. & T. Clark.

—— (2000) *Karl Barth: An Introduction to Hid Early Theology 1910–1931*, Edinburgh: T. & T. Clark.

—— (2001) *Reality and Scientific Theology*, Eugene, OR: Wipf and Stock.

Tyrrell, George (1963) *Christianity at the Crossroads*, London: George Allen & Unwin.

Vermes, Geza (1995) *The Dead Sea Scrolls in English*, Sheffield: Sheffield Academic Press.

Volf, Miroslav (2006) *The End of Memory: Remembering Rightly in a Violent World*, Grand Rapids, MI: William B. Eerdmans.

Voysey, Charles (1879) *Fragments from Reimarus*, London: Williams & Norgate.

Ward, Graham (1995) *Barth, Derrida and the Language of Theology*, Cambridge: Cambridge University Press.

—— (1996) *Theology and Contemporary Critical Theory*, London: Macmillan.

—— (2000) *Cities of God*, London: Routledge.

Warnke, Georgia (1987) *Gadamer: Hermeneutics, Tradition and Reason*, London: Polity Press.

Watson, Francis (1994) *Text, Church and World: Biblical Interpretation in Theological Perspective*, Edinburgh: T. & T. Clark.

—— (1997) *Text and Truth: Redefining Biblical Theology*, Edinburgh: T. & T. Clark.

—— (2004) *Paul and the Hermeneutics of Faith*, London: T. & T. Clark.

Webster, John (2010) 'Gunton and Barth', in Harvey, Lincoln (ed.) *The Theology of Colin Gunton*, Edinburgh: T. & T. Clark, pp. 17–31.

Webster, John, Tanner, Kathryn and Torrance, Iain (eds.) (2007) *The Oxford Handbook of Systematic Theology*, Oxford: Oxford University Press.

Wenham, David (1995) *Paul: Follower of Jesus or Founder of Christianity?* Grand Rapids, MI: William B. Eerdmans.

Wenham, Gordon J. (1979) *The Book of Leviticus*, Grand Rapids, MI: William B. Eerdmans.

—— (1987) Genesis *1–15, Word Biblical Commentary*, vol. 1, Nashville, TN: Thomas Nelson.

White, Hayden (1973) *Metahistory: The Historical Imagination in Nineteenth-century Europe*, Baltimore, MD: Johns Hopkins Press.

Whitehead, Alfred N. (1979) *Process and Reality*, New York: Free Press.

Williams, Bernard (2002) *Truth and Truthfulness: An Essay in Genealogy*, Princeton, NJ: Princeton University Press.

Williams, Rowan (2001) *Arius: Heresy and Tradition*, London: SCM.

Wittgenstein, Ludwig (1968) *Philosophical Investigations*, Oxford: Blackwell.

—— (1969) *The Blue and Brown Books: Preliminary Studies for the 'Philosophical Investigations'*, Oxford: Blackwell.

—— (1975) *On Certainty*, Oxford: Blackwell.

Wolterstorff, Nicholas (1996) *John Locke and the Ethics of Belief*, Cambridge: Cambridge University Press.

Wright, Andrew (2004) *Religion, Education and Post-modernity*, London: Routledge.

—— (2007) *Critical Religious Education, Multiculturalism and the Pursuit of Truth*, Cardiff: University of Wales Press.

—— (2008) 'Contextual Religious Education and the Actuality of Religions', *British Journal of Religious Education*, 30(1), pp. 3–12.

—— (2011) 'In Praise of the Spiritual Turn: Critical Realism and Trinitarian Christianity', *Journal of Critical Realism*, 10(3), pp. 332–358.

Wright, Nicholas T. (1992) *Christian Origins and the Question of God, Volume One: The New Testament and the People of God*, London: SPCK.

—— (1996) *Christian Origins and the Question of God, Volume Two: Jesus and the Victory of God*, London: SPCK.

—— (2000) *The Challenge of Jesus*, London: SPCK.

—— (2003) *Christian Origins and the Question of God, Volume Three: The Resurrection of the Son of God*, London: SPCK.

Wright, Terry J. (2010) 'Colin Gunton on Providence: Critical Commentaries', in Harvey, Lincoln (ed.) *The Theology of Colin Gunton*, Edinburgh: T. & T. Clark, pp. 146–164.

Wyschogrod, Edith (1998) *An Ethics of Remembering: History, Heterology, and the Nameless Others*, Chicago, IL: University of Chicago Press.

Zizioulas, John D. (1985) *Being as Communion: Studies in Personhood and the Church*, Crestwood, NY: St Vladimir's Seminary Press.

—— (2006) *Communion and Otherness: Further Studies in Personhood and the Church*, London: T. & T. Clark.

Index

Made in the USA
Las Vegas, NV
21 September 2026

0c04d069-98e7-4448-a6d5-0012bdc375b8R01